PERSPECTIVES
on Human Sexuality

PERSPECTIVES
on Human Sexuality

Anne Bolin
and
Patricia Whelehan

State University of New York Press

Published by
State University of New York Press, Albany

© 1999 State University of New York

All rights reserved

Printed in the United States of America

Three photos of fetal development © Lennart Nilsson/Bonnier Alba AB, *A Child is Born,* Dell Publishing Company.

For information, address State University of New York
Press, State University Plaza, Albany, N.Y., 12246

Production by Diane Ganeles
Marketing by Fran Keneston

Library of Congress Cataloging-in-Publication Data

Bolin, Anne.
    Perspectives on human sexuality / Anne Bolin and Patricia
Whelehan.
        p.   cm.
    Includes bibliographical references and index.
    ISBN 0-7914-4133-4 (hbk. : alk. paper). — ISBN 0-7914-4134-2
(pbk. : alk. paper)
    1. Sex.   2. Sex customs.   3. Sex (biology)   4. Hygiene, Sexual.
I. Whelehan, Patricia.   II. Title.
GN484.3.B66   1999
306.7—dc21   98-38210
                                                               CIP
                                                               AC

10 9 8 7 6 5 4 3 2 1

This book is dedicated to Jose Andres Gonzalez del Valle (August 4, 1963–September 9, 1991) who taught me to love without fear and who enhanced the lives of everyone who knew him. We miss you.

Pat

For my mother, Vivian Bolin, who left this earth March 21, 1998. I miss you so much.

Anne

Contents

List of Illustrations   ix

List of Tables   xi

Preface   xiii

Acknowledgments   xvii

Chapter 1: Introduction: History and Context   3

Chapter 2: Anthropological and Sexological Views   21

Chapter 3: Evolution of Biological Structures
             Related to Modern Sexual Behavior   41

Chapter 4: Introduction to the Hormonal Basis of
             Modern Human Sexuality   67

Chapter 5: Modern Human Male Anatomy
             and Physiology   81

Chapter 6: Modern Human Female Anatomy
             and Physiology   103

Chapter 7: Fertility, Conception, and Sexual
             Differentiation   135

Chapter 8: Pregnancy and Childbirth as a
             Bio-Cultural Experience   159

Chapter 9: Early Childhood Sexuality   177

Chapter 10: Puberty and Adolescence   207

Chapter 11: Topics in Adult Sexuality: Human Sexual
             Response and Birth Control   235

Chapter 12: Topics in Adult Sexuality: Life-Course
Issues Related to Gender Identity,
Gender Roles, and Aging   281

Chapter 13: Sexual Orientations, Behaviors, and
Life-Styles   313

Chapter 14: HIV Infection and AIDS   335

Chapter 15: Conclusion and Summary   379

Notes   383

Glossary   389

Bibliography   407

Index   459

Illustrations

Fig. 1.1   Culture as Architecture   16

Fig. 3.1   The Evolutionary Relationships of Humans
           and Our Closest Living Relatives   44

Fig. 4.1   Sexual and Reproductive Cycles   70

Fig. 4.3   H-P-G Axis Graphic for Male and Female   72

Fig. 5.1   Man's Hairline   83

Fig. 5.2   Woman's Hairline   83

Fig. 5.3   Circumcised and Uncircumcised Penises   87

Fig. 5.4   Pelvic Structures   88

Fig. 5.5   Scrotal Wall   96

Fig. 6.1   External Genitalia   108

Fig. 6.2   Uterine Development and Musculature   118

Fig. 8.1   Fetus at 9 Weeks, 14 Weeks, and 20 Weeks   161

Fig. 8.2   Major Stages in the Birth Process   162

Fig. 8.3   Baby Emerging; Natural Birth   164

Fig. 8.4   Mom and Dad with Twins; Natural Birth   165

Fig. 9.1   The Kinship Diagram   183

Fig. 9.2   Kinship Principles and Groups   184

Fig. 9.3   Matrilineal and Patrilineal Systems   185

Fig. 9.4   Parallel and Cross Cousins   191

Fig. 12.1   Dallas Denny   289

Fig. 12.2   Dallas Denny   289

Fig. 12.3   JoAnn   290

Fig. 12.4   Gitte-Maria   291

Fig. 12.5   Gwendolyn   292

Fig. 12.6   Wendi   293

Fig. 12.7   Wendi   293

Fig. 12.8   Merissa Sherrill Lynn   294

Fig. 12.9   Michelle   295

Fig. 13.1   Gender Role Categories   329

Fig. 14.1   Ethnic Distribution of AIDS   340

Fig. 14.2   Women and AIDS   341

Fig. 14.5   Safer Sex Kit Brochure   358

Tables

Table 4.2   Comparative Anatomy   71

Table 4.4   Hormones Involved in H-P-G
             Axis Functioning   75

Table 7.1   Summary of Anomalies of
             Prenatal Differentiation   148

Table 10.1   Age of Sexual Intercourse   223

Table 11.1   Who Uses What Kind of Contraceptive?   259

Table 11.2   First Year Failure Rates of Birth
             Control Methods   260

Table 14.3   Common STDs   348

Table 14.4   Activities that Cover the Safer Sex Continuum   357

Table 14.6   HIV Continuum   366

Preface

This preface is a welcome to all our potential readers—students, professors, and researchers—of this human sexuality text. For those of you who teach or conduct research in human sexuality, you may be wondering why the need for yet *another* undergraduate human sexuality text, especially since there are several fine texts on the market. We acknowledge that there are some very good texts available dealing with current U.S. sexual behavior and attitudes (eg. Kelly's *Sexuality Today: The Human Perspective;* Allgeirer and Allgeirer's *Sexual Interactions;* Francoeur's *Becoming a Sexual Person*).

We also believe that since most of the current undergraduate texts are *not* written by anthropologists, there are dimensions of human sexuality that are not covered by these texts. Most noticeably, our text does *not* focus on a view of sexuality as anchored in western behaviors; nor do we wish to inculcate a view of nonwestern peoples as "exotic others." Our behaviors and beliefs are as exotic and alien to people in other cultures, or even to people of previous generations in our own culture, as their behaviors are to us. This book incorporates an anthropological perspective that is unique and different from most of the available texts. Our book integrates evolutionary, cross-cultural and bio-cultural dimensions to human sexuality. We examine patterns of sexuality as they occur in a variety of cultures, including our own, as opposed to a conflict, issues approach of many late twentieth century U.S. sexuality texts.

Translated, this means that we look at modern human sexual behavior as having evolved from a primate heritage. It has changed through time and space physically and behaviorally as a means of adapting to our specific needs as a large-brained, upright organism which depends on learning as its primary survival (adaptive) strategy. We compare ourselves to our nearest relatives—the non-human primates and to other human groups to gain a better understanding and insight into what we share as a human species, as well as carryovers from our primate heritage. We integrate, as much as possible, the biological and learned aspects of sexuality through the life cycle from conception through old age.

The late twentieth century ends with us facing daunting challenges as a species, members of groups and as individuals. HIV infection and AIDS are global, pandemic health concerns. HIV infection cross-culturally and in the U.S. is spread primarily through sexual contact. There will not be a cure or widely available preventative vaccine before the twenty-first century, although it remains on the horizon. Thus, prevention of infection through education coupled with behavior and attitude changes which are geared to the specific needs, perceptions, and values of people at risk continues to be important. This effort necessitates a cross-cultural, relativistic anthropological approach.

We are a human community. Late twentieth century technology include computers and supersonic satellite communication as well as western medical technology that can be available on a worldwide basis make diversity, i.e., a variety of value systems, behaviors, and perceptions a reality of U.S. and international life. To survive as a species we need to appreciate, understand, and accept difference and use it to enhance our humanity as individuals, groups, and a species. To meet these challenges also involves a cross-cultural, comparative anthropological perspective. Therefore our approach in this book has both theoretical as well as applied dimensions in trying to understand ourselves as sexual beings.

To do this, however, we have recognized certain limits to this text. One, we do not cover every aspect of modern U.S. sexuality. Two, this is not a text in late twentieth century U.S. sexual behavior. While we include U.S. sexuality as part of human sexuality, we do not focus on it. Simultaneously, we include sexual behaviors and attitudes of nonwestern people through time and space. We try to incorporate sexuality as part of their world view and sociocultural life—as integrated and related to political, economic, and social structures.

Those sexual behaviors and aspects of the life cycle which we discuss are those which are human—i.e., found in all human groups, although arranged, defined, and constructed according to the specific demands of a given group. For example, fertility issues, pregnancy, and childbirth are pan human concerns. We discuss how the U.S. and other societies culturally define and manage these life cycle issues. Childhood sexuality is part of human sexuality. So, we examine how various cultures, including our own, channel and regard childhood sexuality as a prelude to adult sexuality. The transsexual and transveste social identities are two socio-sexual roles found cross-culturally and in the U.S., and are responded to very differently in the U.S. and the nonwestern societies in which they occur.

On the other hand, fetishes and paraphilias (i.e., focusing exclusively on particular objects or body parts as the primary or only means of sexual arousal) appear not to be particularly widespread cross-culturally. They are *not* dealt with in this text. While they are not dealt with specifically in this text, we do provide an important framework for furthering the understanding of such western based categories, or "culture-bound" syndromes. We are aiming for a better understanding of human sexuality—that which we share as a species, and not exclusively what is occurring in middle class U.S. society.

Given this perspective, instructors, researchers, and students may use this text in a variety of ways. It can be used as a basic text with supplemental or recommended readings in U.S. sexuality. Or it may be used as a supplemental text for those most comfortable in dealing with human sexuality as a late twentieth century U.S. phenomenon. It can also serve as a resource for researchers and teachers who want to incorporate evolutionary or cross-cultural data as part of their writing and teaching.

We hope the approach taken here will be of use and value to you in your understanding of this highly varied and data-rich topic of human sexuality.

One final *caveat* is in order that relates to the speed with which science progresses. There is a necessary time lag in the publication of a manuscript, so that when a book comes into print it cannot, by virtue of the production process, contain the most recent findings of the months prior to its publication date. Scientific research surges forward at a rapid rate in this postmodern world in which we live. As a consequence, we would like to turn the readers' attention to several chapters in which new scientific evidence has occurred and which will continue to occur as *Perspectives on Human Sexuality* is in the process of production and publication. We would like the reader to note that there have been recent discoveries related to the discussion of terrestrial adaption in chapter 3 "Evolution of Biological Structures Related to Modern Sexual Behavior." Recent research has seriously challenged the view that bipedalism (walking with erect posture on two legs rather than quadrapedally) occurred as a result of adaptation to savannah life in open terrain as opposed to a forest environment. This is known as the savannah hypothesis. A more recent hypothesis thrown into the bipedalism hopper of theories, argues that while the savannahs may have played some part in the evolution of bipedalism, it is not the primary causal factor. Very recent findings of early human ancestors known as Australopithecines suggests that they lived in a broad spectrum of environments rather than being relegated to only savannah life.[1]

We would also life to call the reader's attention to chapter 11; "Topics in Adult Sexuality," in which we discuss birth control and chapter 14 which focuses on HIV infection and AIDS. We encourage the reader to check with his or her physician and/or contact Planned Parenthood Federation of America, 810 Seventh Avenue, New York, NY 10019, telephone: 212-541-7800 for any new developments in birth control methods in the recent months preceding and encompassing the production and publication of this book.

Treatments for HIV infection and AIDS are proceeding rapidly. In recent months, AIDS clinicians have had very positive results in treating AIDS through the combination of drugs, known as "the cocktail." Between January and July 1996, five of the nine drugs available for AIDS treatment in the United States have been introduced. These new drugs have offered optimism for the HIV/AIDS landscape, in particular the protease inhibitors which suppress a part of HIV's replication process when used with two older AIDS drugs. For the most recent information on HIV infection and AIDS, we recommend contacting the Centers for Disease Control, Division of Sexually Transmitted Diseases, 1600 Clinton Road, Atlanta, GA 30329. In addition, numerous sources for information on HIV/AIDS are available, among them are the National AIDS Hotline 1-800-342-2437, the ACLU AIDS Project (212) 944-9800 (ext. 545), the National AIDS Clearinghouse 1-800-458-5231, American Foundation for AIDS Research (212) 682-7440, and the National Lesbian and Gay Health Foundation (202) 797-3578.

Acknowledgments

While we recognize and accept the importance of formally acknowledging those of you who have helped us write and publish this text, we also know this statement touches the surface of what you have given us. Each of us will thank our respective sources of support.

Patricia Whelehan wants to express her appreciation and gratitude to SUNY Potsdam's administration and the anthropology department for their one semester course reduction during the earlier stages of manuscript preparation; as well as to the Clerical Center during its existence for its early involvement in draft preparation. Specifically I would like to thank: Ms. Kathy Tyler of the Anthropology Department, SUNY Potsdam, for her invaluable help since 1992 with the manuscript; students Joyce Rice who initially volunteered editorial assistance, Margaret LaCommare-Lewis, Betsy Cosentino, and Rebecca Kibbe who helped with the glossary and bibliography; Dan Batcher who retyped several chapters and Suzanne Gates who worked on permissions. I'd also like to thank my daughter, Rachel Galgoul, for her insights and contributions, as well as all of the students of ANTC 150 and 346, Human Sexuality 1 and 2 over the past twenty years. Your participation, involvement, and challenge sparked the initial interest in writing this text.

Jose Andres Gonzalez del Valle, to whom this book is dedicated, provided substantive criticism of the HIV and sexual orientation chapters. To all of you in the sexual subcultures and the HIV community in New York, California, the Dominican Republic, and Lima, Peru, I wish to express my appreciation for your acceptance. I hope I have accurately represented your points of view in this book. Lastly, I'd like to thank SUNY Press for adopting this as one of their publications.

Anne Bolin wants to thank the Elon College Research and Development Committee for course release time in the initial stages of research for this manuscript. I am grateful to Tom Henricks, sociologist and Associate Dean of Social Sciences (during the life-course of this book) for his encouragement of this project and his general support of research and writing as important components of teaching excellence. I am also indebted to my colleagues in the Department of Sociology. Chair Larry Basirico is an inspirational scholar, leader, and great friend. Larry gave me valuable feedback, advice, and gifts of generous joy and laughter. Special thanks are due to Paul Shankman of the University of Colorado who has taught me so much during the course of my career as an anthropologist. His dedication to scholarship and rigor has remained an invaluable touchstone for me.

Linda Martindale of Elon College played an essential role in the life of this manuscript. I offer her my gratitude for her expertise in manuscript preparation, working her computer magic and her editorial skills. The students in my Anthropology of Sex course deserve special mention as well for their encouragement and enthusiasm for this book. Important technical assistance was provided by the Elon College Media Services Department. The Elon College library staff played an essential role in locating resources. Special thanks are given to Teresa LePors for her expertise in library research. Greg Babcock, my spouse, high school sweetheart and partner, has provided me an environment in which to write that is filled with love and harmony. And my family has always been an important source of emotional and intellectual support for my academic efforts. My deepest appreciation is given to my mother, Vivian Bolin, who read and critiqued my drafts of this manuscript.

OVERVIEW

Chapter 1

Introduction: History and Context

This chapter:

1. Introduces human sexuality from a bio-cultural perspective.

2. Discusses how the social control of human sexuality forms the fundamental basis for the formation of human groups and group life.

3. Discusses ethnographic and comparative approaches to the cultural patterning of human sexuality. Highlights anthropologists such as Malinowski, Benedict, Mead, Ford and Beach, Martin and Voorhies, and Frayser.

Chapter 1

Introduction: History and Context

The Anthropological Perspective

In one ruling by the Supreme Court, sex has been declared as "a great and mysterious motive force in human life [that] has indisputably been a subject of absorbing interest" (in Demac 1988:41). As children we have heard our parents speak euphemistically about the "birds and the bees" and as adolescents we may have shared late night discussions with our friends about the "secrets" of intimacy. As adult Americans our concerns are expressed in an array of new self-help books on the subject flooding the market every year. We even have two new disorders of human sexuality that have captured the imagination of the news media: sexual addictions and inhibited sexual desire (lack of interest in sex). Obviously the subject enthralls more than Masters and Johnson (1966) or Dr. Ruth.

Our intent is to offer a unique way of understanding ourselves as sexual beings through the perspective of anthropology. Some of our readers may not be familiar with what anthropology is and might be wondering what exactly "Raiders of the Lost Ark" has to do with sex. For those of you unfamiliar with anthropology, we welcome you to an exciting new viewpoint. For those of you majoring or minoring in anthropology, we hope that our text inspires you to conduct further research on the subject of human sexuality.

Sex as Biology

Confusion about what anthropology is stems from the interdisciplinary nature of the field. An anthropological approach is one that incorporates an understanding of humankind as biological, as well as cultural beings. We use the term bio-cultural to describe our perspective. While bio-cultural approaches in anthropology may not be appropriate for all the subjects anthropologists might research, for a number of topics such a view lends a fuller and more complete understanding. Bio-cultural perspectives are widespread in fields such as medical anthropology, biological anthropology, and the anthropology of sex and gender to name just a few of our many specializations.

The interweaving of biology and culture into a bio-cultural perspective is the distinguishing feature of this textbook and the theme that unifies the multiplicity of anthropological studies of human sexuality. We are not suggesting that these are the only two dimensions for interpreting sex; indeed, sex has a very important psychological component as well. As anthropologists, we regard the psychological component as part and parcel of culture that shapes our personalities in characteristic ways, yet also allows for the diversity of individuals as unique genetic entities.

The term "sex" has many meanings. Sex is part of our biology. It is a behavior that involves a choreography of endocrine functions, muscles, and stages of physical change. It is expressed through the "biological sex" of people classified as male or female (Katchadourian 1979). Despite this physiological component, the act of sex cannot be separated from the cultural context in which it occurs incorporating meanings, symbols, myths, ideals, and values. Sex expresses variation across and within cultures.

An anthropological definition of sex is necessarily a broad one that includes the cultural as well as biological aspects of sex. We shall offer you a definition of sex, but urge you to remember that defining sex is far more complex than our definition suggests. For example, our definition cannot limit sex to only those behaviors resulting in penile-vaginal intercourse, for by doing that we would eliminate a variety of homosexual, bisexual, and heterosexual behaviors that are obviously sexual but not coital. Therefore, we shall define **sex** as those behaviors, sentiments, emotions, and perceptions related to and resulting in sexual arousal as defined by the society or culture in which it occurs. We qualified our definition by referring to cultural definitions of sexual behaviors since these differ a great deal among ethnic groups and among different cultures. For example, petting as we know it in western society is not universal, that is, it is not necessarily considered a form of arousal among all other peoples of the world. As you read this text, you will begin to broaden your horizons of understanding yourself, your own society and the multicultural world in which we live.

Anthropological Perspectives on Human Sexuality

The study of human sexuality is a cross-disciplinary one. Six major perspectives dominate the study of human sexuality. Included are: the biological with a focus on physiology; the psychosocial with an emphasis on the developmental aspects of sexuality and the interaction of cognitive and affective states with social variables; the behavioral that stresses behavior over cognitive and emotional states; the clinical with a concern with sexual disorders and dysfunctions; the sociological, with a focus on social structures and the impact of institutions and socioeconomic status factors on sexual behavior; and the anthropological which includes evolutionary and cultural approaches with emphases on sexual meanings and behaviors within the cultural context. By **culture** we mean: the skills, attitudes, beliefs, and values underlying behavior. These are learned by observation, imitation, and social learning.

In today's global community it is increasingly important for us to incorporate multicultural perspectives in our knowledge base. Since this approach is at the heart of anthropology, we offer a brief historical overview of some of the more well-known cultural anthropologists who have shaped the study of human sexuality. The contributions of anthropologists studying the evolution of human sexuality are discussed in chapter 3.

Anthropology as a discipline developed in the nineteenth century. From its inception, anthropologists have been interested in the role of human sexuality in evolution and the organization of culture. Darwin, most well known for the biological theory of evolution (see chapter 2 for definition and chapter 3 for discussion), also formulated theories on culture that included ideas on human sexuality. These were presented in *The Descent of Man and Selection in Relation to Sex* (1874, orig. 1871). Darwin argued that morality is what separated humans from animals. In his theory of morality, Darwin regarded the regulation of sexuality as essential to its development. According to Darwin, marriage was the means for controlling sexual jealousy and competition among males. In the course of moral evolution, restrictions of sexuality were first required of married females, then later all females and finally males restricted their own sexuality to monogamy. Darwin's approach incorporated notions of male sexuality and assertiveness, and female asexuality. These views reflected Darwin's own cultural beliefs about sex and gender (Martin and Voorhies 1975:147–149).

Other nineteenth century anthropologists also produced theories of social evolution that included the regulation of sexuality. John McLennan (1865), John Lubbock (1870) and Louis Henry Morgan (1870) conceived of societies as having evolved through stages. These stages represented increasing restrictions on sexuality as societies progressed from primitive stages of promiscuity to modern civilization characterized by monogamy and patriarchy (Martin and Voorhies 1975:150). These theories were flawed by thinking that esteemed western European culture was superior and viewed social evolution as an unwavering linear trend of "progress."

The twentieth century brought new approaches to the study of human sexuality as anthropology shifted from grand evolutionary schemes with little rigor to empirically oriented studies. This turn led to a new methodology for which anthropology has gained acclaim. Bronislaw Malinowski is the acknowledged parent of this anthropological research method known as **ethnography**. Ethnography is the research method of participant observation in which the anthropologists becomes entrenched in the lives of the observed. The ethnographic method serves as the basis for an ethnography, the detailed study of the culture of a particular group of people. Malinowski is known for his analysis of sex as part of the ethnographic context. His groundbreaking work entitled *The Sexual Life of Savages in North-Western Melanesia: An Ethnographic Account of Courtship, Marriage and Family Life Among the Natives of the Trobriand Islands, British New Guinea* was first published in 1929. While others were writing in the 1920s on the subject of indigenous peoples and their sexuality, unlike Malinowski theirs was not based on firsthand research but rather missionary and travelers' reports or short-term field projects (Weiner 1987:xiii–xiv). Malinowski's two-year-term living with the Melanesian Trobriand Islanders and his scientific and systematic methods of data collection left an important legacy for the field of anthropology and the study of human sexuality.

Malinowski was interested in the relationship of institutions to cultural customs including sexual behaviors. His perspective stressed the importance of the cultural context and emphasized how social rules ordered sexuality among the Trobriand Islanders. What appeared to Europeans as unrestrained sexuality were in actuality highly structured premarital sex rules and taboos based on kinship classification (Weiner 1987:xvii). Malinowski seriously challenged the dominant nineteenth century cultural evolutionism of McLennan, Lubbock, and Morgan. He rejected the notion that early human life was represented by sexual promiscuity. The Trobriand Islanders illustrated that even the most nontechnologically complex peoples regulated their desires through systems of kinship. Rather than promiscuity as a prior condition, Malinowski focused on the ordering of sexual relations in creating the family (Weiner 1987:xxv–xxvi).

Malinowski was also influenced by another trend impacting anthropology: that of psychoanalysis. He was impressed with the psychoanalytic openness to the study of sex, but was critical of Sigmund Freud's theory of the incest taboo and the Oedipus complex. In a nutshell, Freud's argument is that unconsciously little boys experience a desire to marry/have sex with their mothers and murder their fathers. (See chapter 9 for further discussion). In *Sex and Repression in Savage Society* (1927), Malinowski ". . . argued that Freud's theory of the universality of the Oedipus complex had to be revised because it was culturally biased. Freud based his theory on the emotional dynamics within the patriarchal western family" (Weiner 1987:xxi). This resulted in a heated debate with psychoanalist Ernest Jones. Malinowski again argued that the Oedipus complex was a result of the western patriarchal family complex. The Trobrianders presented quite a different picture from the western nuclear family because the Trobriand culture is a **matrilineal** one; that is, people traced descent through their mother's family. This produced different family dynamics so that Malinowski concluded that the Trobrianders were free of the Oedipus complex. Unfortunately his work did not influence the psychoanalytic position to any great degree.

Ruth Benedict and Margaret Mead loom large in the history of anthropology and in their respective contributions to the study of sex. Both were students of Franz Boas, the parent of American anthropology. Benedict's contribution continues to be felt today. Her perspective, in revised form, is embedded in contemporary anthropology in the concepts of ethos (the "approved style of life") and world view (the "assumed structure of reality") (Geertz 1973:126–141). Benedict's *Patterns of Culture*, published in 1934, offered an approach in which cultures were regarded as analogous to personalities. She stressed how each culture produced a unique and integrated configuration. This was known as the configurational approach (Benedict 1959:42–45).

That Benedict was light years ahead of her time was demonstrated in her concluding chapter where she reiterated points from her paper "Anthropology and the Abnormal" (1934). She was concerned with individuals whose temperaments were not matched to their cultural configuration and the psychic costs to those such as homosexuals who were "not supported by the institutions of their civilization" (1946:238 in Bock 1988:52). She proposed that "abnormality" was not constant but is rather culturally constituted. She suggested what, at that time, was a radical view: tolerance for non-normative sexual practices such as homosexuality. Implicit in her view is that sexuality is no different than any other social behavior, it was culturally patterned. Benedict argued that ". . . in a society that values trance, as in India, they will have supernormal experience. In a society that institutionalizes homosexuality, they will be homosexual" (1934:196 in Singer 1961:25). Benedict challenged prevailing notions of homosexuality as pathology. In 1939, in her "Sex in Primitive Society," she concluded that homosexuality was primarily social in nature, shaped by the meanings of gender and sex roles (Dickermann 1990:7).

For the study of human sexuality, Benedict's major contribution was that sex, which is a part of culture, is patterned, fitting into the larger society, the cultural whole or the **gestalt.** The configurational approach was certainly not without flaws and anthropology has moved well beyond regarding cultures as personalities. But, Benedict has left an important legacy for anthropology in her emphasis on patterning and cultural holism. For the field of sexology, Benedict was bold and unafraid in her perspective on sexual variation.

Margaret Mead was also an important and powerful figure in anthropology and sexology. Before her death in 1978 she was more widely recognized for her work than any other anthropologist in the world. In numerous books and articles, Mead addressed the subject of sex and gender. While her contributions are many, we shall focus on her first book *Coming of Age in Samoa* (originally 1928) investigated when she was not yet 24 years old.[1]

Mead was a proponent of cultural explanations for understanding human behavior. She explained this approach by saying:

> It was simple—a very simple point—to which our materials were organized in the 1920's, merely the documentation over and over of the fact that human nature is not rigid and unyielding, not an unadaptable plant which insists on flowering or becoming stunted after its own fashion, responding only quantitatively to the social environment, but that it is extraordinarily adaptable, that cultural rhythms are strong and more compelling than the physiological rhythms which they overlay and distort . . . We had to present evidence that human character is built upon a biological base which is capable of enormous diversification in terms of social standards (Mead 1939:x in Singer 1961:16).

In *Coming of Age in Samoa*, her commentary addressed female adolescence in Samoa as well as in the United States. She proposed that the turbulence of the American girls' adolescence was not typical of adolescence throughout the world. Mead was responding to a popular biological theory of adolescent stress and storm believed to be caused by the changes in hormones during puberty. Her study of Samoan adolescence provided a very different picture. Unlike American adolescence, for the Samoan youth, this was not a period of turbulance and high emotion. Based on evidence of a carefree Samoan adolescence, Mead reasoned that the conflict experienced by the American teenagers was due to culture rather than hormones. The latter part of *Coming of Age in Samoa* explained the strife of American adolescence as a cultural phenomenon. Mead offered cultural explanations. For example, she identified the importance of rapid culture change in American society as contributing to the adolescent unrest.

In contrast, the Samoan girls' adolescence was conflict free. This was due to Samoan culture which was relatively homogeneous and casual. So casual that according to Mead the young woman:

> defers marriage through as many years of casual love-making as possible . . . The adolescent girl's total interest is expended on clandestine sex adventures . . . to live with as many lovers as possible and then to marry into one's village . . . (Mead 1961:157).

Samoan society was one in which extremes in emotion were culturally discouraged. It was characterized by casualness in a number of spheres including sexuality, parenting, and responsibility. In contrast to western culture, the young Samoan woman's sexuality was experienced without guilt. She concluded that the foundation of this casual approach to sex and painless adolescence could be explained by the following: 1) a lack of deep feeling between relatives and peers, 2) a liberal attitude toward sex and education for life, 3) a lack of conflicting alternatives and 4) a lack of emphasis on individuality. In this work, she established the importance of the study of women when little information was available (Howard 1983:69). She also challenged notions of biological reductionism that even today are all too often used to support status quo politics.

While the approaches of Malinowski, Benedict, and Mead contributed to the creation of the ethnographic study of sexuality with an emphasis on the cultural, Clellan S. Ford and Frank A. Beach's *Patterns of Sexual Behavior* deserves credit in 1951 for offering the first synthetic study that incorporated biological, cross-cultural, and evolutionary considerations. Their work is distinctive for its inclusion of homosexual and lesbian data, a trend continued in Gregerson's 1994 *The World of Human Sexuality: Behaviors, Customs and Beliefs.* According to Miracle and Suggs (1993:3), Ford and Beach's book is "[t]he single most important and provocative work on sexuality to date . . . It also provided the intellectual—if not the methodological—foundation for the subsequent work of Masters and Johnson." (See chapter 11 for discussion of Masters and Johnson). *Patterns of Sexual Behavior* integrated information from 190 different cultures as well as provided comparative data on different species with an emphasis on the primates (humans, apes, and monkeys). Their work includes an encyclopedic collection of sexual behavior cross-culturally. For example, Ford and Beach offer discussion and information on sexual positions, length (time) of intercourse, locations for intercourse, orgasm experiences, types of foreplay, courting behaviors, frequencies of intercourse, methods of attracting a partner, among numerous other topics.

Ford and Beach's study of human sexuality employed the cross-cultural correlational method. This is a statistical method for comparing attributes (variables) in large samples of cultures (Cohen and Eames 1982:419). This approach is valuable for testing hypotheses about human sexuality, establishing patterns and trends, and formulating generalizations. Their study relied on ethnographic data that is collected and coded in the HRAF files (the Human Relations Area Files). HRAF is a rigorous classification scheme for information on the world's societies. Categories of information for over 1,000 societies are now coded and available to researchers.

The cross-cultural correlational statistical method was subsequently used by Martin and Voorhies in *Female of the Species* (1975). Like Ford and Beach, Martin and Voorhies included evolutionary and biological issues. Their focus was broader in that they were interested in the relationship of human sexuality to gender status/ roles, social organization, and type of subsistence (how people make a living). Martin and Voorhies tested hypotheses to arrive at generalizations about the relationships of these factors. In a sample of 51 foraging societies Martin and Voorhies found that 30% of them allowed premarital sexual experimentation (1975:188–189). This pattern was related to matrilineality (where descent is traced through the mother's side of the family) and matrilocality (where the couple resides in the village of the wife's mother). Their studies of horticultural groups also revealed a statistical correlation between matrilineal societies and sexual permissiveness toward premarital sex, while patrilineal (tracing descent through the father's side) societies tended to control female premarital sexual behavior (1975:246–247).

There are many research applications for this methodological approach to sexological research. For example, Schlegel and Barry in *Adolescence: An Anthropological Inquiry* (1991) report that premarital restrictiveness occurs in societies in which a dowry is given (wealth from the bride's family is included in the marriage transaction). They conclude that "[f]amilies guard their daughters' chastity in dowry-giving societies in order to protect their property (dowry) against would-be social climbers and to ensure that they can use their daughters' dowries to attract the most desirable sons-in-law" (Schlegel and Barry 1991:116). Chastity rules guard against a lower status man impregnating a higher status women and thereby making claim on her dowry and inheritance by trapping her into marriage. In this way property exchange and status considerations are factors in restricting premarital sexuality (Schlegel and Barry 1991:117–118). Davis and Whitten report that the general pattern found in HRAF studies such as these is that sexual restrictions tend to be associated with complex societies (1987:74).

Frayser's *Varieties of Sexual Experience* (1985) is in the tradition spawned by Ford and Beach incorporating the cross-cultural correlational approach with biological and evolutionary concerns. Frayser presents an integrated model in which human sexuality is regarded as ". . . a system in its own right, related to but not subsumed by social, cultural, psychological, and biological factors" (in Frayser and Whitby 1987:351). For Frayser, although the cross-cultural record reveals an almost infinite variety in sexual expression, there is continuity with our evolutionary past. In regard to evolution, Frayser examines cross-species sexuality, particularly that of our close relatives, the non-human primates. For example, she points out that human sexuality is distinguished by unique sexual and reproductive attributes; these include the ability for sexual arousal that is not limited to estrus ("heat"), and the evolution of the female orgasm. These capabilities are present in our relatives to a limited extent, but emerge full blown in humans and may be linked to extraordinary amounts of non-reproductive sexual behavior among humans in contrast to other animals.

Frayser has distinguished the social and cultural aspects of human sexuality in terms of the **social system** defined as patterned interactions. The social system is contrasted to the **cultural system** which is defined as the patterned beliefs and meanings that influence sexual expression (Frayser 1985:7). This model is one in which the biological, the social, and the cultural system converge to influence the sexual system. It is a valuable approach for understanding sexual patterning and for recording the continuities and heterogeneity within and between cultures.

Sex as Culture

The regulation of human sexual expression as to when, where, how, and who—may serve diverse socio-cultural goals. George Peter Murdock's pioneering study, *Social Structure* (1949), offers us a classic approach to the different ways that the regulation of sexuality contributes to the organization of cultures. In all societies sexual access among members of a society is regulated. The most obvious example of this is the incest taboo. With an almost universal prevalence, the **incest taboo** prohibits sexual access between siblings and between siblings and their parents. But, even those societies which have allowed incest have regulations surrounding it that are integrated in the wider social organization and belief system. The exceptions include Hawaiian royalty, kings and queens of ancient Egypt, and Inca emperors. These elites were regarded as so powerful and sacred that only their very close relatives had the equivalent status to qualify as a mate and to perpetuate the lineage. Such sexual unions and marriages were not allowed, however, for the population at large (Murdock 1949:13).

Rules for sexual access also extend beyond the immediate nuclear family. **Exogamy** is a rule requiring that people marry outside their group, while **endogamy** specifies marriage within the group (not the immediate family). These rules create kin groups through different kinds of restrictions on sexual access. Rules of exogamy and endogamy are defined by reference to marriage. This illustrates how sexual ideologies are integrated in the social organization of kin groups. One should, however, not make the error that sex and marriage are always equated. This is a mistake often found in the literature on human sexuality, but one seldom made by the people involved in extramarital affairs. "**Marriage** is a publicly recognized union between two or more people that creates economic rights and obligations within the group . . . and guarantees their offspring rights of inheritance" (Crapo 1987:148). It is regarded as an enduring relationship and includes sexual rights (Ember and Ember 1988: 13). Murdock (1949:8) offers clarification:

> Sexual unions without economic co-operation are common, and
> there are relationships between men and women involving a divi-

sion of labor without sexual gratification, e.g., between brother and sister, master and maidservant, or employer and secretary, but marriage exists only when the economic and the sexual are united in one relationship, and this combination occurs only in marriage.

Ford and Beach's pioneering *Patterns of Sexual Behavior* (1951) proposed that sexual partnerships consist of two types: **mateships** defined in the same way as marriages; and **liaisons** ". . . less stable partnerships in which the relationship is more exclusively sexual" (1951:106). Sexologists and anthropologists generally subdivide human liaisons on the basis of their premarital or extramarital character (Ford and Beach 1951:106).

The regulation of sexual partnerships makes it is possible to define groups of people by relationships based on offspring and kinship. These kin relationships are formalized through marriage systems. Sexual prohibitions function to ". . . minimize competition among relatives and to increase the bonds of cooperation and friendship between neighboring groups" (Crapo 1987:61). Because descent is important for a number of reasons such as inheritance, obligations, and affiliations, we can regard sexual unions as having the potential to shape kin group formation. Sexual access therefore defines kin groups. The importance of sexuality is socially recognized through marriage as an institution with sexual rights and obligations. But, it should be kept in mind that there is a great deal of sexual activity that occurs prior to and outside marriage, and this includes sexual activities between people of the same gender, ritual and ceremonial sex as well.

Societies differ as to their tolerance of premarital and extramarital activities and the conditions under which it is acceptable and/or prohibited. According to Broude and Greene's (1976) survey of the cross-cultural record, in 69% of the societies studied, men "commonly" participated in extra-marital sex while in 57% of the societies women did so as well. This leads us to another thorny issue for sex researchers, the contrast between **ideal** and **real** culture. The ideal culture or normative expectation is that in 54% of the societies extra-marital sex is allowed only for men, while only 11% allow it for women. But the data suggests that many more people actually violate this ideal, particularly in the case of women.

In summary, human sexuality is a central force in the origin of kin groups. In Murdock's words: "All societies have faced the problem of reconciling the need of controlling sex with that of giving it adequate expression" (1949:261). The regulation of sexual relations is the basis for descent and inheritance, critical factors for human societies in the maintenance of social groups. Yet, sex and marriage do not necessarily "go together" like a horse and carriage. Sex is not the central factor in the bonding of two individuals through marriage. To think so is to engage in a bias shaped by recent modern U.S. views of marriage. Sex is indeed critical for kin groups and their perpetuation; and while sex is a right and an obligation in marriage, it is not necessarily the basis upon which marriages are made. Economic cooperation emerges as an important factor in marriage both in evolutionary terms and in the cross-cultural record. This will become more evident in our discussion of "The Patterning of Human Sexuality."

The Patterning of Human Sexuality: The Bio-Cultural Perspective

Human sexuality has a foundation in human biology which provides us with certain inherited potentialities. " . . . The inherited aspects of sex seem to be nearly formless." It is only through culture that sex assumes form and meaning (Davenport 1976:161).

Our human biological wiring is very different from what we think of as animal instincts. For example, the drive for food that allows us to survive is fulfilled through learning how to get food; so in some cultures people collect food, some fish and yet others like us go to the grocery store. The drive for sex is also shaped by culture and is very unlike a mating instinct. When a female animal comes into heat, she automatically (through hormonal mechanisms) becomes sexually responsive and follows her mating instinct. Humans, however, may ignore their drives; for example, Catholic priests and nuns deny their sexuality in order to live in celibacy (Scupin and DeCorse 1992:164). Others delay sexuality until marriage which may not occur until their twenties or later.

Human biological predispositions are not " . . . rigidly determined . . . They may orient us in particular directions in pursuing certain goals, but they do not determine our behavior in a mechanical fashion without learned experiences" (Scupin and DeCorse 1992:164). This biological underpinning to our sexuality and other behaviors is part of what is called an **open biogram**, "an extremely flexible genetic program that is shaped by learning experiences" (Scupin and DeCorse 1992:164). Through socialization humans acquire their culture. This capacity to learn and to adapt nvirons is a part of our unique bio-cultural evolution as Ve can say that our biology sustains us as cultural beings ng us with an unusual capacity to learn.

Sexual behavior is culturally patterned; it is not accidental or random but is integrated within the broader context of culture and is intermeshed in a web of other cultural features as we have seen in our discussion of sex, marriage, and kinship. A number of cultural characteristics are associated with patterns of human sexuality. These may include: the level of technology, population size, religion, economics, political organization, medical practices, kinship structure, degree of acculturation and culture change, gender roles, power and privilege (stratification). Consequently, larger cultural patterns are important in shaping reproductive and non-reproductive sexual behaviors and values in a society. Sexuality is patterned across cultures in relation to these variables as well as within a culture. Davenport suggests that sex is molded by the "internal logic and consistency of the total culture. As one sector of culture changes, all other sectors that articulate must undergo adjustments" (1976:162).

Cultures are integrated systems that exist within particular environmental and historical contexts. We have discussed the biological basis of human sexuality, we offer now an overview of the cultural basis of human sexuality. To comprehend how sexuality is embedded in culture necessitates an understanding of the culture concept. We can think of culture in terms of architecture[2] (see figure 1.1). The basement represents our biology as humans; including our evolution and physiology. The floor in figure 1.1 is the foundation for understanding that cultural variation lies in how people have adapted to their environments. This includes how people make a living, their technologies, and economics. There are a number of ways people have found to survive in the world. Anthropologists have classified societies in terms of: foraging, horticulture, agriculture, pastoralism (herding), industrial, and post-industrial adaptations.

Adaptation to the environment impacts the social system including social organization and social structures which may be likened to the frame of a house. The social system is the means that people adapt to one another. It includes social organization and its elements including kinship and marriage, various institutions and structures such as religion and political organization. The social system is influenced by how people make a living through demographics, the relations of work such as age, gender, and kinship; who controls the means of production and the power relations of society. Societies have been classified in terms of their social systems as bands, tribes, chiefdoms, pre-industrial states, and industrial states.

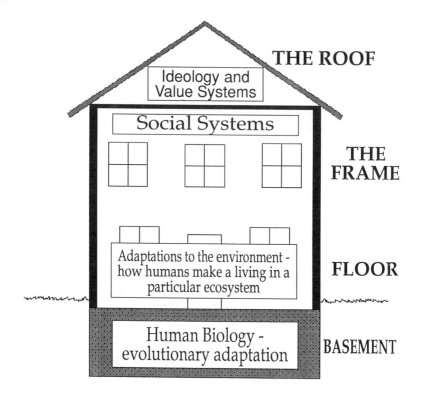

Figure 1.1 Culture as architecture.

The roof of our building may be conceptualized as the ideological value system. This is the system of meanings and beliefs in a culture. It includes expressive elements of culture like art, music, rituals, myths, folklore, cosmology. It is the meanings and beliefs behind and sustaining the patterning of cultures such as marriage norms, gender roles, courtship, etc. The foundation, the frame, and the roof are all interrelated parts of the cultural whole.

Human sexuality is part of that cultural whole. We may first encounter it in the basement in terms of our evolution and our unique human physiology. To grasp human sexuality as part of a cultural matrix we may locate it in any of our architectural levels. For example, in investigating beliefs about human sexuality, we might begin with our roof (ideology and the value system). We may observe that a particular culture has very few restrictions on premarital sex. This culture may regard premarital sex among adolescents as an amusement (Schlegel and Barry 1991:21), as part of an experiential kind of sex education, or perhaps as a way to find a marriage partner. In short, there are numerous meanings and beliefs around premarital permissiveness among cultures which allow and encourage its practice.

In order to see how the meanings behind premarital sex are part of the "internal consistency and logic" of a culture, we will want to investigate how premarital sex relates to the social system. As we saw earlier, Martin and Voorhies (1975) found a correlation between matrilineal social organization and premarital permissiveness, while patrilineal organizations restricted female premarital sexuality. From this information explanations may be proposed.

To be even more rigorous in our investigation of premarital sexual permissiveness requires an analysis of the foundation of culture, adaptation to the environment. For example, agricultural systems are associated with higher populations, stratification, and generally greater complexity. Earlier we noted in the work of Davis and Whitten (1987:74) that the greater the socio-cultural complexity, the more likely there are to be premarital restrictions on sexuality. Since subsistence type is associated with complexity, it has been proposed that foragers and matrilineal horticulturists are more likely to be permissive (Martin and Voorhies 1975).

Our analysis could go even further and include the biological. For example, research into premarital permissiveness among foragers will reveal that **adolescent sterility** may be a variable to be considered (see chapter 11). Adolescent sterility is a period of infertility among young females after the onset of menarche. They are not fertile until their late teens or early twenties. If premarital sex is allowed in societies in which this occurs, young people may explore their sexuality without the consequences of pregnancy and responsibilities of parenthood.

We offer this architectural approach to culture and sex to illustrate that culture is a complex whole in which the parts are interrelated. One can begin anywhere in our bio-cultural architecture and explore human sexuality. Some researchers prefer limiting their research to one area; for example, Masters and Johnson's investigations of human sexual response have focused on the biological. Others, such as the anthropologists cited, may be more interested in the relationship between beliefs and premarital sex practices and how these are related to social organization. Even others may want a bigger picture and explore how premarital sex norms are related to the types of subsistence adaptation.

These are the kinds of opportunities for understanding human sexuality offered by a bio-cultural perspective. We hope this approach will allow our students a greater awareness of themselves as sexual beings, a greater understanding of themselves as cultural creatures, and an appreciation of our evolutionary past and biological heritage.

Summary

Chapter 1: Introduction: History and Context

1. Human sexuality is a bio-cultural experience and phenomenon.

2. Human sexuality is a means used by human groups to achieve socio-cultural goals such as the creation of kin groups.

3. A variety of anthropological perspectives were introduced including: Malinowski, Benedict, Mead, Ford and Beach, Martin and Voorhies, and Frayser.

4. We concluded that human sexuality has two components, one in human biology which provides us with certain potentials and limitations, and the other in culture, wherein our sexuality is learned and integrated in the broader cultural context.

OVERVIEW

Chapter 2

Anthropological and Sexological Views

This chapter:

1. Compares and contrasts psychological, sociological, and biological perspectives of human sexuality.

2. Presents anthropological concepts, terms, and definitions. Specific examples from the fields of physical anthropology and cultural anthropology that are relevant to our understanding of sexology are offered.

3. Provides a definition and discusses the scope of human sexuality.

4. Offers the importance of a relativistic perspective of human sexual expression.

Chapter 2

Anthropological and Sexological Views

This chapter presents an in-depth discussion of the anthropological perspective. Anthropology is contrasted with other disciplines in order to highlight its unique contribution to the study of human sexuality. In addition, key sexological terms and definitions are presented.

Anthropological Perspectives in Contrast

Because of its interdisciplinary nature, the anthropological approach incorporates psychological, sociological, and biological views but is not limited to any one. It is precisely because of anthropology's interdisciplinary nature as well as its interest in spanning great periods of time and vast distances, that it includes, yet may be contrasted with biological, sociological, and psychological approaches to human sexuality.

Biological

The biological perspective focuses on the physiological basis of sexual behavior. Biological perspectives on human sexuality stress **essentialist** views of human sexuality. These views look at instinct as an "essential" attribute of sexuality and regard reproduction as the core of that instinct. Katchadourian and Lunde (1975:2–3) have challenged this perspective of human sexuality and counter that:

> The incentive is in the act itself, rather than in its possible consequences [reproduction]. Sexual behavior in this sense arises from a psychological "drive," associated with sensory pleasure, and its

21

reproductive consequences are a by-product (though a vital one)
... [O]ur sexual behavior involves certain physical "givens," including sex organs, hormones, intricate networks of nerves, and
brain centers.

To reduce human sexuality to an instinct to reproduce ignores
the importance of the symbolic and the affective in motivating sexual
behaviors. In addition, such a perspective ignores the role of the
group and shared cultural meanings in the survival of the species.

The biological view, however, is important for our understanding of human sexuality. The "physiological givens" serve as the
basis for our discussion of human physiology and sex, but even
they must be placed in a cultural context. For example, we might
ask if the sexual cycle of Americans, which is believed to peak in
males at about 18 or 19 years of age and in females between the
ages 35 and 40, is not shaped by cultural factors. Evidence suggests that this is the case (see Hyde 1982:342, 353). The perspective of anthropology is one that regards biology and culture as
tandem developments in human history. Anthropology emphasizes
the importance of cultural systems and learning as central features
in human evolution.

Sociological

The sociological tradition in sex research is characterized by
research that focuses on contemporary western sexuality. It looks
at the importance of "social learning, social rules and role playing"
in the expression of human sexuality (Musaph 1978:84) and stresses
patterns of social interaction. The survey method remains the most
popular sociological research technique for collecting sexological
data (Katchadourian 1985:11). Sociological research has provided a
valuable contribution to sexology by its attention to the intersec-
tion of class, status group, and the sexual experience. This approach is evident in such works as Komarovsky's *Blue Collar
Marriage* (1962) and Rubin's *Worlds of Pain* (1976).

Anthropology and sociology are very compatible perspectives.
There are, in fact, a number of anthropologists whose sexological
interests are primarily on U.S. culture. The anthropologist, in contrast to the sociologist, is trained to maintain a comparative and
bio-cultural view with reference to the cross-cultural record regardless of the research topic; whether it is studying childbirth or middle-
aged women (Jordan 1993; Trevathan 1987; Brown and Kerns 1985).
While the sociological perspective tends to focus on the importance
of social structure and patterns of interaction, the anthropological

one additionally integrates the significance of beliefs in understanding human behavior. This is essential in order to move beyond our own western cultural biases around sex which can creep into research. It is therefore useful in the study of U.S. sexology to sustain a broader frame of reference including structure, meaning, and cultural variation internally and cross-culturally. For the anthropologist this may also include an evolutionary understanding as well. However, generally speaking, anthropology and sociology are very closely related disciplines and it is often impossible to distinguish between the works of anthropologists and sociologists.

Psychological

The psychological perspective, while acknowledging the role of the neurophysiology of emotion, tends to emphasize the importance of the mental and the affective in relation to behaviors in the expression of human sexuality (Katchadourian and Lunde 1975:3). There is perhaps no one psychological perspective, but several different thrusts in a general concern with cognitive and emotional components of human sexuality.

From the Freudian perspective, the sex instinct termed *libido,* was regarded along with the death instinct, as a driving force in human behavior (Hyde 1982:6). Biology was regarded at the root of the individual's psychosexuality. Developmental aspects of sexuality were considered part of our physiological inheritance. As the individual developed, s/he encountered various stages in which sexuality and conflict were characteristic.

In contrast, psychological social learning theories acknowledge the significance of the environment in learning one's sexuality and in shaping its expression (Katchadourian and Lunde 1975:3). For example, developmental psychologists are interested in how sexuality unfolds in the child, while clinical psychologists typically deal with sexual dysfunctions and "pathologies" (Katchadourian 1985:10–11). The scope of psychological studies covers extensive areas including sexual motivation, familial and peer influence, self-esteem issues, and a number of other subjects as far ranging as gender identity and gender differences in sexual response.

While the topics may vary, the approach is usually focused on the psychology of the individual and family dynamics in the development of sexuality. The general trend in psychology is toward a far smaller or micro-level of analysis than that undertaken by anthropologists. While psychological anthropologists may be interested in the mental and emotional structures behind the expression of human sexuality in individuals, the cultural context remains an important feature for analysis. Anthropologists are more likely to be interested in the impact of culture on family dynamics or perhaps the cultural patterning of sexual dysfunction within society. For example, the psychological perspective locates dysfunction within the individual and the family milieu, in contrast to an anthropological one which locates its source in society. In addition, like anthropology, psychology emphasizes the role of learning, but unlike anthropology it does not usually consider it within an evolutionary framework or a cross-cultural one (Katchadourian 1985:11).

Anthropological Concepts Defined with Examples

Having discussed the anthropological perspective in comparison and contrasted with biological, sociological, and psychological ones, we would like to introduce you to several concepts from anthropology necessary for understanding human sexuality in a biocultural context. We have taken as key terms: evolution, the culture concept, ethnocentrism, and cultural relativism. Other anthropological terms are offered in the glossary. Anthropological terms and concepts are discussed in greater depth than other terms because of their importance. The anthropological terms of particular relevance to understanding the bio-cultural perspective are: society, primates, bonding, ethnological, ethnographic, comparative, cross-cultural, holism, emic, etic, and genetic fitness. The interested student should look these up in the glossary.

Evolution

The modern theory of **evolution** challenged the prevailing view of the seventeenth and eighteenth centuries that all species were separate and divine creations. Through his famous travels on the HMS Beagle, Darwin formulated his theory of natural selection. At this same time, naturalist Alfred Russell Wallace independently arrived at a similar conclusion: species are not separate creations but have evolved through a process of natural selection. In 1858 Darwin and Wallace together rocked the meetings of the Linnaean Society of London, and in 1859 Darwin published *The Origin of Species*, documenting and detailing the theory of natural selection (Ember and Ember 1990:14–17).

The central tenants of natural selection are straightforward. Natural selection is a mechanism of evolution whereby those individuals who are better adapted to their environment over long periods of time will be more likely to reproduce offspring who will survive than those who are not. Those individuals who reproduce themselves are more likely to pass on the traits they possess than those who are not so well adapted to their environments. This has been referred to as **survival of the fittest** and is calculated in terms of reproduction, not lifespan of the individual. Since environments do not remain stable over time, different characteristics may emerge as more adaptive so that what was adaptive in one environment at one time is no longer adaptive at another time. **Adaptation** is defined as "a process by which organisms achieve a beneficial adjustment to an available environment, and the results of that process" (Haviland 1989:59).

While Darwin knew that traits were inherited, he could not explain how new variation in populations occurred. It was the Austrian monk, Gregor Mendel, who pioneered the study of genetics. His findings were incorporated in the theories of the scientific community in the early 1900s. Studies of genetics are now an essential component in the study of evolution (Ember and Ember 1990:18–20).

The Culture Concept

The culture concept was developed by the end of the nineteenth century. The first clear definition was proposed by Sir Edward Burnett Tylor, who is considered the parent of anthropology. In 1897 he defined culture as ". . . that complex whole which includes the knowledge, belief, art, morals, law, custom and any other capabilities and habits acquired by man . . . [and woman] . . . as a member of society." By 1952 Kroeber and Kluckhohn, in reviewing the anthropological literature, found 164 different definitions of culture (Lett 1987:54–55).

We offer the following definition: ". . . information—skills, attitudes, beliefs, values—capable of affecting individual's behavior, which they acquire from others by teaching, imitation, and other forms of social learning" by Boyd and Richerson (1989:28). We have selected this definition because of its thrust on culture as cognitive, i.e., what an individual knows about her or his culture. This aspect, borrowing from a linguistic model of language in anthropology, may be thought of as **competence**. It is all the rules you need to know to act like a native of "x" group. Spradley (1987:17) calls this "cultural knowledge" and notes that it has two dimensions: **explicit** and **tacit**. Explicit culture is the knowledge we can easily communicate about, for example, knowing what our genealogies are or that we practice monogamy, albeit serial, in our marriage systems. Tacit culture is what is "outside our awareness." Hall's work on non-verbal communication, *The Hidden Dimension* (1966) has described a number of spatially oriented rules about how close to stand next to someone and when to touch or not touch that are examples of tacit culture. We tend to be aware of these rules only when they are violated (Spradley 1987:22–24) as in the case of when someone "violates your space" or "gets in your face."

But culture is not just floating around in our heads; culture is behavior too. It includes **performance**, our socially acquired lifeways and our patterned interactions; the things humans do and make. For the most part, it can be observed. This is what makes sex research so difficult. Human sex, except in certain cases of public and ritualized religious events, is private and not readily observable. In order to understand observable patterns of behavior or performance we need to know about competence, the values and beliefs underlying the behaviors.

Culture has certain characteristics that anthropologists have delineated. 1) Culture is shared in that it is composed of a group of people who share a common culture although they need not share all the attributes of the culture. 2) Culture is learned and transmitted. The process of learning one's culture in a society is "enculturation" (Ember and Ember 1990:171–172). 3) Culture is symbolic. Thus, culture can be seen as the making of meaning as meaning is arbitrarily assigned to behaviors, events, and the world in general.

Ethnocentrism

According to Bernstein (1983:183), **ethnocentrism** is "unreflectively imposing alien standards of judgment and thereby missing the point of the meaning of a practice." It is "the attitude that other societies' customs and ideas can be judged in the context of one's own culture" (Ember and Ember 1990:510) and ". . . that one's own culture is superior in every way to all others" (Haviland 1989:296). As a discipline, anthropology has reacted against this view as a result of the method of participant-observation in which anthropologists early on came to know that "savages" were as human as western peoples and that their behavior could only be understood as part of their culture (Haviland 1989:296). To fully comprehend the meaning and dangers of ethnocentrism, it is important to adopt the anthropological stance of cultural relativism.

Cultural Relativism

According to Ember and Ember (1990:510), relativism is "the attitude that a society's customs and ideas should be viewed within the context of that society's problems and opportunities." Thus, ". . . there is no single scale of values applicable to all societies" (Winick 1970:454). For anthropologists it is crucial to remain relativistic in order to describe and explain objectively and to discover meaning without western bias. For example, it is obvious that western cultural biases against homosexuality could impact a scientific understanding of the subject. Herdt (1988, 1987, 1981) and Williams (1986) have written about nonwestern male homosexual practices and Blackwood (1984) on nonwestern lesbian behavior. They offer a relativistic and non-judgmental view of the subject. It is evident from their writings that even the terms we use like homosexual and lesbian are so loaded and culturally specific that they cannot be directly translated into the meaning given to, for example, ritualized homosexuality among Sambia of Highland New Guinea (Herdt 1987). The homosexuality of these peoples is simply not commensurable with our western conception of what appears superficially to be a similar behavior.

At some point in adopting a culturally relativistic perspective, you might be faced with a clash of values. How far can you take your cultural relativism and where do you draw the line are questions often asked by students. It is one that concerns anthropologists as well. In fact, *Ethos*, the Journal of the Society for Psychological Anthropology, devoted an entire issue to the question of moral relativism (1990:131–223). The introduction begins with:

> What sort of theory of human values can be devised which encompasses and accords legitimacy to the obvious cultural and historical diversity in moral systems, without being so open that "anything goes"? That is a core problem of moral (ethical) relativism . . . (Fiske and Mason 1990:131).

These issues are significant particularly around topics directly affecting the lives of our students. Often it is easier to be relativistic regarding behavior or beliefs of people in a far and distant land. We, your authors, maintain it is more difficult to be relativistic within one's own culture when certain behaviors clash with the views of dominant society. For example, homosexuality is one of several options for sexuality, yet in several southern states such sexual practices are considered illegal and people are currently prosecuted under such ethnocentric and inhuman laws.

But what about incest, rape, or abortion? Are you asked to find these acceptable with cultural relativism? No, you are not. But perhaps on the last item, abortion, you are not so sure. Rape and incest are in most cases clearly "crimes" in our society. However, abortion which once was a crime is no longer, but rather a choice of the individual until that legally granted choice is removed. Where you draw the line is ultimately your decision on many of these issues. We recommend a simple stance of individual choice as a guiding principle, but we also understand even that may be problematic for some readers. In the case of rape and in most incest situations, the participants are clearly victims of crimes and have no choice. Because we are feminist in perspective, on the abortion issue we take the position that it is the right of the woman to choose. See also *emic* and *etic* in glossary.

Definitions of Human Sexuality

We turn now to defining human sexuality. In keeping with the anthropological perspective, we will use a very wide lens and integrate a relativistic perspective. As we shall see, defining human sexuality is no simple task. It is impossible to set narrow boundaries on what is included within the category we call sex.

Katchadourian (1979:8–34) has described the many meanings of the word "sex." According to Katchadourian's research, the English word, derived from the Latin *sexus*, can be traced back to the fourteenth century (1979:9). The term *sex* has undergone a variety of permutations and grammatical uses including rock musician George Michael's "I Want Your Sex" and Kip Winger and Fiona's "You're Sex'n Me." Sex commonly refers to what people "do," is usually termed "sexual behavior," and often is described as erotic (Katchadourian 1979:11). Barale offers a down-to-earth definition: "Genitals are the given: what we do with them is a matter of creative invention; how we interpret what we do with them is what we call sexuality" (1986:81 in Duggan 1990:95).

Time and Space

Definitions of human sexuality have varied across time and space. Human sexuality is symbolic behavior as much as it is reproductive behavior. It includes self-stimulation (masturbation) as well as copulation and other activities related to coitus. It also includes non-coitally oriented pleasuring as well as sex between partners of the same gender. Questions of where the boundaries are and what to include in the definition of sex are difficult. For example, Katchadourian asks if we can include sexual fantasy in our definition of sex. It is certainly erotic, ". . . but is it 'behavior'"

(1979:11)? Others have attempted to avoid the problems of behavior and emotional arousal by referring to the term "sexual experience." When this array of terms is placed in the even wider context of the cross-cultural record, we are led to ask if there are indeed any sexual universals amidst such a wide range of human sexual expression?

We have chosen the broadest possible scope to explore the many meanings of sex including: species-wide behavior, biological, sociocultural, and behavioral definitions. We will also include other attributes such as the functions of sexuality in the cultural context. Contributing to the complexity of our definitional task is that sexual expression may be part of actual behavior or may be purely symbolic. Or sexuality may be expressed in terms of the metaphorical as is often found in rituals. For example, among the Ndembu as studied by Turner (1969:10–43; Cohen and Eames 1982:250–251), the Isoma fertility ritual is rich with sexual symbolism. In one part of this ritual the infertile woman holds a white chicken which represents semen and good fortune in Ndembu cosmology.

Definitions of human sexuality do not remain stable over time, as is documented by the history of western views of sexuality in the field of sexual science. Scientific studies have shifted their interests from sex as reproduction to perspectives that focus on sexuality and its non-reproductive aspects, including larger issues such as gender variance (Jacobs and Roberts 1989:439–44). Western attitudes in society at large have also changed over time. Medical views have fluctuated in tandem with changes in the wider culture. Thus, the well-known prudish and sexually repressive cultural atmosphere of "mainstream" (i.e., middle class) Victorianism in the period of mid-1800s to 1900 was reproduced in medical views so that masturbation was believed to cause mental illness (Masters, Johnson, and Kolodny 1982:11–12). The following quotation illustrates all too clearly the Victorian view of female sexuality. We quote from Ruth Smythers' 1894 "Instruction and Advice for the Young Bride on the Conduct and Procedures of the Intimate and Personal Relationship of the Marriage State."

> To the sensitive young woman who has had benefits of proper upbring, the wedding day is ironically, both the happiest and most terrifying day of her life ... On the negative side, there is the wedding night, during which the bride must pay the piper, so to speak, by facing for the first time the experience of sex. At this point, dear reader, let me concede one shocking truth. Some young women actually anticipate the wedding night with curiosity and pleasure! Beware of such an attitude. A selfish and sensual hus-

band can easily take advantage of such a bride. One cardinal rule of marriage should never be forgotten: **GIVE LITTLE, GIVE SELDOM, AND ABOVE ALL, GIVE GRUDGINGLY** . . . while sex is at best revolting and at worst rather painful, it has to be endured, and has been by women since the beginning of time, and is compensated for by the monogamous home and the children produced through it (1989:5-7). (Our bold).

This view represents a fundamental change over time in how sex is regarded for women. In today's world, the model of sex is one of "sex as pleasure" rather than sex as duty. Such changes in attitude influence how human sexuality is experienced and integrated within culture.

Species-Wide Behavior

Human sexuality or more accurately the capacity for sexuality is a species-wide behavior. The term *species* merely refers to "a population or group of populations that is capable of interbreeding, but that is reproductively isolated from other such populations" (Haviland 1989:66). While all humans may mate with one another, it is characteristic of cultures to restrict sexual and reproductive activities between people. Sometimes the cultural meaning assigned to certain gene pools (i.e., races) prohibits groups of humans from interbreeding with one another even though they are perfectly able to do so (Haviland 1989:66).

Biological Definitions and Dimensions

Biological definitions include sex in reference to ". . . the two divisions of organic beings identified as male and female and to the qualities that distinguish males and females" (Katchadourian 1979:9). This latter definition is frequently termed *biological sex*. Yet, this definition is problematic as well as ethnocentric. Relying on a model described by Money and Ehrhardt (1972:4–15) and expanded upon in classes by our undergraduates, we can offer a multifaceted view of biological sex and challenge the simplicity of the notion of defining one's sex as male and female. How, in fact, do we reach that determination? How do you know what sex you are? Our list of sex attributes includes: chromosomal sex, gonadal sex, hormonal (endocrine) sex, sex of internal reproductive structures in addition to the gonads, secondary sexual characteristics (including distribution of fatty tissues, hair growth, breast development), gender identity (self-perception as male or female), gender role, sex of assignment and rearing, and legal sex among others.

Obviously, there are many aspects to defining one's sex. For example, male-to-female transsexuals who have undergone sex reassignment through surgery mix these definitions in unique ways. Post-operative transsexuals have female genitals, a female hormonal mix with estrogen and progesterone dominant over testosterone, a legal sex as a female, but a sex of assignment at birth and rearing as male. They have a female gender identity, but "male" secondary sexual characteristics until hormonal reassignment and alteration of hirsuteness through artificial means occurs, and a history of a male gender role which is followed by a female one.

One can easily see how the biological can be mixed up with the cultural when the discussion turns to gender roles and legal sex. Yet, strictly biological determinants like hormones, secondary sexual characteristics, internal and external reproductive features are not so clear either. Biological sex exists on a continuum. There are a number of gender anomalies that illustrate this variety in biological sex (Masters, Johnson, and Kolodny 1982:504).

In testicular feminizing syndrome, the **androgen-insensitive** male produces enough testosterone to make him into the Incredible Hulk, but because of a genetic problem, "he" cannot absorb and process "his" testosterone. As a consequence, the fetus develops a blind vagina (it doesn't lead to a uterus) and female genitalia. At birth, the infant who looks female but who does not have internal female reproductive organs, is usually identified as a girl. At puberty, "she" develops breasts but cannot menstruate and is infertile (Money and Ehrhardt 1972:280).

To suggest to such a young woman at puberty that because she is a chromosomal and hormonal male that she is a male, overlooks the fact that gender is a lived phenomenon (Kessler and McKenna 1978:76–77). In our society there are only two choices. The androgen-insensitive male is reared as a female, and has a gender identity as a woman. She is a woman in the socio-cultural sense although she may not be a physiological female, but a hermaphrodite (having both male and female sex cells).

Not only are there individuals who are biological hermaphrodites, but cultures may recognize more than two genders as well. In some cultures, gender may be an achieved (acquired) rather than an ascribed (assigned) status. The cross-cultural record reveals all these possibilities. Anthropologists have long reported on the existence of the **Berdache** or **two spirit**.[1] It now appears that a variety of kinds of behaviors and variations in gender expression have been lumped under this term. The two spirit or *Berdache* has been referred to as a gender transformed status, as an alternative gender, and/or as a cross-gender role. Nevertheless, it may generally be described as a position in society in which a person takes on some or all of the tasks, dress, and behaviors of the other gender. Rather than just two genders as in the western case, the Mohave recognized four genders: woman, *hwami* (female *Berdache*), man and *alyha* (male *Berdache*). The Chuckchee reported seven genders, three female and four male (Jacobs and Roberts 1989:439–440; Martin and Voorhies 1975:96–99; 102–104). Martin and Voorhies refer to these alternative genders as **supernumerary genders** (1975:94). Jacobs and Roberts (1989:439) have proposed that: "If one uses the criteria of linguistic markers alone, it suggests that people in most English-speaking countries also recognize four genders: woman, lesbian (or gay female), man, and gay male."

The other biological definition of sex focuses on the physiology of sexual arousal and coitus, and on the reproductive biology of humans. This includes changes in the human cycle in both reproductive physiology as well as human sexual response. But as Jacobs and Roberts (1989:441) so eloquently point out "[r]eproduction and sexuality are codependent variables in the human life cycle. But sex and sexuality are much more complex than linking them with reproduction."

Behavioral, Cognitive, Affective, and Symbolic Definitions and Dimensions

Behavioral definitions of human sexuality focus on behaviors and consequences that can be observed and measured. This approach adds a valuable dimension to our understanding of human sexuality by elaborating on the biological.

The work of Kinsey and his colleagues represents one of the most well-known behaviorist studies of human sexuality: *Sexual Behavior in the Human Male* (1948) and *Sexual Behavior in the Human Female* (1953). The Kinsey Reports focused on six sexual outlets leading to orgasm (masturbation, sex dreams, petting, coitus, homosexuality, and sex with animals). They were based on interviews with 5,300 males and 5,940 females. Kinsey was central in the creation of the scientific study of sex. He exemplified how such an emotionally charged subject could be studied with scientific rigor. The Kinsey Reports opened a forum for the public discussion of human sexuality. Gagnon (1978:93) believes that the public furor created by this work ushered in ". . . a major increase in the publicly sexual character of society that occurred in the late 1960's and early 1970's."

The behaviorist definition of sexuality is concerned with the scientifically measurable, i.e., external states. Kinsey was critical of sex research done through the single case method or the method of ethnographic sexology. He advocated the sociological method of the survey of large populations with concern for representativeness and the "statistical sense" without which one was "no scientist" (Gagnon 1978:93).

Behaviorism may be contrasted with the Freudian psychoanalytic perspective with its interest in the internal and the unconscious, the affective and the symbolic. The psychoanalytic approach is less amenable to measurement by conventional scientific techniques based on the natural sciences model of scientific inquiry. For Freud, infant and childhood sexuality was regarded as part of a broader phenomenon of seeking pleasure. Sexual desires were interpreted as potentially in conflict with the demands of civilization, thereby necessitating inhibition and suppression. Actually Kinsey agreed with this view, but felt that ". . . cultural values that oppose biological facts are social errors" (in Gagnon 1978:93).

Socio-Cultural Definitions and Dimensions

As we have seen, biological definitions of human sexuality focus on anatomy and physiology, physical development, and changes in human sexual response throughout the life cycle with an emphasis on the reproductive. Biological definitions of human sexuality also include behavioral dimensions. In contrast, socio-cultural definitions accent the role of customs in shaping human behavior and tend to take a relativistic stance. This position opposes the biological and behavioral definitions that look to physiology to explain alleged sex differences in human response. Socio-cultural definitions regard sexual behavior as culturally constituted and created (Gagnon 1978:95). Gender is a socio-cultural construct in which meanings are assigned to biology. The **sex-gender system** is defined as "'the set of arrangements by which a society transforms biological sexuality into products of human activity'" (Rubin 1975:159 in Vance 1983:372).

Socio-cultural studies have gradually evolved from an early interest in sex as reproduction, e.g., Martin and Voorhies' *Female of the Species* (1975), to studies of sex as institutional, e.g., marriage systems, and finally to sexuality itself (Jacobs and Roberts 1989:439). This approach is amply represented throughout this text, especially in chapters 7–14.

Sex, Gender, Masculinity, and Femininity

In this section, we present a list of terms related to the concept and construct of **gender**. It is important to define gender early in our discussion of masculinity and femininity. In the 1970's research, the terms *sex* and *gender* were collapsed and mixed together, although by the early 1980s successful efforts were made at separation and redefinition (Jacobs and Roberts 1989:439). Jacobs and Roberts (1989:439) offer an excellent definition:

> **Gender** is the sociocultural designation of biobehavioral and psychosocial qualities of the sexes; for example, woman (female), man (male), other(s) (e.g., *berdache*). Notions of gender are culturally specific and depend on the ways in which cultures define and differentiate human (and other) potentials and possibilities.

Kessler and McKenna's (1978:7–16) definitions also serve us well. While it is conventional to define **sex** as the biological aspects of male or female, and to define **gender** as the "psychological, social, and cultural aspects of maleness and femaleness," Kessler and McKenna argue that even the concept of two biological sexes is a social construction (1978:7).

In summary, for purposes of clarity, sex will be used to refer to activities related to sexual pleasure, arousal, and intercourse whether recreational or for reproduction (Jacobs and Roberts 1989:440). Gender will refer more broadly to the cultural aspects of being male or female. Elsewhere, specific usages such as chromosomal, hormonal, or morphological sex will be presented, even though these biological characteristics are always interpreted through a cultural lens (cf Kessler and McKenna 1978:7).

Through an understanding of the "attribution process," how people assign gender to others, we can gain insight into the social construction of femininity and masculinity. Western femininity and masculinity are integrated in a gender scheme whose central tenants are that there are only two sexes, male and female, and that these are appropriately associated with the two social statuses of gender: men and women, boys and girls. ". . Whatever a woman does will somehow have the stamp of femininity on it, while whatever a man does will likewise bear the imprint of masculinity" (Devor 1989:vii). Therefore, masculinity and femininity are associated with sex roles. Our western **gender schema** is a shared belief system about sex and gender. It regards biological sex as the basis for gender status which is the basis for gender role. The actual process whereby people attribute gender to another actually occurs in the reverse to our gender schema; a person's display of masculinity or femininity (gender role) indicates gender which is followed by the presumption of appropriate genitalia which are not readily visible (Kessler and McKenna 1978:1–7, 112–141; Devor 1989:149). Without our portable gene scanners and x-ray vision, daily life consists of encounters in which the biological is clearly mediated by cultural expectation in the attribution process. We do not really see genitals and sex, but gender presentations of feminine and masculine beings.

In summary, masculinity and femininity may be defined as components of gender roles which include cultural expectations about behaviors and appearances associated with the status of man or woman in the western bi-polar model of the sexes. For our purposes the following definition for gender role is provided.

> Everything that a person says and does, to indicate to others or to the self the degree in which one is male or female or ambivalent. It includes but is not restricted to sexual arousal and response. Gender role is the public expression of gender identity, and gender identity is the private experience of gender role ... Gender identity ... [is] ... the sameness, unity, and persistence of one's individuality as male or female (or ambivalent), in greater or lesser degree, especially as it is experienced in self-awareness and behavior (Money and Ehrhardt 1972:284).

Because of the attribution process, gender roles are often confused with sex and biology. Gender role stereotypes include ideas that differences in gender are the result of biology, for example, women are more nurturant, men are more aggressive, women are emotional while men are rational. These differences are rather the result of learned behaviors. Stereotypes such as these are classified by sociologists as **expressive** and **instrumental** gender roles. Boys are socialized into instrumental roles that are associated with acting or achieving while girls are socialized into relationship oriented or expressive roles (Eshleman, Cashion and Basirico 1988:182). That these roles are cultural and are not natural, is amply demonstrated in the cross-cultural record in which a diversity of behaviors and expectations are recorded. Mead's study of the Arapesh, the Mundugumor, and the Tchambuli (they call themselves the Chambri) in *Sex and Temperament in Three Primitive Societies* (1963, 1935 original) offers a classic account of gender role variation in counterpoint to our western conceptions. Among the Mundugumor, both men and women were aggressive and non-emotional, while among the Arapesh, both sexes were cooperative and nurturant. The Tchambuli (Chambri) expressed the reverse of our western gender roles with cooperative caring men and assertive women as the behavioral norm.

Biology and Sex: Political Aspects

We conclude with the interface of biological, behavioral, and socio-cultural definitions of sex with political ends. "Whether the boundaries of women's place in society were erected with the bricks of theology or the cement of genetic determinism, the intention is that the barriers shall remain strong and sturdy" (Burnham 1978:51). This quote serves as a good introduction for our discussion of the politics of biology and sex. Biological/sexual functions have been used to serve larger political purposes in societies and ours is no exception. For example, women's role in reproduction has been interpreted to justify conceptions of female inferiority which support ideologies of gender inequality.

Nineteenth century physicians maintained the view that for medical reasons it was unhealthy for women to be educated. This was based on a medical ideology that the brain and the reproductive organs shared the same biological resources, so that development of one meant the other was deprived. Since women's role was defined as a reproductive one, feeding her brain through education was seen as jeopardizing her reproductive capabilities (Burnham 1978:51–52).

Other researchers have described the historical relations of sexuality to changes based on the development of industrial capitalism in western societies, thereby providing a political-economic interpretation (Ross and Rapp 1983:51–73). Ross and Rapp (1983:51) note that "the personal is political" reflecting our point that what one may think of as private is also public in the sense that it is linked with broader institutions of the western political-economy. These institutions are patriarchal and perpetuate beliefs about the differences between the sexes and their respective sexualities. Such ideologies are then supported by a sexology in which **androcentric** (male biased) views are legitimized as science, aided and abetted by biological reductionism.

Sanday's study of rape-prone societies clearly shows the relationship between male dominance and political power. Rape-prone societies were found to glorify strength, power, and violence. In these societies women have no voice in the political sphere or in religious life and are generally regarded as "owned" by men (Benderly 1987:187). In this way sexual behavior is defined as violent and as natural for men. Rape-free societies, however, challenge this view of brutal men. Rape-free societies are associated with resource stability, the absence of competition, and egalitarian social structures for men and women. Rather than a belief in a male supreme being, rape-free societies acknowledged a male and female deity or a "universal womb" (Benderly 1987:187–188). These are societies where rape does not occur.

In conclusion this chapter has presented terms and concepts necessary for understanding human sexuality from an anthropological perspective. We have elaborated the importance of culture in shaping our human sexuality.

Summary

Chapter 2: Topics in Anthropological and Sexological Views

1. Psychology, sociology, and biology offer useful perspectives for the understanding of human sexuality. These viewpoints are incorporated to various degrees by anthropological approaches.

2. To understand the bio-cultural perspective, it is necessary to define our terms. These include concepts and constructs such as evolution, the culture concept, ethnocentrism, and relativism. Other terms are also discussed.

3. Definitions of human sexuality have varied temporally and spatially.

4. Definitions of human sexuality include areas such as anatomy and physiology, the sexual life cycle, and human sexual response.

5. Sex has many components. These include behavioral, cognitive, affective, and symbolic dimensions.

6. Sex and gender are compared and contrasted.

7. Sex has been used to serve larger cultural ends in societies. We examine sex in the context of power and politics.

OVERVIEW

Chapter 3

Evolution of Biological Structures Related to Modern Sexual Behavior

This chapter:

1. Presents an overview of nonhuman primate evolution and ancestral relations.

2. Discusses the consequences of human arboreal and terrestrial adaptations.

3. Focuses on the development of the grasping hand, stereoscopic vision and grooming behaviors, and how these structures are related to modern sexuality.

4. Presents the topics of bipedalism, brain complexity, infant dependency, reliance on learning, and the profound consequences this has had on hominid evolution and human sexuality as well as reproduction.

5. Introduces the concept of bonding in human and nonhuman primates.

6. Relates the development of the brain to evolution and to human sex and reproductive cycles.

7. Addresses estrus and the loss of estrus as well as its implications for human evolution.

8. Integrates the subject of orgasm into an evolutionary framework.

Chapter 3

Evolution of Biological Structures Related to Modern Sexual Behavior

From an anthropological perspective our biobehavioral sexuality, i.e., how we look and behave sexually, did not just appear. It evolved through time and space. That is, certain characteristics adapted to specific environmental contingencies. There are continuities through time and space among humans and with other animals, particularly nonhuman primates, that are part of our modern human sexuality. In this chapter, we are going to discuss aspects of our primate heritage that some researchers believe serve as a model for early **hominid-hominine**[1] sexuality which affect our modern human sexual behavior.

Biologists have developed a phylogenetic tree that places plants and animals in an evolutionary perspective.

Our tree looks like this:

Kingdom:	Animalia
Phylum:	Chordata
Class:	Mammalia
Infra class:	Eutheria
Cohort:	Unguiculata
Order:	Primata
Suborder:	Anthropoidea
Infra order:	Catarrhinae
Super family:	Hominoidea
Family:	Hominidae
Genus:	Homo
Species:	Sapien

(From: Moore et al. 1980 fig. 3-2: 60)

41

We are of the order Primata, frequently described as the most social of the mammals.

In order to explore evolution as it relates specifically to modern human sexual functioning, we will be covering about 65 million years. We will investigate our primate relatives with a special focus on the **hominoids**—monkeys, apes and humans, as well as our **hominid** ancestors. In this chapter hominid refers to all humans, including our ancestors as well as modern humans. We will look at the influence of **arboreal**, or tree, and later **terrestrial**, or ground, adaptations upon human evolution. Those biological and behavioral features that ensured human survival and developed as a result of these adaptations are part of our biology and integral to understanding modern human sexuality.

Evidence from a variety of sources is used by anthropologists in the reconstruction of **hominid** evolution. This includes the fossil record which incorporates skeletal information from primate ancestors, archaeology, the living apes and modern hunters and gatherers. There are strengths and weaknesses in using **hominoid** analogies, that is using monkeys and apes, in understanding hominid evolution and current behavior. As primates—social, intelligent, tool using and making animals that have sophisticated communication systems and emotional displays, monkeys and apes are more similar to us than other animals. Our econiches have overlapped and been shared at various times. Our separate lines of evolution are relatively recent in geological terms. Finally, our chromosomes and blood types, particularly compared with chimpanzees and gorillas, are more similar than different, although we cannot interbreed or share blood with them (Weiss and Mann 1990). These continuities allow for a strong basis for comparison.

Based on immunological and genetic studies, we endorse the view that the appropriate primate group for behavioral comparison in reconstructing the past is the chimpanzee (Zihlman 1989). While baboons are sometimes used as "ancestor models" (Konner 1982) such as in the work of Washburn and DeVore (1961), we feel the chimpanzee is the more likely candidate for insights into our evolutionary past rather than the baboon, based on Tanner and Zihlman's (1976) and Zihlman's (1989) argument of biological and behavioral similarities. Although some types of baboons are terrestrial, studies of their behavior, which are extrapolated to early hominids, tend to fulfill western stereotypes about gender and evolution. We must remember that the baboon is a monkey and not closely related genetically to humans, unlike the chimpanzee and the other great apes.

Nonhuman primate models may be valuable in several ways. They help to show the continuities with other species of our order Primata. They illustrate the high intelligence and sociability of primates. They can serve as a reality marker to check our own biases and perspectives regarding dominance, division of labor, and sexuality. They provide evidence of the relationship of ecological variables such as food, shelter, and predators to social behavior. They indicate the variety and flexibility of primate behavior and patterns of intra- and inter-gender and group cooperation and competition that may be evolutionarily deep-seated. Primate analogies may serve as a model for reconstructing early hominid behavior. Finally, when viewed from a female perspective, female primates may experience a status and flexibility in behavior that few modern human females enjoy (Blaffer-Hrdy 1981).

However, taken too far, analogies defeat their purpose. We must remember that we have had our own line of evolution for 5–8 million years and have adapted to every econiche on the planet. The development of our cerebral cortex allows for qualitatively different kinds of communications and social relations than other primates. These variables have worked to our advantage and disadvantage relative to the expression of our sexuality. We have developed highly complex, functioning social systems and abused members of our species to a greater extent than any other animal.

Based on research with chimpanzees as well as data from modern gatherers and hunters, we can construct a model of early humans that includes the following characteristics: 1) an omnivorous diet, 2) the making and using of tools (the earliest tools were probably constructed of organic materials), 3) the sharing of plant and perhaps other foods, 4) social organization based on small family groups, 5) a heavy reliance on learning.

Aboreal and Terrestrial Adaptations

By the Eocene (54–38 million years ago), evidence of our first primate ancestors, the *prosimians*, appeared. Profound changes were brought about by "life in the trees" (Ember and Ember 1990). Our primate past reflects an important period of adaptation to an **arboreal** niche. As a result of the forest adaptation, primates developed certain shared physical and behavioral features. When these were adapted by our hominid ancestors as they moved into a terrestrial niche and given a long time span, they evolved and contributed to shaping human sexuality and reproduction. Hominid terrestrial adaptation is believed to have occurred somewhere around eight to five million years ago. This was facilitated by climatic changes as forests in Africa gave way to savannahs and the scene was set for the development of **bipedalism** (walking on two legs). Our relatives, in all likelihood, moved gradually from the fringes of the forest into the grasslands (Tanner 1981).

Kottak (1991:123) has coined the term *Hogopans* to describe the ancestors of humans, gorillas, and chimps who diverged somewhere between eight and five million years ago. It was not until about two million years ago, which is 1.2 million years after hominids first appeared, that the brain enlarged significantly with the appearance of *Homo erectus,* an early human remarkably like modern people in appearance from the neck down.

Shared primate characteristics related to modern human sexuality that evolved from the initial arboreal adaptation included: the **grasping hand**, **touching and grooming**, as well as **stereoscopic vision**. Moving from the tree adaptation to the terrestrial environment of the savannahs facilitated the development of **bipedalism**. This freed the hands for carrying objects and manipulating them. It facilitated the primate trajectory of enhanced **brain complexity** and **infant dependency and reliance on learning**, discussed later.

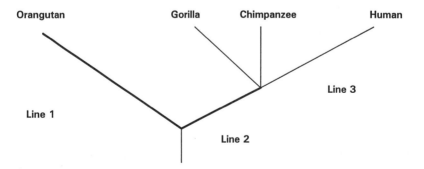

Figure 3.1 The evolutionary relationships of humans and our closest living relatives (Weiss and Mann 1990: 288).

The Grasping Hand, Touching, and Grooming

Being "handy humans" was an extremely important feature of our evolution. Hand coordination interacted with the expansion of the brain as well as with the emergence of culture. According to some theoreticians, manual dexterity is most probably linked to the development of the visual centers of the brain and stereoscopic vision.

In terms of human sexuality, handiness, especially the grasping hand, is related to the importance of touch in general bonding behaviors from the parental to the sexual. To illustrate this point, in 1946 Spitz compared infants in two institutions. In one, the infants were born and reared in an orphanage for two years. Their physical needs were met and they received excellent medical attention, although they had little social interaction with the nursing staff. The other institution was very similar in terms of care with one important exception. The children's mothers, who were institutionalized for delinquency, had a great deal of time to play and interact with their babies. Spitz found that the children in the orphanage fared poorly in comparison to the children who were in the institution with their mothers. Of 91 children in the orphanage, 34% "died within two years of the study, and twenty-one showed slow physical and social development." He concluded that the difference in the health and well-being of the children was a result of attention (Eshleman et al. 1988:167–168). We conclude that touch was a significant factor in this attention. How this came to be is related to our evolutionary past.

For humans the sense of touch is extremely important. Sensitivity to touch is believed to be a result of the primate arboreal adaptation because feeling sensations may have been critical in grasping food such as insects, leaves, and fruits. Primates have pads on their fingers giving them information about what the hand is feeling in contrast to other mammalian species in which hair provides environmental feedback (Haviland 1989). In the course of hominid evolution, the loss of body hair probably enhanced touch as a means of assessing one's surroundings including the social milieu. It is linked with the unique development of primate forelimbs including the hands, precision gripping, tactile sensitivity, and manual motor skills (Ford and Beach 1951).

The grasping hand is associated with primate grooming. Ford and Beach (1951:44) regard grooming as a "basic pattern for primates." "Grooming is an activity that has purposes of cleansing the fur (delousing) as well as in establishing and facilitating a variety of social bonds" (Harrison et al. 1988). Grooming establishes relationships among primates as well as denotes hierarchy. As a social behavior, grooming among sexual partners is very widespread among monkeys and apes and is frequently linked with sexual arousal and activity (Ford and Beach 1951:45). Grooming demonstrates a convergence of biological and social functions including dimensions such as health, sexuality, bonding and social organization. In human sexual activity, grooming is symbolically translated into stroking and patting and may also be found in a variety of cultural activities including cleansing of parasites, combing and arranging hair, and adorning one another's bodies with paints, feathers and/or clothing. All of these involve the sensations of touch and sociality. Ford and Beach's (1951) *Patterns of Sexual Behavior* reports that human grooming activities are frequently a precursor to sexual relations and may be an integral part of foreplay in contemporary human cultures in which sex is treated in a positive manner.

> **ASIDE**: While we have primate propensity for touch, rules, attitudes, and behaviors about touching and body space are culture-specific. For example, our culture has been described by some sexologists and sex therapists such as Domeena Renshaw, M.D. as "touch deprived." We have rather rigid rules about touching and body space and tend to confuse affection with sexual touching. Most dramatically, this can be illustrated by a middle class value on newborns, infants, and children "having their own room" and sleeping separate from their parents and siblings from birth. In contrast, in many traditional societies, women carry their infants with them while engaging in their daily activities and parents and children share a common sleeping space.

Stereoscopic Vision

In addition to the development of the grasping hand, our primate ancestors acquired highly developed visual cortices from their arboreal adaptation that remain with us today. Primate eyes are large, include "stereoscopic, diurnal, [and] colour vision" (Harrison et al. 1988:50). This represents a shift from reliance on smell as a vehicle for information processing to vision (Kottak 1991).[2]

Subsequent terrestrial adaptation to savannahs also encouraged the development of the visual cortices. This was facilitated by a climatic change from heavy forestation to grasslands that also contributed to the development of bipedalism. Hypothesized causes of bipedalism in our early chimpanzee-like ancestors include the ability to see over tall grass, to gather food, energy efficient locomotion, predator avoidance, and to free the hands (Lewin 1989:68).

The evolution of vision and bipedalism are both related to modern sexuality in terms of the signalling of sexual interest. Prior to the loss of **estrus**, visual signalling among our ancestors was probably apparent in the brightly hued purple swelling of the female genital area as seen in modern chimpanzees and some other primates. Estrus is a period of penetrative sexual receptivity among mammals that converges with fertility. After the loss of estrus, discussed later in this chapter, we speculate that the concealment of ovulation led to a greater reliance on symbolic cues between hominids for indicating sexual interest. We can only presume that symbolic cuing and communication of sexual interest continued to elaborate and snowball evolutionarily.

> **ASIDE**: Perper, a biologist and anthropologist, has discussed the importance of visual signals and closing of body space as factors in both human and nonhuman primate attraction and sexuality (1985). Visual cues, as with touch, are culture specific. In the U.S., female breasts are seen as a sexual cue, while in most parts of the world they're seen as a form of nutrition. The controversy (1992) in New York State over allowing women to appear topless on public beaches is an example of this. In Japan, women's necks traditionally were seen as a visual cue, and among Tantrics, people who practice a sensual form of Yoga, the entire body is seen as a potential erotic zone (Gregersen 1983, 1992; Garrison 1984).

Bipedalism

The development of **bipedalism** in humans marks them as distinctive among the primates, although other primates such as chimpanzees have the capacity for short bursts of bipedalism. As we shall see, the subsequent evolution of the human brain is intimately linked with bipedalism, embellishing characteristics already developed in ancestral primates such as a large brain to body size. Bipedalism had consequences for the evolution of the hand and manipulation of tools, the elaboration of the motor areas of the brain as well as memory and thinking.

While the dates may vary, it is generally accepted that humans became bipedal approximately five million years ago during what was a transitional period of human evolution (about 8 to 5 million years ago) from a hypothesized ape-like ancestor of the late Miocene to an early human form, most likely some variety of **aus-**

tralopithecine. In fact, two australopithecines from Laetoli, Tanzania, left their footprints in the sand as they walked side by side, testimony of their bipedalism 3.6 million years ago (Ember and Ember 1990:75).

Bipedalism which entails an upright posture and a **striding gait**—what you use in long distance walking—had a profound impact on the evolution of human sexuality. A number of skeletal and muscular changes accompanied upright posture. One of the major changes for hominid sexuality include a tilting forward of the pelvis; it was shortened and flared as well. The genitalia was moved forward with the female genitals less exposed and more hidden than the males. This affected females a great deal since no longer would the sexual swellings and brightly hued coloration accompanying estrus be visible and hence functional. Without the obvious sexual swellings to indicate sexual receptivity, other ways of communicating were developed. These were probably related to the expansion of communication skills in general, an essential component of culture (Beach 1973).

Changes in the pelvis also affected childbirth. Compared to our primate relatives, humans are born in a very immature state. This trend was escalated by the evolution of bipedalism and the growth of the human brain. In order for a species with such large heads to be born through a rather narrow birth passage, natural selection favored the development of **neotony**, immature physical characteristics, so that the relative size of the head could be reduced. Neotony was associated with overall physical immaturity and dependency which necessitated adult caretaking. This in turn led to a greater reliance on learning and hence longer dependency for hominids.

With the shifting forward of the hominid female genitalia, face to face sex was a possibility and perhaps a probability. We are not necessarily referring here to the "missionary position" where males are on top, since this is not even the most preferred position cross-culturally. Face to face sex included positions in which the female is on top of the male or side by side. The location of the female genitalia in a more forward location certainly contributed to the human potential for a wide array of sexual positions.

Bipedalism also facilitated to the expansion of the brain in a bio-cultural system of escalating complexity. The actual expansion of the brain in terms of size did not occur until about 1.2 million years after the first hominids appeared. This is not to say that size alone indicates high levels of cognition. We may assume that the small brains of the earliest hominids were also relatively complex ones. After all, chimpanzees are very intelligent creatures. The adaptive strategies utilized by the earliest hominids may have promoted a certain amount of cognitive complexity prior to the actual physical expansion of the brain. Generally, however, hominid trends indicate a correlation between increases in the size of the brain and complexity.

As primates became bipedal, hands and arms could be used for carrying. We may regard this stage as the dawn of human culture. With freed hands, early hominids could carry their babies and other objects. Based on contemporary gatherers and hunters, hominids probably used a strategy of carrying food back to some sort of base camp rather than just eating it on the spot. Perhaps the females shared food with their babies first and then with other kin and friendly unrelated males. The earliest stone tools appeared in association with evidence for an expanding brain. These were probably predated by tools and implements made of organic material that related to a food gathering strategy. Digging sticks and some sort of basket or net for carrying food items could have been included. We have no remains of these tools because they were made of organic materials that decompose rapidly. We posit that both males and females gathered plants and protein and possibly used their expanding brains to hunt or chase small prey. Communal and group hunting by the Mbuti of the Ituri forest is well documented. It is not necessary to propose that only men hunted unless the prey was some variety of large game. There is ample evidence of women hunting small animals (O'Kelly and Carney 1986:12–21).[3] Among the Agta of the Phillipines, women hunt larger game including wild pigs and deer and participate in spear-fishing as well (Estioko-Griffin 1993: 225–232).

Generally, the pattern of men engaged in large game hunting and women involved in smaller game or communal hunts is related to the dangers inherent in hunting and adult sex ratios needed for **reproductive success**, i.e., species continuation. Hunting is potentially dangerous. Men can be wounded or killed in the hunt by either wayward spears in motion or the charges of a wounded animal. Secondly, big game hunting can involve exploring new frontiers—whose dangers, human or nonhuman, are unknown. In either of these situations, the evolutionary reality is that as adults, men are more expendable, i.e., a population can lose more adult men than adult women and still have the group continue. For example, only one post-pubertal male is needed to inseminate a number of women. However, women can only be pregnant with one male's child at a time. In addition to the loss of the reproductive women in hunting (or warfare), would be the loss of all their labor as well. For example, in gathering and hunting and horticultural societies, women do a great deal of the work and supply most of the food among many groups.

> **ASIDE:** The present debate in the U.S. military and among some civilians concerning whether or not to allow women in the military to engage in direct combat has interesting evolutionary consequences. In the short run (proximal evolution), it may be "politically correct" to put women on the front lines. In the long run (distal evolution), it makes little survival fitness sense to lose women in combat. For to lose them also means to lose their reproductive (pregnancy, lactation, caregiving) and labor functions and contributions.

It also means that women are much less dependent on men than men are on women for reproductive success in modern societies. While women need sperm to become pregnant, they do not need the man per se. Despite recent reproductive technology, men however, still need women to impregnate, carry the fetus, give birth, and in many societies, lactate to ensure the survival of the child. Now that we have technology to create babies, such as in vitro fertilization, chromosome selection, sperm banks and artificial insemination; technology for infant feeding, such as bottles and infant formula; and technology to secure our food and fight wars, there may well be a shift away from the need for women to act as primary child caretakers, even though they are still needed as childbearers at this time.

Brain Complexity

The development of a large and complex brain as well as our sociality are intimately involved with our sexuality. The large human brain with elaborate centers that include memory, language and symboling, to name just a few, are directly related to our sexuality. Human sexuality is experienced and mediated through a complex web of cerebral functioning which includes the capacity for elaborate fantasy, dreams, verbal and non-verbal thinking and images. Our human heritage as social beings facilitates how we communicate our sexuality, which is experienced as much in our "heads" as it is in our genitals. In addition, because we rely on learning so much, sex, just like other parts of our culture, is also learned. We learn how to experience our sexuality, including appropriate courtship behaviors, gender roles, and related norms and values. The biological capacity for sex is therefore intricately intertwined with a matrix of cultural constructions that shape our perceptions, experiences, and expressions of human sexuality.

"The proportion of brain tissue concerned with memory, thought, and association has increased in primates. The primate ratio of brain size to body size exceeds that of most mammals" (Kottak 1991:91). It has been suggested that this may be a result of reliance on fruit in the diet during the arboreal period of primate adaptation. In such a situation memory is important in locating fruit, particularly seasonal varieties. Selection for a large brain in early primates may also be correlated with a high metabolic rate and increased oxygen needs of the brain (Ember and Ember 1990).

The early foraging strategies associated with an omnivorous diet (if it was smaller than they were, they probably ate it), expedited the expansion of the brain in terms of complexity. In seeking food and carrying it someplace, the human capacity for **"equilibration"** or thinking/decision making skills grew. To take something from one place to another requires brain power and certainly the capacity for displacement—to think about something that is not present in one's immediate environment. To make tools and baskets requires the articulation of cerebral centers with motor skills. The cognitive task of remembering locations of sites and sources for food, some of which were seasonal, also required a high level of cerebral functioning. In short, bipedalism intensified a very important trend for hominids. Similar to other primates, the human brain is large vis-à-vis body size. While larger brains are associated with an increase in body size, " . . . the human brain is roughly three times its expected value for a hominoid of that body size" (Harrison et al. 1988:49).

Throughout the course of hominid evolution, we can assume that humans continued a pattern of dependence on learning and hence culture as a primary means of survival. The size and complexity of the human brain reflects the increased reliance on learning as a means of adaptation. We adapt to our environment primarily through culture. Remember culture is learned, shared, patterned behavior, including symbols and beliefs that are expressed between and within generations, individuals and groups. We need to interact with others of our own kind regularly in order to survive and be functional members of society. Harrison refers to these as "open" as opposed to "closed behavioral programmes" which puts the focus on behaviors that may be "modified" during the life course (Harrison et al. 1988:53). Primates are known for their open behavioral patterns.

Infant Dependency and Reliance on Learning

Early humans were generally similar to other primates who usually give birth to a single infant and whose maturation and survival requires a period of learning (Martin and Voorhies 1975). Reproduction occurs relatively late in primate lives. Primates generally have long lives for mammals, including the capacity for sexuality throughout. Sexual maturation is typically slow (Harrison et al. 1988).

Human young are totally dependent on adult caregivers longer than any other primates and need this contact with adults and their peers in order to survive. It is during this long period of primate dependency that a great deal of learning occurs. This includes the development of motor skills as well as social and, in humans, cultural skills (Ember and Ember 1990). In humans, dependency is even longer. For example, the human infant's period of reliance is twice as long as the chimpanzee's. While apes are dependent for four or five years, the human child's extends to "over a decade" (Haviland 1989). When the human infant is born, its brain is only one-quarter of its adult size, indicative of the evolutionary tradeoff that resulted in neotony and an even longer period of caregiving.

The whole social group, not just their mothers, is very important to young nonhuman primates for the development of their learning skills (Ember and Ember 1990). It is within the group that they learn how to communicate in a variety of ways including the vocal, gestural, and auditory (Harrison et al. 1988). Because the primate group is critical for socialization and survival of the young, we must assume that factors favoring traits for group survival were also selected for in humans along with characteristics for cooperation and bonding. As a consequence, the young human must rely on adults, parents, kin and the broader social group for survival. Thus, human cooperation would ultimately have favored infant survival.

Bonding

Bonding is an anthropological term that refers to enduring, social-affective, or emotional ties between members in primate and human societies. When applied to modern humans and in the current literature on U.S. male-female relationships, bonding is expressed in terms of intimacy and courting behaviors (Perper 1985; McGill 1987; Hite 1987). The form and duration of primate bonds generally relate to securing food and protection from predators, both of which are important to reproductive success. The bonding between mother and infant associated with infant dependency on the mother is part of our primate heritage. This bonding serves as the basis for the establishment of all our other relationships (Money 1986).

Bonding, as we have defined it, is a hominid characteristic that reflects continuities from our nonhuman primate heritage (Blaffer-Hrdy 1981; Jolly 1985). We will review and illustrate the major forms of primate bonding as they relate to human sexuality. As stated, adult female-child bonds are probably one of the oldest, deepest, and most primal forms known. This is a bio-cultural phenomenon related to conception, pregnancy, childbirth, and lactation. Many human societies recognize this process as a critical marker of adulthood, what Rapheal labels "**matrescence**" for females and "**patrescence**" for males (1988). This bonding extends beyond the biological mother and offspring and does not imply a female "instinct" for childbearing. A worldwide pattern exists in which children are primarily raised by the female kingroup, not the individual mother.

> **ASIDE**: In societies that practice various forms of **infanticide** or killing the young at birth, it is usually female babies whom are killed, and they are generally killed by females in their kin group (Murdock 1964; Frayser 1985). Scheper-Hughes (1992) also details a behavior of "**differential investment**" in her research of maternal-child care in the extremely poor shanty towns or *favelas* in Brazil. Women know that most of their children will not live to adulthood because of poverty, resulting in malnutrition and disease. Through a complicated religious rationalization system and behaviors, they invest time, energy, and resources in those children who seem to have the best chances of surviving, who have "a will to live."

Female-child bonds are supported by female-female bonds which can be either cooperative or competitive. Cooperative female-female bonds are generally kin-based or take on kin terminology if *not* of biological (**consanquineal**) or marital (**affinal**) relations. These nonkin relations are called **fictive kin** and can be exemplified by relationships such as sorority (or fraternity) "sisters" or "brothers" or by phrases such as "she is like a sister to me." Cooperative female-female bonds provide psycho-emotional support, and socialize females into "female" behavior. Competitive female-female bonds in both the human and nonhuman primate worlds generally are adversarial over males. In humans this can be expressed by jealousy among co-wives in **polygynous** societies where men have more than one wife; or current U.S. culture where one female pursues another female's man, i.e., the "other" woman (Brown and Kerns 1985; Fernea 1965).

Female-male bonds may also be cooperative or competitive. Cooperative female-male bonds tend to be nonsexual and take on familial characteristics—siblings or fictive sibling relations—"He is like a brother to me." They can provide a great deal of socio-emotional support, protection, and friendship. In **matrilineal** societies where descent is through the female line and a woman's brother fills the social role of "father" to his sister's children, brothers and sisters have a lifelong supportive relationship that can supercede their marital relationships (Kluckhohn and Leighton 1962). Competitive male-female bonds often are sexual and revolve around issues of trust, intimacy, and sexual exclusivity. In a U.S. movie, "When Harry Met Sally," the central theme was whether "straight" men and women could be platonic friends (Sally's position) or if sexual tension would always exist and affect the relationship (Harry's position). The conclusion—see the movie! This concern over adult male-female relations is culturally widespread. In much of the world, adult male-female contacts outside the kin group are carefully circumscribed due to the belief that leaving unsupervised adult men and women together would "naturally" lead to sex (Faust 1988; Whelehan's field notes; Fernea 1965; Mernissi 1975; Schlegel 1977).

Adult-male child bonds generally occur between males and those children the men believe to be their offspring. In **patrilineal** societies where descent passes through the male, and **bilineal** or **bilateral** descent societies where descent passes through both males and females but paternity is necessary for a child's place in the kinship system, knowing the biological father of the child is important. This knowledge is culturally secured by creating sexual double standards for males and females and placing female sexual behavior under restriction to ensure **paternity**, i.e., known fatherhood. Examples range from calling a female in the U.S. who has a number of sexual partners a "slut" or "whore" to "madonna-whore" complexes in many agricultural societies where a woman is seen as virtuous as a sister and mother (i.e., nonsexual) and a whore as a wife where her sexuality is expected and obvious, to "*machismo/marianisma*" complexes of Latino cultures that emphasize male sexual prowess and female faithfulness among other characteristics (see chapter 14: HIV Infection and AIDS for a more detailed discussion of *machismo/marianisma*).

Analagous to female-female bonds, male-male bonds can be both competitive and cooperative. Competitive male-male bonds in the human and nonhuman primate worlds center on dominance—status, power, position in a hierarchy—and sexual access to women. This is exemplified cross-culturally by ongoing warfare found in **horticultural** societies (a form of farming discussed in detail in chapters 7 and 8). In these societies all resources—land, water, food, women—are in scarce supply. Ongoing warfare serves to forge political alliances and is a source of women. Wives are found among the warring factions and foster alliances between different groups. In U.S. society, male-male competition can be found in amateur and professional sports as well as for positions in the labor force. Competition for women is well known. In U.S. culture, males use their power, status, and economic success to attract women (Farrell 1974, 1986; Goldberg 1976, 1979, 1980; Zilbergeld 1978, 1992). Cooperative male bonds, exemplified by fraternities, men's groups, and men's houses in horticultural societies provide a sense of male solidarity and support, often to the exclusion and derogation of females (Buckley and Gottlieb 1988; Murphy and Murphy 1974).

Estrus/Menstrual Cycles

Nonhuman and human sexual expression are similar in that they are linked to an intricate choreography of biochemical responses (see chapter 4 for discussion). However, unlike modern humans, sex and reproductive behaviors of our primate relatives and perhaps some of our hominid ones were tied to **estrus**. As previously introduced, estrus includes: sexual receptivity among females, mating-linked behaviors, and a connection to ovulation (Harrison et al. 1988). It may also be associated with a sex skin and sexual swellings in some of the primates such as the chimpanzees and with the release of chemical odors, called **pheremones**.

Pheremones, sometimes referred to as "olfactory hormones" are under the direction of the sex hormones and have a similar chemical structure (Money 1986; Katchadourian and Lunde 1975:559). As a class of "smells," pheremones are defined by Money (1986:292) as: "a volatile odorous substance or smell that acts as a chemical messenger between species and serves as a foe repellant, boundary marker, child-parent bonding agent, or lover-lover attractant." It is the latter that concerns us here. Pheremones are known to cause a response in the recipient of the odor which may be behavioral, as in attempts at mounting, and/or physiological, i.e., bio-chemical (Haynes 1994:441–442; Katchadourian and Lunde 1975:559). Pheremones have been likened to natural aphrodisiacs. This information has not escaped the interests of perfume manufacturers.

For nonhuman primates the release of chemicals or smells associated with estrus is important in sexual interactions, although more research is needed (Harrison et al. 1988; Haynes 1994). Human sexuality may indeed involve pheremones, although it has been speculated that humans and probably our hominid ancestors relied less on scents in their sexual responsiveness and more on visual cueing and communication. As a consequence of our arboreal adaptation, the visual centers were developed at the expense of the olfactory system. This may help account for the reduced role of pheremones in human sexuality, continuing a trend derived from the primate's tree dwelling adaptation. However, several studies have provided evidence of the role of pheremones and scents in modern human sexual and reproductive behaviors.

While it is established that pheremones are involved in nonhuman primate sex and reproduction, the importance of pheremones remains to be firmly established in human sexual functioning. In fact, Offir (1982) has pointed out that this is a very neglected area of research in the study of human sexuality. However, two lines of evidence are provocative; both are associated with the **menstrual cycles** of females.

Scents or smells in which the male's attractiveness to the female is enhanced (in both human and nonhuman species) are termed *copulins* and these are found in the female vagina (Offir 1982). The human apocrine glands, scattered over the body in areas associated with hair growth, are also known for the odors they produce (Doty 1981 in Frayser 1985:61). Biologists Winnifred Cutler and George Preti of Monell Chemical Senses Center and the University of Pennsylvania School of Medicine, conducted research in which the perspiration from men and women was collected systematically over a three month period. The pheremone mixture from male volunteers was placed on the upper lips of six women with irregular menstrual cycles. To avoid confounding the experiment, women were selected who were not sexually active at the time. As a consequence of this lip swatching with sweat, the cycles of the women were regularized to 29.5 days. No effect was found in a control group whose upper lips were rubbed with alcohol. Additional findings of the female pheremone mixture were also intriguing. Ten women whose upper lips were scented with the female pheremone mixture began to synchronize their menstrual cycles after only a few months (cf. Haynes 1994:442).

McClintock (1971) has reported on the phenemenon of **menstrual synchrony** in which it was found that women who live with or near one another, such as roommates and best friends, had menstrual cycles that had become synchronized. It is hypothesized that pheremones have an important role to play in this synchrony (Katchadourian and Lunde 1975:99n).

Estrus and Orgasm: Theoretical Approaches

As mentioned earlier, the evolution of human sexuality has been debated in terms of several significant events that mark the human female as quantitatively, not qualitatively, different from her nonhuman ancestors. The human female is distinguished by concealed **ovulation**; that is, human females do not go into "heat" and display visible signs of their fertility including coloration and swellings. Such signals of estrus ensure that sex will occur when the female is fertile. It is posited that our nonhuman primate ancestors developed concealed ovulation through the process of natural selection as they made the transition to human status, became bipedal, and the genitalia shifted forward. Concealed ovulation has important implications for human sexuality. It has aroused a great deal of anthropological interest, conjecture, and debate.

One controversial school of thought is exemplified in the work of Donald Symons, author of *The Evolution of Human Sexuality*, 1979. Symons offers a **sociobiological** interpretation. Generally speaking, the classic sociobiological arguments maintain that sexual behavior patterns are genetically coded and based upon different reproductive strategies for males and females. While there is great diversity in sociobiological arguments, Symons presents one view of the loss of estrus in hominids and how this event was related to reproductive strategies for both males and females.

Two possible scenarios are provided by Symons to account for evolution. Both scenarios are focused on early humans making the transition from pre-human (proto-hominid) to human (hominid) status, and both include a discussion of the role of pair bonding and "male control over female sexuality" (Caulfield 1985:348).

The first scenario exemplifies exchange theory in which ancestral males exchanged meat for sex with females. Symons proposes that during estrus ancestral females displayed their fertility and receptivity by vivid coloration and swelling of the genitals. According to Symons, those females whose sexuality was not tied to estrus were able to offer sex to males in exchange for protein foods. The male strategy was to control his partner's sexuality, in order to ensure that he was the father. This is referred to as **paternity certainty**. In this scenario, the male receives "proof of paternity," by "acquiring permanent sexual rights to a female or females" and the female and her offspring benefit from additional protein by this economic exchange for sex. Accordingly, marriage and the family develops from this situation of exchange (Symons 1979:139; Caulfield 1985:348–349).

The second scenario is a "cheat'n primate hearts" theory. In this scenario, mateship and the family evolved before loss of estrus, rather than after as in the other scenario. Ancestral females used their concealed ovulation to cheat on males who were already in a marriage with them, thereby preserving their choices in selecting the biological father of their children yet benefitting from protein resources provided by the *social* father. Hidden ovulation and loss of estrus, according to Symons, may have evolved in just such a marital situation in order for the female to select the biological father of her child while married to another male. This male would become (unknown to him of course) the *social* father (Symons 1979:141; Caulfield 1985:349). According to Symons (1979:141) "By hiding ovulation, females may have minimized their husbands' abilities to monitor and to sequester them, and maximized their own abilities to be fertilized by males other than their husbands."

Symon's view has been criticized as a gender biased position that reproduces traditional western stereotypes of men and women, so that male sexuality and female nurturing are emphasized, while female sexuality and male parenting are not. This brand of sociobiology may be contrasted with theoretical positions that see sexual behavior as "characterized by its overwhelmingly symbolic, culturally constructed, non-procreative plasticity" (Caulfield 1985:344). In addition, Symon's argument rests on male hunting and provisioning. It obscures the importance of protein sources garnered by women (Bell 1983) and the role of women in cooperative hunting and small game hunting (Estioko-Griffin and Griffin 1981; Goodale 1971; O'Kelly and Carney 1986). It also assumes that pair bonding, marriage, and the nuclear family are the ancestral forms of social organization (Caulfield 1985).

The Orgasm in Evolutionary Perspective

We would like to turn now specifically to the issue of the evolution of orgasm, and the significance of **orgasm** for reproductive success. The traditional theoretical stance is that it is not necessary for women to have an orgasm to become pregnant, but for males the contractions that expel the ejaculate into the vagina are clearly central in the process of reproduction. The view that female orgasm is not necessary for reproductive success, while the male orgasm is, is at the root of a major controversy surrounding the role of orgasm in evolution.

Two of the central figures in this debate are Sherfey and Symons. In *The Nature and Evolution of Female Sexuality* (1972), Sherfey has rewritten Freudian psychoanalytic interpretations of female sexuality. Her work reflects evolutionary considerations as well as the findings from Masters and Johnson's research (1966). After analyzing this evidence as well as that from the nonhuman primate record, Sherfey (1972) concluded that the human female was ". . . insatiable in the presence of the highest degree of sexual satiation" (Sherfey 1972:112). She clouded her argument with the idea of a period of matriarchy in human evolution. She provided little evidence to support either of her contentions: insatiability and matriarchy.

Sherfey (1972:51) proposed that female nonhuman primates showed an "extraordinarily intense, aggressive sexual behavior and an inordinate orgasm capacity" that is inherent in human females but that is shaped and suppressed by culture. This view was indeed a subversive one. It challenged western notions of a passive female sexuality embedded in sexological theorizing. Her arguments attempted to show that female orgasm was closely linked with reproductive success. For example, she hypothesized that pregnancy facilitated orgasmic response by enhancing the capacity for vaso-congestion and strengthening the uterus during orgasmic contractions.

Symons in *The Evolution of Human Sexuality* (1979:90–95) took issue with Sherfey and felt that the insatiability of females would actually be a deterrent to reproductive success. In his argument, an insatiable sexuality was more likely to detract from child rearing and subsistence activities. In Symon's scenarios, females who mated too promiscuously would actually decrease the likelihood of choice in selecting the father of their child through sheer volume of mates. According to Symon's "[t]he sexually insatiable woman is to be found primarily, if not exclusively, in the ideology of feminism, the hopes of boys and the fears of men" (1979:92).

For Symons (1979), female orgasm is a by-product of the male orgasm. Symons maintained that orgasm is "necessary" for male reproductive success because it is the vehicle through which men inseminate women. Therefore, orgasmic ability in women is merely an artifact of this capacity in males. Symons can be critiqued for confusing ejaculation with orgasm which are physiologically distinct processes. For example, prepubertal boys can achieve orgasm, but do not ejaculate. For a discussion of these anatomical differences see chapter 5: Modern Human Male Anatomy and Physiology.

If one defines reproductive success only in terms of fertilization, then erection with ejaculation and orgasmic ability is an intricate part of male sexuality and hence reproductive success, while it isn't necessary for the female's fertilization. However, such a view is gender biased and overlooks some very important recent evidence. Reproductive success includes far more than impregnation. Crucial in the process is the period of incubation and parenting. Parenting is perhaps the most important feature in the reproductive success of males as well as females.

Symons' perspective reflects western concepts of passive females and overlooks the evidence of actual female sexual functioning, such as the capacity for multiple orgasms in women. Orgasm in human females is more probably an extension of the pleasurable sensations associated with coitus in primate females generally. In all our discussions of reproductive success, we must keep in mind that the pleasure of orgasm is a powerful motivation for sexual behaviors. Without estrus marking visible fertility, increasing the sheer quantity of sex in humans is an ideal strategy for offsetting the ambiguity of concealed ovulation. That it feels good is a crucial element in the motivations of individuals, be they males or females. For Beach (1973) the argument is that with the loss of estrus, something had to happen so that reproduction would not suffer in the absence of obvious visual cues such as genital swellings and perhaps pheromones. Hominids achieved this through sheer quantity, at its simplest modeling a stimulus-response pattern of behavior. We learn how to have an orgasm; it feels good and we repeat the behavior.

In addition to these arguments, the research of Baker and Bellis (1993:887–909) offers the most serious challenge to the Symonesque view that female orgasm is a by-product of the evolution of the male orgasm. In addition, this research also suggests that the traditional perspective that the female orgasm is irrelevant to conception is an overstatement and in need of revision.

Baker and Bellis' research found that women who had orgasms any time between one minute before the male ejaculated and up to 45 minutes after retained 70 to 80% of the sperm. While it cannot be concluded that the female orgasm is necessary for conception, this research suggests that the female orgasm is functional in influencing conception (1993:883).

Summary: Sociobiology and the Feminist Interpretations

Some of the sociobiological explanations of the loss of estrus and the role of the female orgasm may be criticized as confusing issues of sex with those of reproduction (Cohen 1986:2). Sex for humans is distinctive because of the sheer breadth of its non-reproductive dimension and its cultural elaboration. In fact, the majority of human coital sexual acts are non-reproductive given the narrow window of female fertility within the monthly cycle. The female is fertile approximately only two days of her cycle. The sociobiological focus on sex as a reproductive strategy is counter to the evidence it seeks to explain since most sex acts are non-reproductive. Caulfield (1985:343–363), Cohen and Mascia-Lees (1989:351–366) and Fausto-Sterling (1985) among others offer feminist accounts which instead point to the importance of sex as symbolic and cultural rather than a purely genetic and reproductive act. It is the meanings humans give to it which has ultimately led to successful adaptations. In these feminist approaches, sex as a method of social reproduction through marriage and kinship takes precedence in our human evolution, as does non-sexual affiliation and cooperation in the survival of humans.

The feminist perspective, rather than focusing on competition between males for females, and/or between females for status and reproductive advantage, addresses human "sociability and cooperation" (Haraway in Caulfield 1985:352). Feminist perspectives such as that of Tanner and Zihlman (1976:585–608) shift the focus from the sociobiological emphasis on dominant males to a view that recognizes the importance of females in human evolution and their role in the selection of less aggressive and more sharing kinds of males. An estrus-free sexuality is seen as encouraging communication between the genders regarding their sexual interest (Caulfield 1985:353). Caulfield (1985:353) suggests that " . . . loss of estrus would have come about not as a result of individual 'reproductive strategies,' but as part of the larger process of the development of shared and learned behavior and symbolic communication." It could well be that reliance on symbolic sexuality and communicating interest was "advantageous in the period when estrous cycles were extinguished in the species" (Caulfield 1985:354).

In summary, in the feminist perspective the shared understandings embedded in symbolic sexual cues has provided an adaptive advantage to human groups, and it is those human groups on which natural selection operates (Caulfield 1985:354). In contrast, sociobiology has proposed a biological basis for male dominance and control over females and female sexuality and a passive female sexuality. The sociobiological perspective may be seen to reflect and support contemporary western patriarchal cultural notions and gender stereotypes about the nature of male and female sexuality.

Human Evolution: A Synthesis

In this section, we describe and offer an interpretation that synthesizes the biological with the cultural for understanding issues such as the loss of estrus and the orgasm in human evolution. In such a manner we hope to move beyond the critical.

With the loss of estrus and the accompanying visible genital and chemical cueing, came the capacity for sexual communication through other (non-verbal and verbal) means. If human couples copulated a great deal, then the likelihood they could be "genetic winners" in the conception game was enhanced, so selection favored sheer quantity of sex as discussed earlier (Beach 1973). The more copulation occurred the more likely it was for copulation to occur on those days within the cycle when the female was fertile. Without the cues and biological triggering, humans developed sexiness as an evolutionary compensation. In terms of the sexes, Kelso has pointed out that these evolutionary "changes have synchronized male and female sexualities more closely in men and women. The synchrony is far from perfect but much closer than between the genders in the African apes" (1980:83).

These changes in human ovulation and sexuality are related to the increased reproductive success of humans. Reproductive success consists of two interrelated factors: "reproductive capacity and ability to care for ... [the young] ... once they are born" (Kelso 1980:84). Reproductive capacity is also referred to as the "**fecundity rate**" (Martin and Voorhies 1975). In this way we can clearly see the biological basis for the unique trajectory of human sexuality. It is rooted in principles of selection and adaptation. Such abstract factors should not be confused with the immediate cultural factors influencing an individual's sexuality within specific cultural contexts. The reproductive capacity of human females exceeded that of our closest relatives. The human female has a reproductive life of 37 years with about 13 chances for conception per year, compared with 24 years for the gorilla and 23 for the chimpanzee. There is a minimum birth interval of 14–27 months for humans compared with 36 months of the gorilla and 36–60 months of the chimpanzee (Schaller 1963; Reynolds and Reynolds 1965, and Goodall 1971 in an earlier version of Kelso ND:1–12).

In addition, **reproductive efficiency** depends on the ability to care for offspring. An important part of our success as a species has been our ability not just to bear more young, but to raise those young to a reproductive age. Here success in individual terms is aided by survival and success of the same group. In this case, we have argued that early hominid groups may have been matricentered or female centered and based on the original family unit of mother and infant. Somewhere along the line this group probably expanded to include more grown-up male kin and unrelated friendly males.

It is not possible to place with certainty the exact time when early hominid females lost estrus, developed concealed ovulation and when the development of some sort of matricentric or female-centered family incorporated non-related males. These factors could well have been related to changes in human populations associated with *homo erectus* around a million and one-half to perhaps even two million years ago as suggested by recent evidence (Kelso 1980:88; Lewin 1989; Lemonick 1994).

Homo erectus is generally regarded as an "intermediate" form between early hominids and *Homo sapiens* (Harrison et al. 1988:125; Poirier et al. 1990). *Homo erectus* is known for being a gypsy, with fossil evidence found from Africa into Asia (Harrison et al. 1988:125). Kelso (1980) has argued that this widespread migration could be linked to changes in sex and reproductive evolution. Enhanced reproductive efficiency (associated with loss of estrus and the capacity for continuous sexual arousal) would be an advantage in situations of migration where high mortality is a likely concommitant (Kelso 1980:88). The migration could have supported trends which had begun much earlier with bipedalism and that were still in the process of selection or that were already set in hominid females.

Naturally this has some implications that are of keen interest. Without estrus as a signal, humans have developed through culture an incredible array of ways of communicating sexual interest to one another—visually, verbally and non-verbally (Beach 1973). As we shall see as we focus on sex and reproduction through the life cycle, culture plays a significant role in shaping modern reproduction and sexuality.

Summary

Chapter 3: Evolution of Biological Structures Related to Modern Sexual Behavior

1. The grasping hand, stereoscopic vision, and grooming are adaptations related to primate arboreal and subsequent terrestrial environments.

2. These adaptations have consequences for modern sexual behavior including the importance of touch, feeling, and vision as important components in sexual attraction.

3. Bipedalism played a critical role in enhancing trajectories begun in association with adaptation to an arboreal niche. Bipedalism fostered equilibration. In addition, bipedalism probably began trends for the loss of estrus and escalated reliance on communication and learning. Bipedalism and the consequent enlargement and development of complexity in the hominid brain was involved in the evolutionary tradeoff that selected for infant dependency.

4. Human reliance on learning is a significant aspect of our sexuality associated with the expansion of the neocortex.

5. For humans the social group is vital for survival.

6. As a continuation of nonhuman primate behavior, humans form a variety of socioemotional ties with one another known as bonds. One of the most basic is adult female-child. Male-male, male-female, and female-female bonds can be both cooperative and competitive. Adult male-child bonds often occur between men and children they believe to be their biological offsprings.

7. The human brain is actively involved in human maturation, reproduction, and sexuality.

8. The loss of estrus for humans enhanced their reproductive success in comparison with other primates.

9. Sherfey and Symons offer opposing views on the role of orgasm for female evolution. Orgasm is important as a reinforcing mechanism because it is pleasurable. This is critical in a species without estrus. Human females are unique in being able to be sexually receptive throughout their cycle with nothing more than interest to excite them. With humans, because ovulation was concealed, sheer amount of copulation would enhance the chance for impregnation. Orgasm is important in facilitating this.

OVERVIEW

Chapter 4

Introduction to the Hormonal Basis of Modern Human Sexuality

This chapter:

1. Introduces the hormonal basis of human sexuality.

2. Describes the H-P-G axis.

3. Defines and compares analogous and homologous structures.

4. Distinguishes the sexual and reproductive cycles.

Chapter 4

Introduction to the Hormonal Basis of Modern Human Sexuality

An evolutionary, interspecies perspective for our **hominid** sexuality has been presented in chapter 3. This perspective establishes outer parameters for our biobehavioral sexuality as modern humans by connecting us as a species to our primate heritage through time and space. This connection has been illustrated in discussions of the shift from **estrus** to a **menstrual cycle**, and the role of such interspecies behaviors such as **bonding** and **orgasm**. Modern adult sexual and reproductive anatomy and physiology emphasizing hormones are introduced in this chapter and discussed in more depth in chapters 5 and 6. Normative physiology and anatomy are assumed unless specifically stated otherwise. This discussion of hormones serves as a transition to exploring male and female anatomy.

A basic awareness of anatomy and physiology is important for the following reasons. Our modern anatomy and physiology is panhuman. It provides a baseline for understanding human sexuality as a biobehavioral, psychocultural and social phenomenon. Since our anatomy and physiology are shared as members of the human species, we are able to engage in intra- and intergroup sexual and reproductive experiences. While the biological aspects such as **gamete**—the egg or sperm—production are species-wide, their definition, role, and use are culture specific. Different cultures give different meanings to the reproductive and sexual processes. The behaviors, affect, and values attributed to our anatomy and its functions are culturally defined. For example, female breast development is a universal secondary sex characteristic. In current U.S. culture, it is an erotic symbol or **sex signal.** In many nonwestern cultures, breasts are not eroticized, but are seen as functional structures for nursing the young. This is obvious when one examines rules about breast coverage cross-culturally. In most places in the U.S., it is illegal for a woman (but not a man) to bare her breasts in public because of the sexual connotations. In tropical areas cross-culturally, women frequently do not cover their breasts, while their genitals, in contrast, are covered. Previous generations of many adolescent boys in this culture have poured over *National Geographic* pictorals of bare breasted indigenous women. The ethnocentrism behind this phenomenon is another issue altogether.

> **ASIDE:** On June 1, 1994 New York State passed legislation that stated women could legally breastfeed their babies in public places such as malls, restaurants, and stores.

Another important reason for having an elementary understanding of anatomy and physiology is that many biobehavioral aspects of our sexuality are only comprehensible with a knowledge of basic anatomy. This includes such components of our sexuality as sexual response, pregnancy and childbirth, and birth control. In addition, in order to comprehend how recent technological innovations that affect sexuality and reproduction impact us as a species, members of a group, and as individuals, it is necessary to have at least a baseline awareness of anatomy and physiology. Innovations such as penile implants, drugs, and prostheses to alleviate some physiologically-based erectile problems, **invitro fertilization** (IVF), **embryo transplants, gender sex predetermination,** and **sperm banks** have the potential to radically alter what has evolved as human sexuality.

Definitions of Auxiliary Terms and Concepts

The following terms are fundamental to a knowledge of sexual and reproductive functioning. You will find this list of terms useful for reference as you read about male and female anatomy and physiology.

Adrenal Glands—Two small glands, located on top of each kidney. They are responsible for most of the other gender sex hormone production in males and females (testosterone in females; estrogen and progesterone in males).

Analogous—Structures that share a similar function, such as the **ovaries** and **testes** in gamete (sperm and egg) production.

Anatomy—Refers to a specific body part or structure.

Androgens—The collective term given male sex hormones, of which **testosterone** plays a major role in sexual and reproductive development and functioning.

Endocrine glands—These glands release hormones directly into the bloodstream. Sex hormones are released by endocrine glands.

Estrogen—The term given to a group of female sex hormones found in post-pubertal males and females. It is largely responsible for primary and secondary sex characteristic development in girls.

Gonadotropin—(sex hormones) The generic term given to those hormones involved in primary and secondary sex characteristic development and functioning.

Gonads—Ovaries in the female, testes in the male. Comprises one-third of the H-P-G axis. Primary source of testosterone in males, estrogen and progesterone production in females.

H-P-G axis—Hypothalamus—pituitary—gonad axis which monitors much of the biochemical aspects of human sexuality.

Homologous—Structures formed from similar embryonic tissue, such as the ovaries and testes or the penis and clitoris.

Hormones—Substances released by the endocrine (ductless) glands that affect anatomical development and functioning.

Hypothalamus—An evolutionarily old brain structure found in many species including humans, which monitors a variety of body functions.

Physiology—Refers to the function of a part of the body.

Pituitary—(Master Gland) An organ in the brain which, as one of its functions, releases hormones necessary for sperm formation and egg development.

Primary Sex Characteristics—Those structures in both the male and female directly involved in sexual and reproductive functioning (e.g., the penis and uterus).

Progesterone—A female sex hormone found in post-pubertal females and males. It is responsible for the development of certain primary sex characteristics in the female such as uterine tone.

Secondary Sex Characteristics—Those structures in both the male and female indirectly involved in sexual and reproductive functioning, but frequently used as cultural markers of physiological sexual maturity and gender signals. (e.g., pubic hair in boys and girls, female breast development, beard growth in the male.)

Testosterone—A male sex hormone found in post-pubertal males and females which is primarily responsible for the **libido** or innate sex drive in both genders, and primary and secondary sex characteristic development in the male.

Sexual/Reproductive Structure Distinction

A given structure may serve **sexual** and/or **reproductive functions.** The **vagina** and **penis** are examples of organs which function both sexually and reproductively. **Sexual structures** are involved directly or indirectly in sexual response, without having to serve reproductively as well (e.g., **clitoris**). **Reproductive structures** are directly or indirectly involved in reproduction (e.g., the **vas deferens**).

Sexual and reproductive structures correlate with the **sexual** and **reproductive cycles** as proposed by Frayser (1985). These phenomena can be presented diagrammatically as

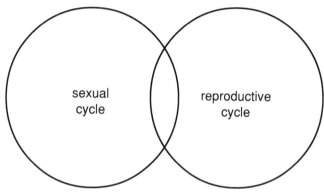

Figure 4.1

The sexual cycle relates to human sexual response, attractiveness, and sexual orientation. The reproductive cycle refers to puberty, fertility, conception, pregnancy, birth, lactation, and menopause. There is overlap in sexual and reproductive cycles and structures, e.g., the **penis** and **vagina** are both sexual and reproductive structures; penile-vaginal (P-V) intercourse has both sexual and reproductive functions.

Much of our sexual and reproductive anatomy is **analogous,** defined as similar in function and **homologous,** defined as similar in embryologic origin, with one another. On a hormonal, physiological and anatomical level, men and women are more similar than different. For example, the **gonadotropins,** the male and female sex hormones, are found in both post-pubertal men and women. The following charts list analogous and homologous structures with their functions.

Table 4.2
Comparative Anatomy

Male	Homologous	Analogous	Function	Female
penis	X		Sexual pleasure in both. Transport of urine and ejaculate in the male.	clitoris
testes	X	X	Primary source of same gender gonadotropins. Responsible for production of sperm in male, maturation of eggs in female.	ovaries
vas deferens	X	X	Transport of sperm in male, of the egg in female. Site of fertilization (conception) in female.	fallopian tubes
scrotum (scrotal sac)	X	X	Holds the testes and spermatic cord in male. Covered with pubic hair and sexually sensitive in both male and female.	labia majora
penile skin (foreskin)	X	X	Sexual pleasure. Covers glans of penis, clitoris in male and female.	labia minora, prepuce or clitoral hood at juncture with mons
glans penis	X		Sexual pleasure in both. Site of urinary meatus in male.	glans clitoris
penile shaft	X		Sexual pleasure in both. Contains internal reproductive/sexual structures in male.	clitoral shaft
urethra	X		Transport urine in both males and females. Transports ejaculate in male.	urethra
Cowper's gland	X		Lubricates urethra and neutralizes urethral acidity in male. Function in female currently not well understood.	Bartholin's gland

The following formula represents the biochemical basis of human sexuality: CC + H-P-G axis = biochemistry of sexuality.

CC = **cerebral cortex**
H = **hypothalamus**
P = **pituitary**
G = **gonads**

While the cerebral cortex can dominate the functioning of the H-P-G axis (e.g., perceived unresolved stress can effect the menstrual cycle), for this discussion we are going to examine the H-P-G axis as it usually works. The H-P-G axis operates as a **negative feedback cycle,** similar to a thermostat, where fluctuations in one part of the axis induce hormone releases in other parts of the axis.

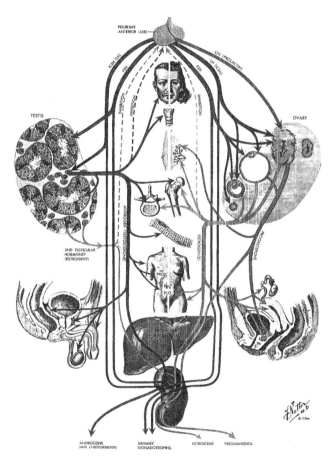

Figure 4.3   H-P-G axis graphic for male and female (Netter 1974).

While the hypothalamus and pituitary monitor a number of body functions, we will focus on their role in sexual and reproductive processes. The **hypothalamus**, located in the parietal or side area of the brain, releases a hormone, **GnRH** (gonadotropic releasing hormone), which triggers the functioning of the pituitary. In humans, the hypothalamus monitors the onset of **puberty** in both genders, the release of **pheremones**, those sexual scent cues, the release of **follicular stimulating hormone (FSH), luteinizing hormone (LH)** and **luteotropic hormone (LTH)**, and **erogenous zone sensitivity**, or those parts of the body which produce sexual arousal when stimulated. It also is part of the limbic system of the brain which influences our emotions.

The **pituitary** or **master gland** monitors a number of body functions. Relative to male and female sexuality, the pituitary gland releases **FSH**, follicular stimulating hormone, **LH**, luteinizing hormone, and **LTH**, luteotropic hormone, in response to stimulation by the **hypothalamus**. FSH and LH in turn stimulate the **gonads**: the **testes** in males, and the **ovaries** in females to produce and release their respective **gonadotropins** in each gender. **Progesterone** and **estrogen** are released in females, **testosterone** in males.

In examining the functioning of the H-P-G axis specifically, concepts of **tonicity** and **cyclicity** are applied. Tonic, which is generally applied to male H-P-G axis functioning refers to the ongoing nature of male hormone production and release. While Ramey (1973) and others clearly document the cyclic nature of testosterone over a 24-hour-period in noting that testosterone levels in males generally are highest in the morning, the general belief is that men, relative to women, are tonic. Male hormone release patterns are ongoing and continuous from puberty until death, although there is a gradual decrease in testosterone production in aging men.

> **ASIDE:** Some researchers believe that in the course of one generation, U.S. men may produce up to 50% less semen and sperm (ejaculate) than in 1950. Environmental pollutants are seen as the possible cause for this decrease (Glenmullen 1993:170).

In contrast, hormone release patterns in women are described as more cyclic, following a rhythmic flow which approximates a 28 day or lunar cycle. There is, however, variation in this pattern from individual to individual and within a woman's cycle. For example, very few women have menstrual cycles which are consistently the same length each month, making birth control options such as

rhythm, cervical mucous checks and **basal body temperature** monitoring less reliable than many other options. (See chapter 11 for a discussion of contraception). The female cycle is defined as a **negative feedback loop,** in which fluctuations in one part of the H-P-G axis influence hormone release in another part of the axis. For example, a drop in the pituitary-based FSH level in the follicular phase (described in chapter 6) of the menstrual cycle triggers the release of **estradiol,** an **estrogen,** from the ovary.

When discussing the release of hormones in the **H-P-G axis,** particularly those released by the **gonads (ovaries** and **testes),** and the **adrenals,** we are referring to **endocrine gland** functioning. Endocrine glands are ductless, which means that hormones are released directly into the bloodstream and can be measured through tests on blood samples. Excess amounts of endocrine hormones are deposited in the urine. Therefore, excess amounts of **androgens,** male sex hormones, which are part of the steroids (a group of sex and other hormones) that some male and female athletes use to rapidly increase muscle size and strength, can be detected in the urine.

The hormones involved in H-P-G axis functioning that we will discuss include the **pituitary hormones FSH, LH** or **ICSH** (interstitial cell stimulating hormone), and **LTH** (prolactin); and the **gonadotropins—androgens** commonly referred to as male sex hormones, and **estrogen** and **progesterone,** frequently referred to as female sex hormones. Each of these hormones is briefly discussed in table 4.4 on the following page.

The **hypothalamic** hormone **GnRH** (gonadotropin releasing hormone) stimulates the frontal lobe of the **pituitary** or master gland to release **FSH, LH** or **ICSH,** and **LTH** in both males and females. FSH, follicular stimulating hormone, is released in the **follicular phase** of the **menstrual cycle** during which time immature eggs develop in their ovarian follicles or sacs. This is discussed in depth in chapter 6. In men, FSH stimulates **spermatogenesis** or sperm formation in the **seminiferous tubules** in the **testicles.**

LH stands for **luteinizing hormone.** It is the same hormone as **ICSH** or **interstitial cell stimulating hormone.** LH functions in both the **follicular** and **luteal phases** of the **menstrual cycle** which will be discussed in depth in chapter 6. In the follicular phase, LH serves to stop egg maturation and helps to release the mature egg from the ovary through triggering androsterone, a male sex hormone. In the luteal phase, LH stimulates the release of **progesterone** from the **follicle** or sac that released the mature egg.

Table 4.4
Hormones Involved in H-P-G Axis Functioning

Hormone	Comparative Function Male	Gonadotropins Primary Release Male	Comparative Function Female	Primary Release Female	Source
Follicular stimulating hormone (FSH)	Stimulates spermatogenesis.	Pituitary	Egg maturation, includes estradiol (an estrogen).	Pituitary	GnRH stimulation from Hypothalamus
Luteinizing hormone (LH)-interstitial cell stimulating hormone (ICSH) in male	Maintains interstitial cells of testes.	Pituitary	Stops egg maturation, releases mature egg from ovary, induces release of androgens at ovulation.	Pituitary	GnRH stimulation from Hypothalamus
Luteotropic hormone (LTH) (prolactin)	Unknown at present.	Pituitary	Uterine tone, stimulates lactation, promotes production of progesterone.	Pituitary	GnRH stimulation from Hypothalamus
Testosterone	Primary and secondary sex characteristics, libido.	Testes	Libido, complement to female primary and secondary sex characteristics.	Adrenal glands	LH stimulation in Pituitary
Estrogen	Skin tone, reduces osteoporosis risk, complement to primary and secondary sex characteristics.	Adrenal glands	Primary and secondary sex characteristics, e.g., menstrual cycle.	Ovaries	FSH stimulation in Pituitary
Progesterone	Possibly anti-aggressor agent.	Adrenal glands	Primary sex characteristics.	Ovaries	LH stimulation in Pituitary

In men, luteinizing hormone (LH) is also referred to as **ICSH** or **interstitial cell stimulating hormone**. ICSH, which clearly describes the functioning of LH in the male, maintains the cells of the **testes** which produce **testosterone**, the primary male sex hormone. These interstitial or **leydig cells** are necessary for testosterone production which is responsible for primary and secondary sex characteristic development in males as discussed in chapter 5.

LTH or **luteotropic hormone** (prolactin) is another pituitary hormone involved in the **H-P-G axis**. It is involved in **lactation** or the nursing process, in the maintenance of uterine tone, and in the production of progesterone. While found in the male, currently its function is unknown.

This chapter is a brief overview of the hormonal basis of human sexuality. The specific hormones introduced here will be discussed in more detail in the next two chapters. Those chapters will integrate these hormones into male and female sexual reproductive anatomy and physiology.

When the limbic system, which is the center of our emotions and includes the hypothalamus is added to the CC + H-P-G axis formula (see page 72), we have a complete biochemical basis to human sexuality. The interaction of these brain functions form the cognitive, affective, and biochemical foundation of human sexuality. This basis is expressed through the physiological maturation and development process, e.g., attainment of puberty, and the learned, culturally specific behaviors, values, norms, and beliefs we as a species believe and act on verbally, nonverbally and symbolically. The latter include each culture's shared definitions of masculinity and femininity, appropriate gender role behavior, speech, demeanor, and affect as well as rules concerning what constitutes sexual, "normal," and reproductive behaviors. Hormones and behavior affect each other and this interaction will be illustrated in the next few chapters.

Summary

Chapter 4: Introduction to the Hormonal Basis
of Modern Human Sexuality

1. The cerebral cortex plus the H-P-G axis comprise the biochemical and behavioral bases of our sexuality.

2. The hypothalamus, an evolutionarily old structure, and the pituitary gland make up the biochemical regulators of our sexuality.

3. The H-P-G axis comprises a negative feedback system which influences the onset of puberty, the release of gonadotropins, the release of pheremones, and erogenous zone sensitivity.

4. Many of our sexual and reproductive structures are both analogous and homologous. Men and women share a hormonal system. This means that biochemically, men and women are more similar than they are different.

5. The sexual and reproductive cycles are distinct, but overlapping systems.

6. Men's hormonal functioning is frequently described as tonic, women's as cyclic.

7. The major sex hormones are testosterone, estrogen, a group of hormones, and progesterone.

8. Culture and biology interact in the expression of our sexuality.

OVERVIEW

Chapter 5

Modern Human Male Anatomy and Physiology

This chapter:

1. Applies the formula of the CC + H-P-G axis = biochemical behavioral aspect of human sexuality to males.

2. Discusses the role of FSH (follicular stimulating hormone) and LH (luteinizing hormone) or ICSH (interstitial cell stimulating hormone) as a tonic process in males.

3. Discusses the external and internal anatomy and physiology of the male sexual and reproductive systems.

4. Discusses male primary and secondary sex characteristics.

5. Introduces the concept of the libido and relates it to testosterone levels.

6. Discusses the effect of alcohol and marijuana use on testosterone levels.

7. Introduces HIV infection and AIDS in men.

8. Introduces male sexual response.

Chapter 5

Modern Human Male Anatomy and Physiology

In this chapter and the next one on modern female anatomy and physiology physical normalcy is assumed unless specifically stated otherwise. A discussion of male anatomy incorporates the CC + H-P-G axis formula presented in the previous chapter.

Applying the CC + H-P-G axis formula to males involves a hormonal exploration of **FSH** (follicular stimulating hormone), **LH** (luteinizing hormone), **ICSH** (interstitial cell stimulating hormone), and the **gonadotropins**, particularly the **androgens**, most specifically **testosterone**. Having an H-P-G axis which is frequently referred to as **tonic**, men tend to have a more continuous release of H-P-G axis hormones in their bodies than do women, whose more rhythmic release is described as cyclic. Men's hormonal patterns continue from puberty until death. In men, the hypothalamic release of GnRH (gonadatropic releasing hormone) triggers pituitarian FSH and LH (ICSH) activity. FSH and LH activate testicular functioning and the production of **androgens**, male sex hormones. FSH aids in **spermatogenesis** or sperm production which occurs in the **seminiferous tubules** of the testicles. LH (ICSH) maintains and promotes the integrity of the **interstitial** cells of the testes, the major source of testosterone production. Testosterone is the primary androgen referred to as the male sex hormone. It is the one on which we will focus.

Testosterone is produced in the interstitial cells of the testicles in men. Another name for the interstitial cells is the **leydig** cells. In women, most of the testosterone is produced in the **adrenal** glands. Testosterone is a crucial sexual cycle hormone in both genders and a reproductive cycle hormone in males. On a hormonal basis, testosterone is responsible for the **libido** or sex drive in both men and women. The amount of testosterone required to maintain the libido in men and women is referred to as the **threshold** level, and it exists in roughly the same amounts in both men and women. As long as this threshold level is maintained, the hormonal aspects of the libido are present in both men and women. Men and women produce testosterone from puberty until death. Thus, both men and women can maintain a hormonal basis for a libido from sexual adulthood, i.e., puberty, through sexual and reproductive aging (e.g., postmenopause).

Testosterone as a major male sex hormone is actively involved in the expression of primary and secondary male sex characteristics. For this to occur, men continuously produce from puberty until death about 10 times as much testosterone on a tonic basis as do women (Greenberg, Bruess, and Mullen 1993). Generally, the tonic release of free circulating testosterone in the male suppresses or binds the release of estrogen and progesterone, the female sex hormones, in the male. The primary sex characteristics are those directly related to sexual and reproductive functioning. In the male, they include the growth and development of the internal and external penis, the testes and scrotal sac, and the auxiliary reproductive structures such as the vas deferens, seminal vesicles, and epididymis.

The secondary sex characteristics are those features less directly involved in reproductive functioning, but probably highly involved in sexual functioning. Secondary sex characteristics involve structures that often are culturally defined as visual sexual and gender cues and indicators of sexual adulthood. Since spermatogenesis is invisible as contrasted to the visibility of menstruation, the appearance of secondary sex characteristics in the male can be used culturally to define sexual and reproductive adulthood in men. This illustrates how physiology can be culturally integrated and interpreted on a behavioral and attitudinal level. For example, the production of testosterone results in beard growth that can be used to define masculinity and manhood. In this culture, a boy's physical ability to produce facial hair is symbolically and behaviorally recognized as a sign of becoming a man. Shaving or plucking male facial hair can serve as part of grooming and hygiene in U.S., European, and native American cultures.

Secondary Sex Characteristics

Testosterone-induced male secondary sex characteristics include a number of features. It is important to remember that these are general patterns and a lot of normal individual variation exists within these patterns. These features are also relativistic, not absolute in comparison to women. For example, as in the primate world, generally men are taller, more muscular, and have more facial and body hair than do women. Sex hormones also interact with genetic and cultural variables to result in adult characteristics, e.g., height has a genetic, hormonal, and cultural basis. Genetic tendencies to be tall or short are reinforced by hormonal release that promotes bone growth and later closes the epiphyses, the ends of the bones, which are further influenced by such cultural practices as nutrition.

Blonds tend to be the hairiest of people, while Africans, Asians, and Native Americans have less facial and body hair. Even within specific kin groups there are individual variations. The pattern and distribution of facial hair, overall body hair, and pubic hair are a function of testosterone. Beard growth and the male hairline shape, but not the amount of head hair,[1] are the functions of testosterone. Men tend to have a scalloped shaped hairline as contrasted to the women's which is more ovoid.

Figure 5.1   Man's hairline.

Figure 5.2   Woman's hairline.

ASIDE: For some forms of reproductive cancers in women, androgens were given as part of the chemotherapy. Side effects of the androgens given to these women included secondary male sex characteristic development in these women. In part, the effectiveness of treatment was measured by beard growth in the women.

In addition, the diamond shaped pubic hair patterning in men as compared to the inverted triangle pattern in women is related to **testosterone**. As with women, men's pubic hair varies in color and amount with the individual and is subject to the aging process. Generally, it is curly, soft, and sensitive to sexual stimulation.

While also a function of genetics and cultural practices regarding nutrition, exercise, and bone development, height and bone growth are related to testosterone. Generally, men are taller than women, with longer, heavier, and denser bones. This also allows them to have greater physical strength and speed than women.[2] The combination of testosterone and estrogen also puts men under 70 years old at lower risk for **osteoporosis,** a degenerative bone disease common in older women.

The enlargement of the larynx that deepens the male voice is the one nonreversible secondary sex characteristic. The enlargement of the larynx and changed voice are permanent, even if testosterone ceases to be produced. In male castration or orchidectomy the testicles are removed. This happens, for example, as a treatment for some prostate cancers, for testicular cancer, or as part of male-to-female transsexual surgery. While other secondary sex characteristics are lost, the deeper voice remains. This has resulted in some male-to-female transsexuals who have deep voices taking voice lessons to soften and raise their voices. In contrast, female-to-male transsexuals benefit from androgen therapy since their voice "naturally" deepens. The male *castrati* who sang in choirs during the middle ages were castrated prior to puberty. Castration also meant that their overall bone size was smaller. Their chest bones were less well developed than non-*castrati*, which contributed to a higher voice. *Castrati* had smaller penises and did have active sex lives. While testosterone is responsible for the sex drive (interest in sex) and reproductive male sex characteristics such as the production of sperm and semen (ejaculate), it is not responsible for erections or feeling sexual pleasure per se. If this is confusing, remember that fetal and prepubertal boys are capable of erections. Prepubertal boys are also capable of masturbating, feeling sexual pleasure, and having an orgasm; they do not, however, ejaculate. *Castrati* tended to lead charmed lives. According to Henderson (1969), the fortunate few boys selected as a *castrati* would have the choicest food, clothes, homes, and women at their disposal for the length of their careers. *Castrati* were not perceived with the same kind of horror that we perceive them in retrospect (Davis personal communication; Bullough 1976; Rice 1982). Eunuchs, also castrated, reflect feminized fat deposition.

> **ASIDE**: Contrary to some popular beliefs, the rock star Michael Jackson sings in a learned falsetto. Scandal sheet reports of castration or female hormone usage sell tabloids, but do not reflect reality.

Men generally also have more muscle mass,[2] are leaner, and have a lower body fat to overall body mass ratio than women. This contributes to men's overall greater physical strength. In current U.S. standards of aesthetic leanness for men, the range of body fat is from about 8% for athletes to about 14% for the "average" male. Athletes in training, for example, football players, may try to achieve a 4–6% body fat content during the playing season. They may be muscularly "bulky," but they are not soft-tissue fat. Body fat distribution and storage in men tends to be upper body fat. Some physical aspects of men's comparative leanness and body fat distribution are that men tend to carry "spare tires," "love handles," or "pot bellies" of excess body fat around their midsections; they may be more prone to coronary heart disease; and they float less easily than women.

Testosterone has an effect on men's skin. Men's skin tends to age more slowly, has fewer wrinkles, and is more prone to acne than women's. Men tend to have more severe acne than do women because testosterone can stimulate sebaceous (oil) gland secretions which contribute to acne. Skin smoothness and aging are also related to their estrogen levels that men retain from puberty till death, as well as cultural practices such as shaving, which removes dead skin cells.

Libidinous functioning or having an interest in sex has attributes of both primary and secondary sex characteristics. As stated previously, a comparable amount of testosterone known as the threshold level is required in both males and females in order to have an interest in sex. Again, this baseline physiological level integrates with cultural values and beliefs about how, when, where, with whom, and how often the sex drive is expressed.

Estrogen and progesterone in men are primarily produced in the adrenal glands and generally are suppressed or bound by the testosterone. Unbound estrogen can produce secondary feminizing sex characteristics such as **gynecomastia** or breast development, loss of facial and body hair, or reduced sex drive.

Progesterone, which does not produce feminizing secondary sex characteristics in men and is given in various forms to some convicted sex offenders as therapy, acts as an antilibido hormone. It not only acts to diminish libido in both males and females, but may mitigate aggressive feelings as well. Silber (1981), a researcher in this area, has administered progestin-based drugs to some convicted sex offenders who have chronically elevated testosterone levels.

> **ASIDE**: A urologist named Silber who investigates male sexuality has developed a theory about certain kinds of sexual behavior related to hypertestosterone levels. The normal level of free circulating testosterone found in men allows for primary and secondary sex characteristic development and a sex drive. A few men, however, have chronically elevated levels of testosterone well outside the range of normal. These chronically elevated levels of testosterone coupled with strongly internalized and culturally supported values on aggression and violence as a means of expressing anger and frustration or resolving conflict may be involved in some of the more dramatic sex crimes. Silber hormonally tested and interviewed a number of men convicted of sex crimes which also involved extreme forms of violence (e.g., rape and body mutilation or dismemberment). In this sample, he found chronically elevated levels of testosterone, a psychological connection between thoughts of sexual violence and heightened arousal, and acceptance of physical violence as a means of expressing anger or frustration and resolving conflict. When these men in his prison sample were given progestin-based drugs, similar to those being given to some convicted sex offenders in this country, their testosterone levels lowered to within normal limits. The testosterone levels remained within normal limits as long as the progestin-based drug was taken. If the drug regimen stopped, the testosterone levels increased to their previously elevated chronicity. On the drug, the prisoners reported less of a connection between sexual arousal and violence (Silber 1981). This is fascinating and potentially powerful research which has tremendously controversial and social and legal implications.

Drug usage definitely affects male sex hormones, particularly testosterone. Two commonly used drugs that affect testosterone levels are alcohol and marijuana. Extensive, chronic alcohol and marijuana abuse can suppress testosterone levels below the threshold level. This can result in loss of libido and the appearance of feminizing secondary sex characteristics such as gynecomastia, or breast enlargement, increase in overall amount and redistribution of body fat, body hair loss, and beard softening. These effects are reversible if the alcohol or marijuana drug abuse stop.

Steroids also affect secondary sex characteristics, particularly muscle size and the lean muscle mass to body fat ratio. Steroids contain androgens that can rapidly increase muscle mass. Their use for this purpose is illegal in most formal athletic situations, e.g., the Olympics. Steroids are stored in the body for at least six weeks and excess amounts are secreted in the urine. This phenomenon explains the mandatory urine tests for steroids in competition-based athletes and for some employees. Steroids also can cause general metabolic problems. They are powerful and potentially dangerous drugs. If abused they may damage the kidneys, liver, and heart, or even result in the user's death. Injection of them through shared needles also puts the user at risk for hepatitis B, for HIV infection, the virus which is related to AIDS, and for other infections.

Primary Sex Characteristics

Primary sex characteristics are those features in men and women that are directly involved in sexual and reproductive functioning. In men, they are hormonally controlled by the androgens or male sex hormones. Testosterone is the major gonadotropin involved in the growth and development of these characteristics.

The **penis,** which increases in size at puberty, is composed of internal and external structures. The external penis includes the **glans**, **shaft**, and **crura** or **root**, structures which are **homologous** to the female's **clitoris**. The glans of the penis, which is

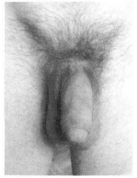

Figure 5.3 There are many variations in the shape and size of the male genitals. The penis in the right photo is uncircumcised; © 1978, Justine Hill.

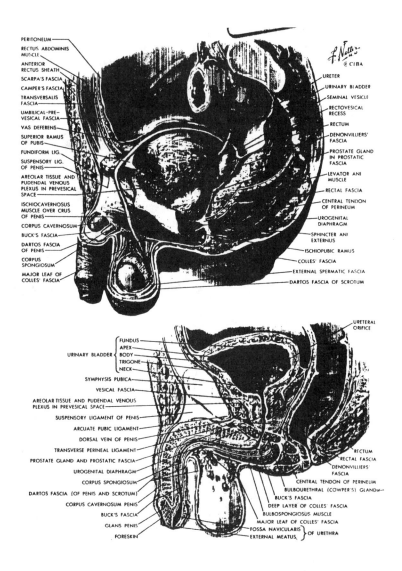

Figure 5.4   Pelvic structures.

acorn-shaped, is formed by the internal **corpus spongiosum**. The corpus spongiosum also contains the **urethra,** which ends in the glans at the opening called the **urinary** meatus. The glans or head of the penis is hairless and is extremely sensitive to sexual stimulation. It is covered with penile skin called the **foreskin**. In some cultures, including the U.S. since about 1850, males have the foreskin surgically removed in a procedure known as **circumcision**. Male circumcision is not universally practiced. Nonwestern societies that practice circumcision usually do it for social and symbolic reasons relating to status changes in males. U.S. society originally performed circumcisions to reduce masturbation. Currently, over 85% of boy babies in this culture are circumcised often without anesthesia within two days of birth for social reasons which have been given medical and hygienic explanations. These include reduction in body odor from accumulation of **smegma**, possible reduction for risk of HIV infection, increased sexual sensitivity, and to be "like the other boys." Currently in the U.S., circumcision is a controversial practice whose medical rationale is certainly suspect.

At the base of the glans penis where the head of the penis and the shaft meet, is the **frenum** or **frenulum**. This small structure is somewhat triangular in shape, sensitive to sexual stimulation, and is the place where the foreskin attaches to the glans. At either side of the frenulum are the preputial glands, which secrete **smegma**. Smegma is a waxy, lubricating substance that allows for smoother retraction of the foreskin over the glans. With circumcision, **smegma** no longer collects under the foreskin. In contrast to Westerners, some Arab groups and Polynesians are meticulous about male genital cleanliness and see men in Europe and the U.S. as dirty by comparison (Marshall and Suggs 1971; Bullough 1976).

The **shaft** is the body of the penis that is covered with relatively loose, hairless skin. The shaft contains the corpus spongiosum and **corpora cavernosa** or cavernous bodies. The shaft increases in size and firmness during arousal as the cavernous bodies engorge with blood to create an erection. The shaft is sensitive to sexual stimulation. Its size, erectile ability, connection with sexuality, and fertility are of great cultural interest in the U.S. and cross-culturally. In this culture, penis size is of major concern to men regardless of sexual orientation. In some other cultures, genital surgery on the shaft such as **subincision,** an incision on the underside of the penis, or **superincision**, an incision on the ventral (top) side, is performed to heighten men's sexuality, their masculinity, or as an indication of status change from boyhood to manhood. These procedures will be discussed in more detail in the chapter on adolescence. The **crura** or root forms the base of the penis and can be felt externally as a ridge at the point where the penis attaches to the body at the lower abdomen.

The **scrotal sac** or **scrotum** is a multilayered pouch of loose skin located behind the penis that contains the **testicles, epididymis,** and **spermatic cords**. The scrotal sac generally is hairless and sensitive to sexual stimulation. It is homologous to the **labia majora** in females.

From the base of the genitals to the anus is an area of skin in men and women called the **perineum;** in men the perineum extends from the scrotal sac to the anus. The perineum is soft, generally hairless, and sensitive to sexual stimulation. Stimulation of the prostate gland through the perineum can be highly erotic for many men (Ladas, Whipple, and Perry 1983).

The **testes** or **testicles**, derived from the Latin, *testare,* to testify, are two spherical spongy bodies located in the scrotal sac. They are **homologous** and **analagous** to the female's ovaries. They are mobile in the scrotal sac. Testicles can be elevated or lowered in the scrotal sac in response to such factors as temperature, surprise, or fright. The ability of the testicles to move in the scrotum contributes to reproductive success. Retraction of the testicles in response to a threat may preserve them! Since sperm can only exist at temperatures of about 32° C, having sperm produced outside the body cavity helps to ensure their survival. In addition, retraction and extension of the testicles in response to external temperature maintains the proper temperature for sperm production and survival. Many men have experienced this when diving into a cold body of water and felt their testicles retract toward their body. Most simply, the testicles, which may rest unevenly in the scrotum, are the site of **spermatogenesis**. The internal construction of a testicle is composed of tightly coiled tubes called **seminiferous tubules,** and a cellular arrangement resembling either a sponge or honeycomb. This cellular arrangement is referred to as the **interstitial** cells or **leydig** cells. They are the source of **testosterone**, whose production is influenced by the release of FSH and LH from the pituitary gland. LH acts directly on the leydig cells to maintain their integrity so that testosterone can be produced and secreted. FSH acts on the formation of sperm in the seminiferous tubules. Testosterone release is continuous from puberty until death, as is sperm production, though both diminish in quantity as men age. Millions of sperm are produced in the seminiferous tubules each day from puberty until death. They are produced in an immature state and are matured outside the testicle, in contrast to the mature egg released by the ovary.

ASIDE: Testicular cancer is most commonly found in men in their twenties. As a means of early diagnosis of testicular cancer and as a means of maintaining a man's overall andrological health, regular testicular self-exams (TSE's) are encouraged. They need to be performed regularly and consistently, similarly to breast self-exams (BSE's) in women. TSE's are most effective after a warm shower, when the scrotal sac is relaxed and the testicles are descended. Visual examination to detect changes in color, size, shape, or to note the appearance of lumps or growths is done first. Then the man is encouraged to gently palpate (feel) the testicles and spermatic cords, initially to familiarize himself with his own anatomy, and secondarily to feel for any unusual lumps or changes. NB: For many men, one testicle rests lower in the scrotal sac than the other. This is an anatomically common occurrence.

From the seminiferous tubules, the newly produced immature sperm transverse the testes to the **epididymis**. The epididymis are two, crescent shaped, grayish structures which curve around each testicle. They house the immature sperm for approximately 4–6 weeks until they are sufficiently mature to be released into the **vas deferens**. It is important to remember that the manufacture and maturation process of sperm is continuous. Millions of sperm are produced and matured daily.

The **spermatic cords**, located on the side of each testicle and extending to the pubis, contain several structures such as the **cremasteric muscle**, **vas deferens**, blood vessels, and nerves. The spermatic cord itself functions to raise and lower each testicle in the scrotal sac. The difference in the length of the spermatic cords is the reason that one testicle may hang lower in the scrotal sac than the other testicle. As stated earlier, sperm are heat sensitive and need a temperature of about the 32° C in order to survive. The elevation and lowering of the testicle in response to changes in temperature, fright, stress, or sexual arousal is a function of the cremasteric muscle. The actual response of the cremasteric muscle is called the **cremasteric reflex**. This reflex can be triggered spontaneously by running the side of one's thumb quickly along the inner thigh of an unsuspecting male. The testicles will spontaneously contract. Triggering this response is not recommended unless you know the male well!

Another structure located in each spermatic cord is the **vas deferens**. The vas deferens, analogous and homologous to the **fallopian tubes**, transport the mature sperm from the epididymis out of the pelvis to the **seminal vesicles** and **ejaculatory tracts** or **ducts**. Release of mature sperm by the epididymis into the vas deferens is continuous. However, during times of intense sexual arousal, an average of 200–400 million sperm are released into the vas deferens. The vas deferens begin as external structures in the spermatic cord and then proceed internally to loop around behind the bladder until they join with the ejaculatory tract on each side of the man's body. The vas deferens are the site of a **vasectomy**, the most common form of voluntary male sterilization.

After looping around behind the bladder, the vas deferens connect with the **seminal vesicles**. There are two of these structures as well, on each side of the man's body, adjacent to the bladder. The seminal vesicles produce the majority of **semen**.

After the vas deferens loop around the bladder joining the seminal vesicles, the vas deferens become the ejaculatory tract. The juncture of the vas deferens and the seminal vesicles is known as the ejaculatory duct. The sperm carried by the vas deferens now becomes part of the ejaculate when the seminal vesicles release semen into the ejaculatory tract and the semen mixes with the sperm.

At the point that the vas deferens becomes the ejaculatory tract, the tract now contains **ejaculate**. Ejaculate is composed of semen and sperm. By volume, ejaculate is about 98% semen and 2% sperm. Semen is a pearly-colored, sticky, viscous fluid that leaves a white stain on material such as clothing or bedding when dried. It will, however, wash out of clothing or bedding. Semen is essential to sperm transport and survival. Semen is an alkaline or basic substance with a pH of about 9.5. Sperm needs an alkaline environment in order to survive.

Semen is composed of a number of ingredients including albumen, the same substance found in egg whites that gives semen its slippery texture; sugars-glucose and fructose; bases, which give it its salty taste; and proteins. Semen has several functions. Its composition nourishes the sperm. Semen also is a transport medium for the sperm, aids in sperm motility, and lubricates the urethra. The amount of ejaculate per expulsion averages about two teaspoons, though it may feel and appear to be by the male and his partner(s) a great deal more. Ejaculate contains from 20–37 calories each expulsion. About 23 samples of ejaculate equal the calories in a piece of lemon meringue pie!

> **ASIDE**: There is an infamous story about the prostitute who went to a doctor to be put on a diet. She followed the prescribed regimen carefully, but still did not lose weight. Frustrated, she returned to the doctor who told her to record everything she swallowed. Her special service to her clients was fellatio, oral sex, or giving head. She swallowed sufficient amounts of ejaculate in a day to prevent her from losing weight.

To note, this true story occurred in the pre-AIDS era. Currently, with the very real concern and problem with HIV transmission through semen and vaginal fluids, oral sex on either a man (**fellatio**), or woman (**cunnilingus**), is *only* a *safer* sex activity when using a condom on the penis or vaginal dam barrier over the vulva, the external genitalia, of the female. The HIV virus is found in semen; its effect on sperm is currently being studied (Padian 1987; IXth International AIDS Conference 1993). On the risky sex continuum from riskiest to least risky sex, HIV exists in sufficient quantities in semen to infect a partner through unprotected penile-anal, penile-vaginal, or oral sex (IXth International AIDS Conference 1993). Therefore, properly used latex barriers such as condoms and vaginal dams need to be used consistently to reduce the risk of infection (see also chapter 14).

Semen, part of the ejaculate, an important substance in male sexuality and reproduction, also has socio-cultural dimensions. In many cultures it is recognized as a vital life substance. There are a variety of beliefs about its functions, quality, and quantity. Barker-Benfield (1975) has coined the term "**spermatic economy**" to connote some Mediterranean groups and nineteenth century British attitudes toward semen (sic) i.e., ejaculate. In these cultures, ejaculate is seen to exist in finite supply and judicious caution against "spending" ejaculate freely; e.g., masturbation, is advised. Among the Sambia in New Guinea, prepubescent and adolescent boys ritually engage in fellatio (oral sex) in order to build strength and physical reserves of ejaculate so that they do not run out of it in adult heterosexual relations (Herdt 1982). Among the Sambia, women are seen as sexually powerful and voracious. Ejaculatory contact with women is carefully regulated so as not to use up all of a man's vital life essence (see also chapter 10).

In our own culture, we have mixed views on the wisdom of frequent ejaculation or "spending." One philosophy promotes a "use it or lose it" approach: the more orgasmic (ejaculatory) one is, the more one will continue to be (Masters and Johnson 1974). The other approach, exemplified by college athletic coaches from the late-nineteenth to mid-twentieth centuries encouraged a "spermatic economy" perspective. Male athletes were advised not to engage in ejaculatory sex before an event so that they would "save" their strength and energy. Currently, most U.S. college coaches recognize this piece of folklore, and most do not pass it on as serious advice to their players (Gordon 1988).

The ejaculatory tract is essentially an extension of the vas deferens. The ejaculatory tract transports the ejaculate, i.e., sperm and semen, to the **urethra**. The urethra is surrounded by the

prostate, which produces the balance of the semen to deposit a full ejaculation in the urethra.

The prostate is a walnut shaped, spongy organ that lies below the bladder. The urethra runs through it. The prostate can be felt through the perineum and by finger insertion into the rectum. Perineal and rectal stimulation of the prostate can produce intense levels of sexual arousal. The prostate produces semen that contributes to the ejaculate carried in the ejaculatory tract. The prostate is a common site of both minor irritation and major problems. In younger men, prostate trouble can be due to either localized or systemic infection or irritation and is known as **prostatitis.** This can generally be easily remedied through antibiotics. In older men, enlargement of the prostate due to either atrophy as part of the natural aging process or due to prostate cancer commonly occurs. Prostate cancer currently is most reliably diagnosed by a combination of a PSA blood test and digital rectal exam (Oesterling 1991:24). It occurs in geometric proportion to age: in 50-year-old men, there's a 40% chance of enlargement, in 60-year-old men, a 50% chance and so forth. It's pretty much a given that the older a man lives, the greater the chances are that he will have problems with his prostate. One of the more serious immediate concerns of an enlarged prostate either due to irritation, atrophy, or cancer, is that the enlargement constricts the urethra. Constriction of the urethra makes urination painful, difficult, or impossible. In fact, painful, slow, or incomplete urination is frequently a sign that prostate problems exist. Treatments include antibiotics, or in the case of enlargement or cancer, surgical removal of the prostate. For benign, noncancerous prostate enlargement, either drug treatments, laser therapy, or a TURP is performed. TURP stands for transurethral resection of the prostate. A man does not ejaculate after a TURP, but should retain erectile and orgasmic ability. For prostatic cancer, more radical surgery is performed. Currently, an orchidectomy, or removal of the testicles may be performed in some cases of prostatic cancer to avoid testosterone feeding the cancer.[3]

The male's urethra runs from the base of the bladder through the corpus spongiosum, the underside cylinder of the internal penis, ending in the urinary meatus at the glans of the penis. The male urethra has two functions. It transports urine from the bladder to outside the body. It transports the ejaculate, deposited in the urethra during the emission phase of male ejaculation, from the ejaculatory tract and prostate to outside the body. Both urine and ejaculate leave the body through the urinary meatus.

As stated previously, sperm survive in an alkaline or basic environment. Urine is acidic and the urethra can be acidic from transporting urine. To counteract the acidity of the urethra so that sperm can survive, two phenomenon occur. There is a sphincter or small closure between the bladder and urethra. This sphincter closes during arousal and ejaculation so that urine does not leak into the urethra and damage sperm. The common belief in U.S. culture that one may swallow urine when swallowing ejaculate, "cum," during oral sex is therefore erroneous. Again, it is important to remember in this age of AIDS, that oral sex is risky without using a condom from beginning to end, in which case oral sex becomes safer.

> **ASIDE**: Some yogis and Robert Noyes, the founder of the Oneida Community, a religious sect started in the nineteenth century near Oneida, New York, claim to be able to attain conscious control over the urinary sphincter. They use this discipline as a part of their birth control. By concentrating intently during the emission phase of ejaculation when the ejaculate enters the urethra, they open the urethral sphincter and force the ejaculate into the bladder instead of out through the urinary meatus. This process is called retrograde ejaculation which also occurs as a side effect of a TURP, discussed previously. It may also occur as a side effect of taking some major tranquilizers or severe alcohol abuse. When the man urinates, the ejaculate is expelled, causing the urine to have a milky-white coloration.

To further counteract the acidity of the urethra, secretions from the **Cowper's glands** neutralize acid levels and lubricate the urethra for the passage of ejaculate. Cowper's glands and ducts are located just beneath the prostate on either side of the urethra. They are homologous to Bartholin's glands in the female. They release a clear, slippery fluid known as **preejaculatory** fluid or precum into the urethra. This fluid flows through the urinary meatus immediately prior to ejaculation and may be used to lubricate the glans and increase stimulation. Since this fluid may contain sperm or traces of semen, it is important not to swallow it or to have it come in contact with either the woman's genitals or the anus of either gender in order to avoid possible HIV infection or conception in the case of heterosexual genital contact.

The internal penis is composed of three cylindrical or corpus bodies: the **corpora cavernosa**, Latin for cavernous bodies, and the **corpus spongiosum**, Latin for spongy body. These bodies appear like this:

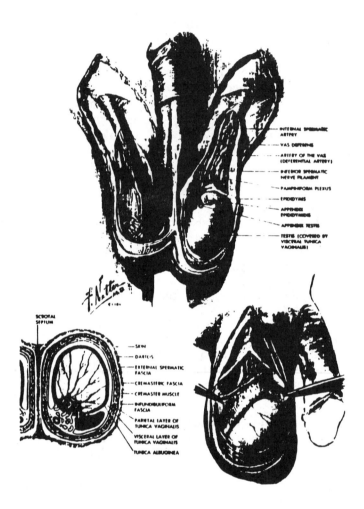

Figure 5.5 Scrotal wall.

The corpora cavernosa are the top two cylindrical bodies of the penis. They are composed of spongy tissue and a rich vascular or blood supply. During sexual arousal, it is primarily these two structures that engorge with blood to create an erection. In addition to neural responses, an erection is achieved and maintained vascularly as long as the blood flow into the corpora cavernosa occurs faster than the blood flow from it; this process is helped by sphincters which close to keep the blood in the cavernous bodies. Human males do not have a penis bone or other structure to maintain an erection. One drug known to have an effect on the vascular structure of the corpora cavernosa is nicotine. Nicotine constricts blood vessels. Since free circulating blood is physiologically important in achieving and maintaining an erection, smokers and chewers may impair full erectile ability. People who have stopped smoking or chewing tobacco and whose bodies are nicotine-free report quicker, fuller, firmer erections (Buffum 1982). Lack of nicotine allows the blood vessels of the penis to open more completely. Please see chapter 11, Human Sexual Response and Birth Control, for a discussion of other drugs which can effect erectile ability.

The third cavernous body of the internal penis is the corpus spongiosum. The corpus spongiosum forms the glans penis and is the structure through which the urethra runs ending in the urinary meatus at the tip of the glans. Men who have been subincised, have had the penis slit open as part of initiation, as discussed in chapter 10, do not urinate through the urinary meatus in the glans. Urine is released farther back along the urethra.

Male internal and external genitalia comprise his sexual and reproductive structures. The external genitalia, in contrast to the female's, are highly visible. Both internal and external structures operate dramatically during male sexual response. The distinctions between conscious and unconscious (out-of-awareness) responses and among erection-ejaculation-orgasm are introduced here and will be discussed in more detail in chapter 11.

Sexual response in general is an interaction of conscious and unconscious mechanisms. The conscious awareness involves the cerebral cortex, the limbic system (feelings), and to some extent the hypothalamus. It includes the perceived, learned triggers of arousal, and awareness of excitement or erogenous zone activity. The unconscious, out-of-awareness mechanisms involve the hypothalamus, the neural responses such as triggering the reflex arc on the spinal column, hormonal release (H-P-G axis functioning), and vascular responses. Of these two mechanisms, the conscious may dominate, defining pleasure, sensuality, sexuality and the perception of the intensity of arousal and orgasm. In the male this relates specifically to perceptions of erectile firmness, "staying power" (ability to maintain an erection), ejaculatory force, sensation, and amount. These are learned, culturally patterned responses. They can override physiological response and ability as evidenced in the sensual-sexual arousability and pleasure experienced by people with spinal cord injuries.

Erection, ejaculation, and orgasm are physiologically distinct processes, although they may be perceived as being the same, particularly male orgasm and ejaculation. The fact that postpubertal males often achieve orgasm and the expulsion phase of ejaculation concurrently reinforces the belief and sensation that orgasm and ejaculation are the same in males.

> **ASIDE**: The voluntary control over erection is dramatically illustrated by the true story of Swami Rama. Swami Rama was debating with a group of U.S. graduate students about the voluntary aspects of erection. To illustrate the conscious control in erection, he had a graduate student fill a 5 gallon bucket with water. Before his incredulous students, he spontaneously generated an erection without overt stimulation of his penis and proceeded to suspend the bucket of water from his erect penis! The man must have done thousands of kegels, an exercise which strengthens the muscles surrounding the crura. This is the most impressive case of the integration of conscious and unconscious mechanisms in erection that this author knows.

Summary

Chapter 5: Modern Human Male Anatomy and Physiology

1. The CC + H-P-G axis discussed in chapter 4 is applied to males.

2. FSH and LH are involved in the tonic process of spermatogenesis.

3. Androgens, particularly testosterone, are involved in male primary and secondary sex characteristic development. The role of estrogen and progesterone in men is presented.

4. Male internal and external sexual and reproductive anatomy and physiology is discussed relative to normal functioning, the libido, and effects of alcohol, nicotine, and marijuana use on male libido and sexual response.

5. It is possible for men to contract HIV infection through unprotected penile-anal intercourse, unprotected penile-vaginal intercourse (particularly if the women is menstruating), and either unprotected fellatio (oral

sex on a male), or unprotected cunnilingus (oral sex on a female), particularly if she is menstruating.

6. Cultural responses to male sexual and reproductive functioning include such practices as circumcision, subincision, and superincision, as well as cultural beliefs about sexuality.

7. Common problems of the prostate such as prostatitis and enlargement of the prostate can occur in men across the life cycle and increase as he ages.

8. Erection, ejaculation, and orgasm are physiologically distinct processes.

OVERVIEW

Chapter 6

Modern Human Female Anatomy and Physiology

This chapter:

1. Applies the formula CC + H-P-G axis to female sexual and reproductive anatomy and physiology.

2. Describes the cyclic, negative feedback aspects of the female's H-P-G axis.

3. Presents the menstrual cycle's analogies and differences to spermatogenesis.

4. Discusses the female's primary and secondary sex characteristics including those that are homologous and analogous with the male.

5. Describes the four phases of the menstrual cycle as well as cultural responses to it. Describes menstrual cramps, menstrual synchrony, and PMS.

6. Introduces cultural responses to female anatomy and physiology.

7. Introduces models of female sexual response.

8. Introduces conception and recent western technologies that increase the chances of conception and gender selection.

9. Summarizes the importance of knowing basic sexual and reproductive hormones, anatomy, and physiology.

Chapter 6

Modern Human Female Anatomy and Physiology

The discussion of female anatomy and physiology parallels that for the male. On a hormonal basis, the formula and systems are analogous for both genders: the cerebral cortex (**CC**) + **H-P-G** axis is involved. On a relative scale women are described as **cyclic**. In the course of a period of time, frequently measured in monthly or lunar cycles, a woman completes one round of hormone release through the H-P-G axis. By comparison, the male's relative tonicity means his pattern of hormone release occurs over twenty-four hours.

To introduce the H-P-G axis in women is to discuss it as a **negative feedback system**. The release of **LTH**, particularly during lactation, **FSH** and **LH** from the anterior lobe of the pituitary is related to fluctuating ovarian hormones, i.e., estrogen and progesterone. Analogous to the male H-P-G axis, Gonadotropin releasing hormone (**GnRH**) is released from the hypothalamus which stimulates the production of pituitary hormones. Follicular stimulating hormone (**FSH**) helps to mature eggs in the ovary during the follicular phase of the menstrual cycle. Luteotropic hormone (**LTH**) not only helps to maintain uterine tone and promotes progesterone production, but is directly involved in the lactation process. Luteinizing hormone (**LH**) which is synonymous with interstitial cell stimulating hormone (**ICSH**) in the male, helps to release the mature egg from the ovary, induces **androgens** at ovulation, and induces progesterone production and release in the **luteal phase** of the menstrual cycle. The cyclic release of these hormones in the female creates a system of balance and regularity. Contrary to popular lore in U.S. culture, female hormone patterning is not "raging," "erratic" or "uncontrolled." It is interesting to note the

103

level of cultural concern regarding women's hormone release vis-à-vis the relative lack of concern towards the male. For example, Martin notes that even medical texts refer to menstruation, menopause, and female hormonal patterns in negative or injured terms—"degenerative, deteriorated, weakened, repaired." Analagous processes for spermatogenesis or other body functions are labelled more neutrally or even positively—"shedding of the stomach lining; phenomenon of spermatogenesis" (1987:47–50). This will be discussed in more detail when the menstrual cycle and its biobehavioral dimensions of **menstrual cramps, menstrual synchrony,** and **PMS** are presented.

As with males, females produce their own gonadotropins (sex hormones), as well as those of the other gender, e.g., testosterone. In females, androgens or the male sex hormones, are largely produced by the adrenal glands. Androgens, particularly testosterone, are produced from puberty until death in the female. Androgens are released at the end of the follicular phase of the menstrual cycle in order to help expel the mature egg from the ovary. The libido hormone, testosterone, is produced in roughly the same amounts in men and women to ensure the threshold level necessary for the sex drive. Women continue to produce this threshold level postmenopausally. Thus, on a hormonal basis, women retain their libido and capacity for sexual response after menopause.

Specific female sex hormones, **estrogen** (actually a group of hormones) and **progesterone**, are primary ovarian hormones produced from puberty until menopause. Their production and release follow an H-P-G axis pattern analogous to the male. FSH induces estrogen; LH induces progesterone. As with the male, the use of recreational drugs such as alcohol, marijuana, cocaine, and some prescription drugs can affect this release pattern, influencing not only the libido and sexual response cycle but the menstrual cycle as well (Buffum 1982).

Estrogen and progesterone are responsible for the development of primary and secondary sex characteristics. As with the male, these characteristics cover a wide spectrum of individual variation, are relative when comparing men and women, and interact with genetic and cultural variables of custom, nutrition, and health in their expression. Estrogen appears to be more operative than progesterone in the development of many of the secondary sex characteristics.

Secondary Sex Characteristics

Secondary sex characteristics include hair patterning, skin quality, bone integrity, breast development, muscle mass, and amount and distribution of body fat. Estrogen levels influence the ovoid shape of a female's hairline, axillary or underarm hair and **pubic hair** growth, and the inverted triangle shape of her pubic hair. As with the male, pubic hair can be sexually sensitive to tactile stimulation or serve as a visual sexual cue. It is curly, soft, generally darker than other body hair, and tends to form an inverted triangle from the pubis down to the groin, or upper inner thigh area. The color, amount, and vagaries in distribution of pubic and axillary hair are individualized and tied to cultural and genetic factors. For example, women in some Arab societies pluck or shave their pubic hair as part of female hygiene practices. In many societies outside late twentieth century U.S. culture, women do not shave their axillary and leg hair. In the U.S., shaving or not shaving one's underarms and legs may be a political statement, a gesture of femininity and aesthetics, or a custom of hygiene.

One of the functions of estrogen is to keep skin supple and soft, and promote collagen production, a substance which helps to maintain the integrity of skin cells. Estrogen plus reduced levels of testosterone also help to inhibit acne in women relative to men. Loss of estrogen in menopausal women, lack of facial hair, and not shaving promotes faster skin aging in women than men. In U.S. culture, this is compensated for by a highly profitable market in facial scrubs, emollients, and even plastic surgery.

In general, women's bones are shorter, lighter, and less dense than men's. In part this is due to heredity and cultural factors related to diet and exercise, but it is also due to the ratio and release of estrogen relative to testosterone in a woman's body. The loss of estrogen at menopause also increases a woman's chances of broken and more slowly healing bones as she ages, as well as the risk of **osteoporosis**, a degenerative bone disease. While osteoporosis cannot be reversed or cured once it develops, it can be prevented. Prevention includes early life cycle attention to diet, particularly sufficient calcium intake through food, not supplements, and reduction in animal protein; exercise, reduced intake of alcohol and nicotine, and hormone replacement therapy (HRT), a controversial treatment due to its possible links with certain cancers. Osteoporosis may be a culture specific disease of middle class western or westernized women. It is not reported in China or in certain Latin American peasant societies where diet or exercise may serve as preventative agents (Benyene 1989; "American Health" Oct./Nov. 1989).

Breast development, not breast size or firmness, is another estrogen-related secondary sex characteristic. Breast development includes the growth of the nipple and **areola**, the pigmented area surrounding the nipple, and the development of the milk ducts within the breasts. Breast size, shape, and firmness are a function of heredity and exercise, in that exercise can strengthen the supporting pectoral muscles so that breasts sag less. Breast tissue is highly fatty. In the past two generations our culture has placed a great deal of emphasis and measure of female attractiveness based on breast size and shape. It is ironic that in a culture currently obsessed with female thinness, fatty, i.e., large breasts are viewed so positively. In contrast, most nonwestern societies value women's breasts as a life-sustaining source of nourishment for the young, rather than primarily as an erogenous zone.

Women's and men's bodies, not culturally manipulated through extreme dieting, use of steroids, or lifting weights, contrast markedly in relation to muscle mass and fat ratios relative to overall body mass. Again, for women, on a hormonal basis this is primarily due to the estrogen-testosterone ratio. Women overall, tend to have less muscle mass and more body fat than do men. Not only do they float more easily than men but they tend to have more endurance as well. This is exemplified in their roles as gatherers with kids in tow, and their ability to go through labor and childbirth—the supreme mobilization of prolonged energy expenditure. Women tend to be slower, lighter, and physically weaker than men and tend to carry lower body fat, i.e., fat which is distributed primarily over their hips and thighs. Since fat is a metabolic insulator, it is one way of keeping reproductive pelvic organs (uterus, fallopian tubes, and ovaries) both warm and protected. It also means that women vis-à-vis men tend to be less prone to coronary heart disease (CHD), which currently is correlated with upper body fat, more often associated with men.

From an evolutionary perspective, the higher fat to body mass ratio in women and its distribution pattern could be adaptive for the developing fetus and neonate. The fat insulates and protects the fetus. Given the reality of periodic food shortages and famines which have occurred throughout hominid evolution, it is estimated that the fat reserves a woman carries could sustain her and a fetus/neonate for about 18 months (Ember and Ember 1990). In nonculturally induced famine situations, this would probably allow a group sufficient time to locate new food resources without seriously jeopardizing the survival of a large number of their young, and their reproductive age females.

Females also need a minimum of 15% body fat in order to reach puberty, establish and maintain a regular menstrual cycle, including regular ovulation. This is in marked contrast to men. By current U.S. standards of leanness, a male with 15% or more body fat would be considered "fatty," but not necessarily fat relative to poundage. This also means that in many nonwestern societies due to mobility patterns which build up muscle and aerobic fitness, and the amount of dietary fat she consumes, girls may reach puberty later than in the U.S. and other western societies where sedentism is more common and the diet is higher in fat and overall calories. In a few African societies for example, prepubescent girls undergo a period of fattening which accomplishes several goals. It increases their body fat to the point they achieve puberty and fattens them so they are culturally defined as attractive and eligible for marriage and pregnancy. The common occurrence in nonwestern societies of adolescent sterility may actually be a function of insufficient body fat for puberty and regular menstrual cycles to occur. See chapter 10 on adolescence for further discussion.

In this culture, current standards for a lean female are 18-21% body fat, the average is about 24%. Given our current interest in female thinness, many younger women in particular diet and exercise themselves into fashionable leanness and irregular menstrual cycles. One clear sign of **anorexia nervosa**, a severe and dangerous eating disorder that primarily is found in middle class adolescent females, is cessation of menstruation. Women athletes whose body fat is less than 15% also can experience menstrual and ovulatory irregularities depending on the quality of their diet. These irregularities are reversible upon increasing the body fat ratio beyond 15%. Our present standards of female beauty in essence would have women look like lean males with breasts, and reinforce a cultural pattern of potentially serious eating disorders for a sizable number of our population.

It is also interesting that a female athlete's price for competitive leanness may be altered menstrual cycles. By comparison, extremely lean males who have less than 6% body fat may have reduced production of sperm (Wheeler et al. 1984).

These secondary sex characteristics can be culturally interpreted not only as indicators of sexual adulthood, but also as visual sex cues exemplified by features such as breast development, appearance of pubic hair, or widening of the hips. In contrast to males where spermatogenesis is hidden, menstruation is a visible and clear marker that primary sex and reproductive characteristics have been achieved. The development of the primary sex characteristics is related to the H-P-G axis and release of estrogen and progesterone from the ovary.

Primary Sex Characteristics

Estrogen and **progesterone** overlap in their effects on the primary sex characteristics which include the growth and development of the external genitalia and the internal reproductive structures. A female's external genitalia is referred to as the **vulva**.

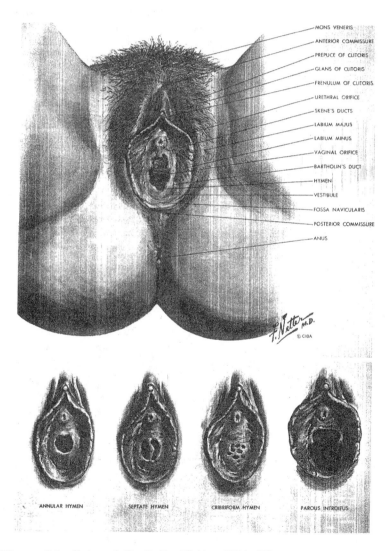

Figure 6.1   External Genitalia (Netter 1974: 90).

The **mons**, **mons pubis**, or **mons veneris** is a fatty pad of tissue covering the pubis. It is sensitive to sexual stimulation and is covered with the upper part of the triangle of pubic hair. The **labia majora**, or outer lips, are homologous to the scrotum. They are fatty pads of tissue covered with pubic hair. Sensitive to sexual stimulation, they can engorge with blood during sexual arousal. They extend from the bottom part of the mons to the base of the exterior vagina or **introitus**. The **labia minora**, analogous and homologous to the foreskin and shaft skin of the penis are hairless, fatty pads of tissue which are surrounded by the labia majora. They form the **prepuce** or **clitoral hood** at their top and the base of the introitus at the bottom. Sensitive to sexual stimulation, they vary in size, shape, and degree of pendulousness with the individual. In some societies they are seen as symbols of beauty and erogenous zones. For example, the Mangaia have more words to describe the aesthetics of the vulva and clitoris than do western societies, where many of the terms carry negative connotations (Marshall and Suggs 1971). Calling a woman a "cunt" in the U.S is not a compliment. In other societies such as in the Sudan, the labia minora are ritually surgically removed in order to preserve a woman's modesty, virginity, and chastity. Currently, this female genital surgical procedure known as **excision** is generating tremendous controversy. It and other forms of female genital surgery will be discussed separately.

The clitoral hood or prepuce, formed by the juncture of the labia minora loosely covers the **clitoris**. The prepuce is analogous and homologous to the foreskin, particularly where it covers the glans of the clitoris. **Smegma** also collects under the clitoral hood as it does under the foreskin and functions as a lubricant in both instances. The friction of the prepuce over the clitoris can produce intense sexual stimulation.

The clitoris, composed of a glans and shaft, is supported by a dense pelvic musculature and has more nerve endings in its glans than the glans of the penis. While homologous to the penis, it is not analogous (Freud aside!). The only known function of the clitoris is sexual pleasure. It is the only organ in the human body whose only function is sexual pleasure. This characteristic functionally distinguishes it from the penis which has four functions: sexual pleasure, a transport mechanism for ejaculate, an organ of reproduction in heterosexual genital behavior, and a transport mechanism for urine.

Controversies in western societies surrounding the function of the clitoris are long standing, widespread, and ongoing. Within the twentieth century alone, there have been the Freudian clitoral vs.

vaginal model; Masters and Johnson's model with physiological centering of female sexual response in clitoral stimulation; the Singer model of sexual response, and the G-spot model of sexual response (see chapter 11 for a more detailed discussion of female sexual response.) The clitoris is a source of cultural interest and definition in nonwestern societies as well. In both western and nonwestern societies this interest has been expressed at times in various forms of clitoral surgery, including **circumcision** which removes the prepuce, and **clitoridectomy**, removal of the glans or shaft. In the U.S., these surgeries began in the mid-nineteenth century to curb female masturbation and cure female insanity due to sexual problems and persisted into the 1930s. In the nineteenth century, women's sexual and reproductive structures were seen as the source of their behavior, affect, and attitudes. Female sexual and reproductive surgery was performed on women who did not conform to their culturally defined roles as good wives and mothers. Medical residents learned these procedures on slave, lower class women, and emigres and performed them on middle class women (Barker-Benfield 1975; Ehrenreich and English 1978; Martin 1987). A gynecologist performed a variation of clitoridectomy in the 1970s and 1980s by moving the clitoris closer to the introitus in order to "ease" vaginal orgasms for some women in the U.S.

Cultural-psychological and physiological studies of sexual response in U.S. females consistently indicate that clitoral stimulation of some form is a necessary and important aspect of female arousal and orgasmic ability (Kinsey 1953; Masters and Johnson 1966, 1970, 1974; Hite 1976, 1987). The clitoris is a physiological center of sexuality for many women.

Between the clitoris and the introitus is the **urinary meatus**. This opening is the end point of the urethra which transports urine from the bladder to outside the body in both men and women. **Urinary tract infections** (UTIs) and **cystitis** are two common problems which occur more frequently in women than men. This is due to women's shorter urethras and the kind of friction a woman experiences during penile-vaginal intercourse (PV-I) which more readily leads to irritation of the female's urinary meatus and urethra than the male's.

The external **vagina** is below the urinary meatus. It is called the introitus and is surrounded by a ring of muscle known as the **pubococcygeus** muscles which help the introitus to open and contract. These are the same muscles surrounding the **crura**, or root of the penis. The introitus is the final passageway for a fetus and menstrual blood, and is the entry point for penile-vaginal intercourse and the beginning of reproductive sex. The introitus as the external part of the vagina takes on culturally defined physical, sexual, psychological, and reproductive connotations.

The prepuce, clitoris, and introitus are all subject to cultural curiosity, definition and interpretation in both western and non-western societies. Western interest has been discussed briefly. Nonwestern has been introduced. **Circumcision, clitoridectomy**, and **infibulation** are forms of genital surgery presently practiced in some nonwestern societies, most notably in Moslem Sudan and some other sub-Saharan Moslem African groups. Infibulation entails extensive genital surgery.

Infibulation can include circumcision, excision of the clitoral glans, shaft, and labia minora, and closure of much of the introitus through excising and sealing the labia majora. These procedures have received a great deal of attention recently by varied groups including the media, United Nations, and researchers. Introduced here, the practices which follow are also discussed in chapter 10. These indigenous procedures have been affected by **acculturation** or culture contact. They are sources of intense controversy concerning their sexual, reproductive, and psychological effects on the women, their partners, and the kin groups involved (Ohm 1980; Lightfoot-Klein 1989, 1990). Issues of ethnocentrism, cultural relativism, and indigenous cultural integrity are involved. For example, among the Gbaya, a horticultural group in central Africa, removal of the glans clitoris in pubescent girls is a **rite de passage**, a symbolic and real marker of her transition from girlhood to womanhood. The girls go through this ritual together as part of a larger social group. Gbaya interpret the clitoridectomy as making her less male and more female. She emerges from the ritual as a recognized adult. Pubescent boys, circumcised at a younger age, undergo a similar procedure in which their glans is nicked so that they can participate in their female peers' ritual. Do we as westerners have a right, or obligation to interfere with these practices? How well do we socialize our adolescents into adulthood? What painful physical and psychological costs do we extract from our young? Are people being mutilated physically, socially, psychologically? Who makes this determination? These are extremely difficult, highly emotionally charged behaviors within cultures, let alone between cultures. Resolution of their controversial aspects will take a long time to achieve.

Another anatomical structure which generates much cultural interpretation and concern is the **hymen**. The hymen is most likely a vestigial organ with no currently known function. It is semipermeable and partially or completely covers the introitus in many women. Its presence and condition is of particular concern in patrilineal (descent through the male line), bilineal (descent through the male and female line), and patricentered (male-centered) descent groups where paternity must be known for offspring to have a legal, social, economic, political, and ritual place in the society.

The condition of the hymen is used to define a female's virginity and chastity. **Virginity**, a physiological state is attributed to someone, regardless of gender, who has never had penile-vaginal intercourse.[1] **Chastity** is a sociocultural condition in which an individual, regardless of gender, lives by the culturally appropriate sexual code of behavior, affect, and attitude as defined by the group. For example, a woman who is a "good wife and mother" by her culture's definitions and standards is no longer a virgin, but can still be chaste. How? By following her culture's rules concerning such behaviors as appropriate dress, speech (she "watches her language" in front of her children), or sexual behavior (she responds to her husband's advances, but does not initiate sex). In contrast, in the 1950s and early 1960s, in the U.S. a "girl" could be a virgin, but not chaste. These "girls" sometimes known as "(cock) teases" would do "everything" except P-V intercourse with boys or would arouse a boy ("get him hot") and then not "give" him an orgasm ("leave him hanging" or "not come through").

While chastity and virginity are gender-free, their cultural interpretation in bilineal and partrilineal descent groups usually results in a double standard of sexual behavior. Promoting non-chastity and non-virginity for males demonstrates their potential fertility and actual virility, both necessary for reproductive success. In contrast, repressing female sexual behavior in these societies except under careful culturally controlled situations such as strictly enforced monogamous marriage accomplishes two goals. It achieves reproductive success since the woman does not have to be orgasmic or ejaculatory in order to conceive and give birth. Secondly, controlling her sexual behavior and severely limiting her sexual outlets and number of partners helps to "know" the paternity of the child. He is probably her one and only sexual partner. This double standard is culturally widespread and persists in U.S. society. For example, recent surveys of U.S. college student sexual behavior indicate women still are subject to greater disapproval than men for having multiple sexual partners, particularly when emotional attachment does not exist (Whelehan and Moynihan 1984; Reiss 1986). In some Greek peasant villages, chastity and virginity are affirmed on a bride's wedding night when blood, either hers or from a vial of sheep's blood given to her as a gift by her female relatives, appears on the sheet after her first intercourse. During the middle ages *ius pramae noctis* or first night rite was a common ritual performed on brides on their wedding night. They were deflowered by the owner of the fief in order to establish proprietary rights over the serfs— men, women, and children, as well as the land (Bullough 1976).

Returning to physiology, on each side of the base of the introitus are two small glands, **Bartholin's glands.** These glands generally are unnoticed unless they are infected, in which case they can swell painfully. Bartholin's glands secrete a clear fluid; their function is unknown presently. Bartholin's glands are homologous to Cowper's glands in the male (see chapter 5).

The **perineum** is homologous in males and females and extends from the base of the introitus to the anus. Soft, generally hairless, and sensitive to sexual stimulation, the perineum stretches during vaginal childbirth to allow the baby to pass through the introitus. As with many other aspects of female sexual and reproductive anatomy, the perineum is subject to cultural scrutiny and controversy. The ability of the perineum to stretch during childbirth is culturally manipulated.

In most nonwestern societies women give birth semi-upright which relaxes and stretches the perineum. In addition, many societies further stretch and soften the perineum through massage or the application of warm compresses or oils. These efforts reduce the chance of perineal tears during childbirth (Arms 1975; *Boston Women's Health Collective* 1976, 1984; "Medical Anthropology" 1981, v. 1).

In the U.S., the most common way of stretching the perineum during childbirth is through an **episiotomy**, a surgical incision in the perineum. This is believed to reduce the possibility of perineal tears, a high probability given the dorsal-lithotomy position (lying on one's back) used in over 90% of U.S. births (Arms 1975; Masters, Johnson and Kolodony 1982; Davis-Floyd 1992). Episiotomies are one of several current controversies in U.S. childbirth practices. These controversies will be dealt with in more depth in chapter 8 on childbirth.

Internal sexual and reproductive structures include the **vagina, uterus, fallopian tubes, ovaries,** and **broad ligament**. The internal vagina, a tubular, muscular organ extends from the introitus, back about 4–5 inches in a curved manner, ending in a blind pouch or cul-de-sac known as the Pouch of Douglas. As a structure, the walls of the vagina rest on one another. Androcentrically, i.e., from a male perspective, the vagina is often described as a potential space. It is a passage way for menstrual blood, sperm, and the birthing fetus. Its expansion capacities are remarkable. In a non-filled state, the walls of the vagina rest on each other. From a state of collapsed walls to being able to accommodate a penis or a full term baby indicates a high degree of flexibility. As a sexual structure, its orgasmic function is a source of cross-cultural interest, definition, and discussion (e.g., the *Kaama Sutra* of Vatsayana in Gregersen 1983; Garrison 1964; Freud 1920, [1953]; Kinsey 1953; Masters and Johnson 1966; Hite 1976, 1981, 1987). For many U.S. women, it is a psychological center of sexuality (see Hite 1976, 1981, 1987). In addition to its sexual functions, it is a reproductive organ.

It is a sexual and reproductive structure with a pH of about 4.5–5, making it slightly acidic. The acidity of the vagina and its natural flora keep it clean and healthy. It is the cleanest orifice in the human body. Regular, frequent douching is unnecessary and can upset the pH levels and irritate the mucosa, leading to irritation and infections. Disruptions in its pH balance or flora can be caused by antibiotics, STDs, stress, or illness. A healthy vaginal mucosa is pinkish, firm, springy, and moist. Vaginal lubrication or exudate, which occurs during sexual arousal, passes directly through the vaginal mucosa (lining). A healthy vaginal mucosa is maintained largely through estrogen, low stress, a balanced diet, and general hygiene.

Since a healthy vagina is slightly acidic, there is a possibility of incompatibility with a male sexual partner's semen, which is basic. In some instances this may cause fertility problems. Also, popular fads to alter the pH of the vagina to increase the chances of conceiving the desired gender are generally a waste of time, money, effort, and may damage sperm or irritate the vaginal mucosa.

The vaginal mucosa changes with menopause. The cessation of estrogen production can result in drying and thinning of the mucosa which can lead to painful intercourse. The irritated mucosa can be soothed through the use of water based lubricants such as K-Y Jelly, Probe or Forplay, or through estrogen replacement therapy (ERT) either as a topical ointment, or in pill or tablet form (Stewart et al. 1979; Seaman and Seaman 1977).

Located on the anterior wall of the vagina, under the pubis about one-third of the way in from the introitus, is the location of the alleged "**G spot**" or **Grafenberg spot.** "Alleged" is used because the very existence of the Grafenberg spot is challenged by some sexologists, e.g., Masters and Johnson (1985). Those who accept the existence of a G spot state that it is a source of orgasmic potential and associate it with female ejaculatory ability (e.g., Ladas, Whipple, and Perry 1983). Under penile penetration with the woman on top, or direct deep finger pressure, the G spot, an area of soft tissue, increases in size and produces a sense of sexual pleasure. Continued stimulation may result in orgasm as well as the ejaculation of fluid from **skene's** or the **paraurethral glands**, located on each side of the G spot. The chemical composition of this fluid, sometimes referred to as female ejaculate, is debated (Ladas, Whipple, and Perry 1983; Mahoney 1983). Since some researchers believe this fluid is similar in composition to prostatic fluid, skene's gland can be referred to as the "female prostate." This fluid is deposited into the urethra and expelled from there during G spot orgasmic response. The expulsion of this fluid out the urethra has led some women, their partners, some researchers, and physicians to believe that these women experience urinary stress incontinence (USI) or involuntary leakage of urine. Unless these individuals are into "water sports" (playing with urine, which in the age of AIDS may be a risky behavior), this belief has created orgasmic problems for some of these women, distress for some of their partners and led a few physicians to perform surgery on these women for USI (Ladas, Whipple, and Perry 1983). The intensity of the controversy surrounding the entire G spot phenomenon as opposed to accepting it as a possible normal variation of sexual response is another indication of our continued discomfort with sexuality in general and women's sexuality specifically.

The **broad ligament** is a band of connective tissue across the woman's lower abdomen which supports the uterus, fallopian tubes, and ovaries. As a supportive structure, it is one reason why women have fewer abdominal and pelvic hernias than men.

The **ovaries** are homologues and analogues of the testes. Pearly gray and almond shaped, they are located in the lower pelvis, slightly lower and adjacent to the fallopian tubes. The ovaries are the female gonads, responsive to stimulation by the pituitarian hormones FSH and LH. As primary sources of estrogen and progesterone, they are palpable under gentle stimulation during a pelvic exam. As an introduction, the ovaries contain the eggs, or ova in tiny sacs called follicles. A woman is born with all the ova she will ever have. She matures and releases about 400 eggs during her reproductive life cycle. Egg development and number are in some ways a degenerative process in contrast to spermatogenesis. A female is born with about half the eggs she had as a fetus. While several eggs start to mature with each menstrual cycle, usually only one achieves maturity and is released. The other ripening eggs during that cycle are reabsorbed by the ovary.

The ovaries have hormonal, reproductive, sexual, and cultural functions. They produce the female gonadotropins, estrogen, and progesterone, and mature and release eggs in preparation for fertilization. By releasing estrogen and progesterone they help maintain uterine tone and functioning, and maintain a healthy vaginal mucosa both of which act in reproductive and sexual response. They are given cultural attributes related to femininity, maternal behavior, and appropriate gender role behavior. In fact, oophrectomy, removal of the ovaries or female castration was commonly performed in the U.S. during the 19th century to "cure" female insanity or to enforce "gender appropriate behavior" (Barker-Benfield 1975; Klee 1988; *Boston Women's Health Collective* 1976, 1984; Ehrenreich and English 1978). This practice is another example of our culture's negative attitudes toward female sexuality.

The **fallopian tubes** extend from the fundus of the uterus for about 5–7 inches to approach the ovaries. They do not cover the ovaries. Hollow, with hair-like cilia projections along their inner walls, they are about the size of a broom straw or thin strand of spaghetti. Homologous and analogous to the vas deferens, they end as a fibrillated or tendril-like organ. At ovulation, when an egg is released by the ovary, the fibrillated ends of the fallopian tube draw the egg into the tube. Generally the adjacent tube draws in the egg. Fertilization usually occurs in the distal or far end of the tube called the **ampulla**, where the tube curves. The egg, regardless of whether it is fertilized is moved down the tube by the cilia. It takes the egg several days to reach the uterus. If fertilized, it may implant in the corpus section of the uterus and begin to develop into an embryo, then a fetus. If unfertilized, the egg passes through the uterine cavity.

The fallopian tubes are the normal site of fertilization and the uterus the normal site of implantation. Occasionally, however, a fertilized egg embeds outside the uterus. This is called an **ectopic pregnancy**. The most common site of an ectopic pregnancy is a fallopian tube. An ectopic pregnancy is a life threatening medical situation. The embedded egg will grow in the tube until the tube bursts. Initially mimicking a normal pregnancy (e.g., cessation of menstruation, possible breast tenderness, and nausea), an ectopic pregnancy can be diagnosed by a careful and thorough pelvic exam. Early diagnosis is important, as the tube will burst by the end of the first trimester. Treatment currently entails removal of the affected tube. Drug treatments for ectopic pregnancy are being tried (Medical Aspects of Human Sexuality 1991:14). While fallopian tube transplants in humans have been done, (Schenker and Evron 1983) as of yet, successful transplants of ectopic pregnancies to an uterine environment are not possible. Ectopic pregnancies occur in about

2–5% of the pregnancies. Incidence rates of ectopic pregnancies are increasing due to the rise in sexually transmitted diseases such as gonorrhea and chalmydia which scar the fallopian tubes, endometriosis, and pelvic inflammatory disease (PID), all of which can damage and cause blockage of the fallopian tubes.

The **uterus** or womb is composed of several parts and performs both sexual and reproductive functions. The uterus, a pear shaped organ about the size of a large thumb is located in the central lower pelvic area. It extends into the vagina. The uterus is a muscular organ which produces **prostglandins**. Prostglandins are hormones which perform a number of functions including contraction of the uterus. Prostglandins may be involved in uterine contractions that occur during orgasm, those that occur during menstruation and cause menstrual cramps, and as a partial cause of uterine contractions during labor. As one of the strongest muscles in the human body, the uterus can be exercised and toned. Part of the tone of the uterus is due to the release of such hormones as LTH, progesterone, and prostglandins. Uterine tone is also maintained through orgasms, during which the uterus contracts (Sherfey 1972). A healthy, toned uterus allows not only for greater chance of retaining an implanted embryo, but for its development and expulsion nine months later. Female orgasms then, not only feel good, but help to maintain uterine tone.

The uterus is composed of several layers, the myometrium, parametrium, and endometrium. The parts of the uterus to be discussed include the fundus, the corpus, the cervix, the os, and the endometrium.

The **fundus** is the rounded top part of the uterus. As the uterus enlarges during pregnancy, the fundus can be felt externally through the lower abdomen as it rises above the pelvic bone. The **corpus** or body of the uterus is composed of several layers of which only the endometrium will be discussed. The fallopian tubes enter the uterus at the bottom of the fundus and beginning of the corpus. The corpus extends into the vagina from its lower section, the **cervix**.

The cervix extends into the vagina and can be felt by deep insertion of the fingers into the vagina, and seen by the use of a mirror and speculum, an instrument used during a pelvic exam to separate the walls of the vagina. The cervix receives a lot of medical and lay attention. The cervix is the site of the **PAP** test. The PAP test is part of a pelvic exam. Cells are gently scraped from the cervix and analyzed to detect cervical normalcy and abnormalcies, including cancer. Cervical cancer is one of the more common cancers to affect women. Regular PAP smears help to ensure early detection and treatment.

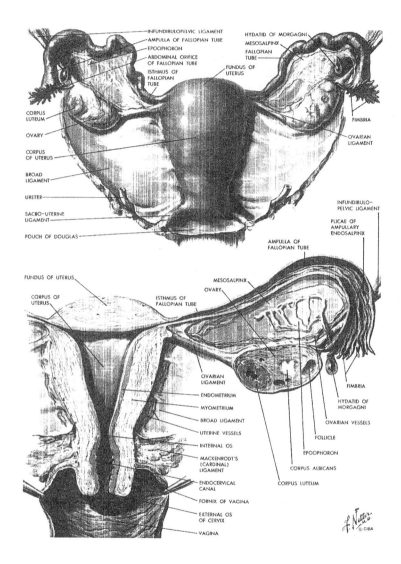

Figure 6.2   Uterine Development and Musculature (Netter 1974: 110).

The cervix changes in color and texture depending on whether or not a woman is pregnant. In nonpregnant women, the cervix is pinkish and cartilaginous in texture, similar in texture to the tip of your nose. In response to hormones released during pregnancy, the cervix softens to a state more similar in texture to your lips, and changes color to a bluish-purplish hue. During sexual arousal and orgasm, the uterus, including the cervix, responds in various ways by retracting, contracting and lowering into the vagina.

The cervix is the site of the **os**, an opening in the tip of the cervix generally about the size of a thin pencil lead. It also changes in color and shape depending on whether or not a woman is pregnant. A nonpregnant woman's os is donut-hole shaped: O and pinkish; a pregnant woman's os is purplish or bluish and more slit-like ⟳. This change in shape is irreversible after a woman has a child. These changes in the cervix and os appear in the first trimester and are used as signs of pregnancy during a prenatal pelvic exam. The os is the passageway for **menses** (menstrual blood), the fetus, and sperm. The os is the structure which dilates or opens during the first stage of labor to allow the baby to pass through. During menstruation the os dilates slightly to allow the shedding endometrium or menstrual blood to pass through.

Most of the month the os is covered with a thickish, sticky substance called **cervical mucous**. This acts as a protective barrier to keep foreign objects, e.g., sperm, douches, contraceptive foams, bacteria, out of the sterile uterine cavity. Just prior to and during ovulation however, the cervical mucous thins and becomes more permeable in order to allow the sperm through to the fallopian tubes where the egg may be fertilized. If the egg is fertilized and implants in the endometrium, another mucoid substance, the cervical plug forms over the os as a protective seal for the fetus against foreign substances entering the uterine cavity. The cervical plug usually is expelled during the first stage of labor and is used as a sign that labor is imminent or has begun.

The **endometrium**, which has been mentioned several times, is the innermost lining of the uterus. Through the H-P-G axis and interaction of estrogen, which helps to build it, and progesterone, which helps to maintain it in the uterus, the endometrium serves as the anchor for the embryo and fetus. The endometrium is a thick, cushiony layer of blood, tissue and mucous which accumulates each month in preparation for a fertilized egg. The fertilized egg implants in the endometrium where it remains attached for nine months as it develops into an embryo and fetus. It is shed, as the menses or menstrual blood, if fertilization and implantation do not occur.

There is a clinical condition called **endometriosis** which is being diagnosed and occurring with increasing frequency among middle class, college educated, career women in the U.S. who are in their late twenties through thirties and who are generally nulliparous (they have never borne a child). Endometriosis involves patches of endometrial tissue found on the ovaries, external uterus, or other pelvic and abdominal organs, and in the fallopian tubes. Endometriosis in the fallopian tubes can cause tubal blockage and scarring, interfering with conception. While the exact cause of endometriosis is unknown, it is theorized that it may be due to a variety of factors including prenatal disposition, hormone fluctuations, and delayed pregnancy (Stewart et al. 1979; American Health 1989).

Symptoms of endometriosis include fertility and menstrual problems, pain during both intercourse, known as dyspareunia, and menstruation, known as dysmenorrhea. Definitive diagnosis of endometriosis is through laparoscopy, a surgical procedure that involves an incision in the abdomen where organs are viewed through a lighted tube or laparoscope. Treatment can be hormonal or surgical depending on the severity of the situation. In a number of cases fertility may be restored (Wade and Cerise 1991; American Health 1989).

The uterus, composed of several structures is a marvelous organ. Its capacity to change in size and function depending on pregnancy is phenomenal. As a reproductive and sexual organ, it is imbued with a range of cultural connotations and definitions. For example, the sexual function of the uterus is only recently being accepted and understood, making its already controversial ritual removal (hysterectomy) post-menopausally in the U.S. an even more debated issue (Klee 1988; *Boston Women's Health Collective* 1984). In general, in western societies for the past several hundred years, women's mental health has been defined as a function of her reproductive organs (Barker-Benfield 1975; Klee 1988). Freud used a psychiatric classification known as hysteria from the Greek word for womb, another term for the uterus. Hysteria is primarily an affliction of women, characterized by over-emotionality, denial, and depression (Freud 1920a [1953], 1920b).

In both western and nonwestern societies the function of the uterus and ovaries are subject to cultural scrutiny. For example, in Tiwi society, an Australian hunting and gathering group, the essence of sexuality is female. Female totems, religious and animistic guardian spirits, which are inanimate symbolic structures that are given lifelike characteristics, play a role in conception. While the Tiwi realize that heterosexual contact is necessary for conception, they need to explain why only certain contact results in conception. Their causal explanation for heterosexual contact which results in conception is that a female totem breathes life into the woman's body. Part of Tiwi contraception then includes appeasement of the female totems to avoid pregnancy (Goodale 1971). Cultural concern over this aspect of human behavior is matched by the degree of concern human groups have with another biobehavioral phenomenon, menstruation. Key concepts related to menstruation, the shedding of the endometrium, are menarche, a girl's first menstruation or period, and the menstrual cycle.

Menstrual Cycle

The **menstrual cycle**, a primary sex characteristic has both sexual and reproductive functions and is reproductively analogous to spermatogenesis, discussed in chapter 5. Both the menstrual cycle and spermatogenesis produce the gametes, the egg or ovum and sperm respectively which are necessary for conception to occur. The menstrual cycle and spermatogenesis function in response to the H-P-G axis. Other similarities include the involvement of homologous structures such as the ovaries and testes, fallopian tubes and vas deferens. The menstrual cycle and spermatogenesis are physiological processes culturally defined as signals of sexual and reproductive adulthood.

There are also sharp contrasts between the menstrual cycle and spermatogenesis. The menstrual cycle is just that, cyclic, roughly taking a lunar, 28-day-month, to follow through a round of H-P-G axis hormone release. The menstrual cycle is rhythmic as opposed to the tonicity of spermatogenesis and male H-P-G axis functioning. The ovary does not produce eggs as the testes produce sperm, rather several immature eggs start developing each month in their follicles or sacs in the ovaries. Generally, only one egg reaches maturity and is released at one point in the cycle, **ovulation**, as opposed to the continuous, numerous—several million daily—production of sperm. The menstrual cycle is highly visible in the menstrual phase when the endometrium is shed through the os and out of the body vaginally. In contrast, spermatogenesis is unmarked. There are no clear primary markers of this process other than the irregularity of nocturnal emissions or wet dreams. Indicators of spermatogenesis often are culturally defined through secondary sex characteristics such as beard growth, voice changes, the growth spurt in height and limb length, or through the imposition of initiation ceremonies or *rites de passage*, rituals that socially take a person from one stage in the life cycle to another.

Another contrast between the menstrual cycle and spermatogenesis is the arbitrary, finite nature of the menstrual cycle. Menstruation begins at puberty, occurs roughly once a month for an average of 4–7 days and ends at menopause. It is bleeding without injury, illness, or provocation. In contrast, spermatogenesis is invisible. Sperm do not appear in other situations such as illness or injury as blood may, and spermatogenesis is continuous from puberty to death. The attributes of the menstrual cycle allow for a range of cultural interpretation and action.

The discussion of the physiology of the menstrual cycle is adapted from Speroff et al. (1978). The menstrual cycle is discussed in four phases: follicular, ovulation, luteal, and menstrual or menstruation. Born with all the eggs she'll ever have, about 700,000, a woman will mature and release a range of 200–400 eggs during her reproductive life cycle, barring illness, injury, pregnancy, or surgery on her reproductive organs (American Health 1994). Based on a lunar, 28-day calendar cycle, she will mature and release about 13 eggs a year. These are averages for healthy western women. In reality there can be considerable variation relative to the length and regularity of the cycle and egg release, depending on variables such as nutrition, amount of body fat, stress, illness, or pregnancy.

> **ASIDE**: The regular, relatively uninterrupted menstrual cycles that many twentieth century, middle class, western women experience is probably a recent, evolutionary anomaly. They are a function of bottle feeding, frequent use of external, chemical or barrier contraceptives, and fewer pregnancies. The female hominid pattern until recently probably was one of fewer regular menstrual cycles, since much of a woman's life was spent lactating for several years, interspersed with pregnancies (Beyene 1989).

The **follicular** phase is the longest and most irregular of the four phases of the menstrual cycle. Its length determines the overall length of the cycle and the regularity from one cycle to the next. The follicular phase ranges 11–16 days in length. Egg maturation occurs during the follicular phase. Based on hypothalamic release of GnRH, the pituitary releases FSH which stimulates multiple egg maturation in the follicles of the ovary. About midpoint in the follicular phase, FSH starts to falter which induces the release of ovarian estrogen, specifically estradiol. Estradiol helps to stabilize the FSH level and induces the release of LH. At this point, the end of the follicular phase is approaching and LH performs several functions. It stops multiple egg maturation, and helps to release the most mature egg from the follicle, which is also aided by the release of androgens, specifically androstriol.

During this time estrogen regulates the levels of FSH and LH, maintaining a delicate balance between them since a surge of FSH would stimulate egg maturation again and a drop in LH could impede the most mature egg's release from the ovary. For this reason, estrogen or estradiol is called the key regulating hormone in the menstrual cycle. At the end of the follicular phase, the egg is ready to be released, FSH is stabilized, LH and estrogen levels are high.

Data is inconsistent as to whether or not women experience an increased interest in genital, heterosexual contact at ovulation. From reports some women are "horniest" at ovulation, others are just prior to or during their periods, others are throughout their cycle, and some not at all (Hite 1976; Masters, Johnson and Kolodny 1985; Masters and Johnson 1966).

Ovulation is the briefest phase of the cycle. It is the release of the egg from the follicle into the lower pelvis where it is generally drawn into its corresponding fallopian tube. Ovulation occurs at midcycle. Generally a woman ovulates once a month, but there can be variations due to stress, intense orgasms, or irregular follicular and ovulatory patterns (Speroff et al. 1978). Thus, since a woman can, though rarely does ovulate more than once a month, it is untrue that she cannot get pregnant during her period, that she's totally "safe" then. In addition, some women's cycles are sufficiently irregular such that ovulation may occur during menstruation. A woman may be aware of ovulation through a cramping or pinching sensation in her lower abdomen from the ovary which just released its egg. This sensation is similar to a stitch in one's side after running. The cramping is called **mittelschmerz.** It occurs as the egg bursts through the surface of the ovary; a small amount of ovarian bleeding, which is absorbed by the body, may occur at that time as well.

The **luteal** phase is the third phase of the menstrual cycle. A lot of activity can potentially occur at this time. A woman with **Premenstrual Syndrome** (PMS) experiences this during the luteal phase. During the luteal phase the egg is either fertilized or not. Each case will be presented. The follicle that just released the egg is now called the **corpus luteum**, Latin for yellow body. Upon stimulation by pituitarian LH, the corpus luteum secretes progesterone. Progesterone maintains the endometrium which was built up by estrogen in the uterine cavity. Progesterone is produced throughout this phase until it receives a signal from the decomposing egg that fertilization has not occurred. LH levels are also elevated to stimulate progesterone release until a "no fertilization" message occurs. If the egg is not fertilized, it starts to degenerate.

In turn, LH levels drop which trigger a drop in progesterone levels. At a certain point, the progesterone level is sufficiently low that the uterine lining, the endometrium, can not be maintained. It is shed as the menses. Concurrently there are drops in FSH and estrogen. When estrogen and FSH are sufficiently low, the hypothalamus is triggered and the hormonal release pattern begins again.

If fertilization occurs in this luteal phase, another set of hormonal patterns occur. Estrogen and progesterone levels, maintained by FSH and LH, remain elevated to keep the endometrium in place. The fertilized egg begins producing its own hormones from the developing placenta, which at this stage is called the **chorion**. The chorion produces **human chorionic gonadotropin**, **HCG**, or "the pregnancy hormone." HCG is called the pregnancy hormone because it is the substance detected by standard at-home and clinical pregnancy tests. HCG is produced until about the 10–12th week of gestation, at which time the placental steroids, another group of hormones, function to keep the placenta attached to the endometrium. Up to the 10–12th week of embryonic development, HCG keeps the developing placenta attached to the endometrium. Excess levels of HCG are secreted in the mother's urine, with the highest levels of HCG secreted early in the morning. Thus a woman uses a "morning sample" or urine specimen for the pregnancy test.

The fourth phase of the menstrual cycle is **menstruation** or menses, on average, the 4–7 days when the endometrium is shed. Menses is composed of blood, tissue, and mucous. It amounts to about $1/2$ cup of liquid, most of which is expelled in the first 48 hours of the woman's period. The entire menstrual cycle as well as menstruation specifically generate widespread cultural interest and reaction.

In nonwestern societies that are **polygynous**, i.e., allow more than one wife; that engage in endemic or ongoing warfare; that are patrilineal where descent is traced through the male line, and that have a high degree of segregation between the sexes, negative attitudes and beliefs about menstruation and women's sexuality are common in men's culture. The fact that men and women in these societies do not interact often and therefore do not know each other well; are joined as adults in marriages that are formed as political strategies due to warfare in groups such as the Yanamamo and Mae Enga; that men fight, bleed, and die for women who bleed spontaneously without apparent injury; and that men must be careful of women's sexual behavior or the lineage will be damaged, probably contributes to these beliefs and their expression. In addition, in many of these societies women control the food supply and men are dependent upon them for food. The sum effect of these intense behaviors can result in strong anti-female ideology by men concerning women's sexual and reproductive functions (Buckley and Gottlieb 1988).

In many of these societies women are further segregated from their peers and men during menstruation by staying in a menstrual hut. Menstrual huts are frequently described in the anthropological literature as the *sine qua non* of female oppression and degradation. However, while in the menstrual hut, women spend time with other menstruating women, eat special, (i.e., restricted) foods, and do not have to assume routine cooking, child care, food preparation, and other work responsibilities. Are menstrual huts oppression, a break from hard work, or an opportunity for women from different areas to bond?

Menstrual taboos are by no means restricted to nonwestern societies. They are alive and functioning in western and industrialized societies, impacting on women at work and at home (Olesen and Woods 1986). In one of the author's human sexuality classes, students are asked to generate two lists. One list describes everything they have heard from peers, parents, media, books, and street talk about spermatogenesis. The other list does the same for menstruation and the menstrual cycle. Very little appears on the list for spermatogenesis and that which does is either neutral or positive. The menstrual cycle list can be extensive. Most of the phrases and terms are negative, a few are neutral, and even fewer are positive. This is sexually sophisticated, late twentieth century, post sexual revolution U.S. culture!

THE LISTS

Spermatogenesis	Menstrual Cycle
being a man	that time of the month up
"cum"	on the rag
ready to have kids	"bitchy"
wet dreams	moody, irritable
	irrational
	being a woman
	moon cycles
	dirty
	don't swim, bathe, shower or have sex

As stated at the beginning of this section, menstruation is a biobehavioral phenomenon. Three specific examples of this include menstrual cramps, menstrual synchrony, and premenstrual syndrome or PMS. Data for these phenomena largely are derived from western cultures. **Menstrual cramps** have a physiological basis in that prostglandins are released by the uterus. Prostglandins cause uterine contractions. Depending on the amount of prostglandins released and the strength, intensity, and frequency of the contractions, they may be experienced as cramps. The woman's pain threshold and toleration for this kind of sensation, in addition to her learned attitudes and behavior towards her body, menstruation, and expressions of pain, all contribute to the phenomenon of cramps. Exercise, orgasms, and aspirin are all reported to be helpful in alleviating cramps.[2] Cramps are no more "all in your head" than they are a complete function of "raging hormones," both of which are popularly held beliefs in U.S. culture.

Menstrual synchrony is a clinically documented, widespread phenomenon. Menstrual synchrony is the eventual synchronization of the menstrual cycles and periods among women who live near one another and are in close contact. Physiologically, menstrual synchrony may be a function of pheremone release. Pheremones, discussed earlier in this text, are sexual scent signals or olfactory cues. It is believed that pheremones may be the hormonal basis in evening out and regulating women's periods. Since menstrual synchrony only occurs among women who are both emotionally bonded and who are in frequent contact with each other, (e.g., they live together and spend time with each other), this would allow for them to key into each other's pheremone patterns. The pattern is broken if either the emotional tie or contact is disrupted (McClintock 1971). Menstrual synchrony could have potential for reproductive success given a regular heterosexual genital sex partner. Regular ovulation and menstrual cycles coupled with continuous sexual receptivity among these women would increase the chances for conception. There is some evidence that women who have regular P-V sex with partners have shorter and more regular menstrual cycles (Jarrett 1984). How this works in polygynous societies would be an interesting study.

Premenstrual Syndrome or **PMS** currently is a controversial medical, social, and legal phenomenon in U.S. and other western cultures (Martin 1987; Buckley and Gottlieb 1988). While some physicians deny its existence (Masters, Johnson and Kolodny 1985), and others debate what it is, the American Psychiatric Association is considering including PMS under the diagnosis of Late Luteal Phase Disorder (LLPD) in the new edition of the Diagnostic and Statistical

Manual (DSM IV) (Lips 1993: 217; Hamilton and Gallent 1990). There are clinics and physicians in the U.S. and other western countries such as Canada and Great Britain who address PMS. PMS is a collection of symptoms which range in type, frequency, duration, and intensity. These symptoms occur during the luteal phase of the menstrual cycle and disappear when menstruation begins. PMS may have a cumulative progression in its intensity, but it reportedly primarily affects women in their twenties and thirties. The cause is being investigated. One current explanation is that PMS is caused by fluctuating progesterone levels during this phase (Martin 1987). It is estimated that the vast majority of women (numbers vary widely) experience at least the milder forms of PMS at some point during their reproductive life cycles. Symptoms range from mild to extreme. Milder PMS includes: headaches, irritability, water retention resulting in clothes or jewelry not fitting well, a feeling of lower body heaviness, lethargy, food cravings, particularly for salt and chocolate,[3] and weight gain that can range from 2–15 pounds and is lost after menstruation. More severe symptoms tend to include emotional ones: mood swings, depression, increase of drug intake particularly alcohol, as well as nausea, and migraines. In its extreme form, women state they experience uncontrollable fits of rage, violence, and depression which may be acted out toward oneself or others in the form of suicide attempts, child abuse, and physical assault toward men they know (Martin 1987; American Health 1989). These behaviors have legal and social consequences which will be discussed later.

> **ASIDE**: While PMS clearly may be culture specific, it may also be further categorized as a folk illness in those western cultures which report it. The controversy within the medical profession in these cultures concerns whether PMS is an actual clinical entity or a vague collection of symptoms. Aside from this debate is the folk perception of PMS. For example, a client of one of the authors labeled every mood change she experienced as PMS, regardless of when in her cycle these mood changes occurred. Increasingly on the college campus where one author teaches, many of the reasons given for "pigging out," being grouchy, or not working are attributed to PMS by both men and women, regardless of the accuracy of the label.

Treatments of PMS include lifestyle modifications and hormone therapy if necessary. A woman who believes she has PMS needs to chart her symptoms over a period of several months to note whether a pattern emerges and to try to control other factors such as stress at work or home. If a correlation appears between the symptoms and the luteal phase, and these symptoms do not appear at other times, a diagnosis of PMS may be made.

Lifestyle modifications refer to nutrition, sleep, exercise, use of drugs, and reduction in stress. Interestingly, the lifestyle modifications resemble those currently recommended as "healthy living" which simulate our hominid gathering and hunting behavior. These include a reduction in salt, saturated fat, refined sugar, caffeine, alcohol, red meats, and an increase in the consumption of complex carbohydrates, lean fish and poultry, grains, fruits, and vegetables. Sufficient sleep and aerobic exercise are encouraged as well as the reduction of the use of recreational drugs. Stress reduction techniques include biofeedback, meditation, guided imagery; anything that relaxes from within.

> **ASIDE**: The extreme behaviors attributed to PMS elicit legal and social responses. These behaviors include suicide attempts, child abuse, and murder. Over the past 10 years, some women accused of child abuse and murder of men they knew have entered a plea of PMS to courts in the U.S., Canada, and Great Britain. This plea has been accepted, and in some cases tried. In one case in Great Britain, the woman was acquitted on the grounds of PMS. Culturally, this evokes a strong response from men, women, feminists, and nonfeminists in support of both sides of the PMS controversy. One set of arguments supports the reality of PMS as a cause of violent behavior and wants judgment and treatments given with PMS as a consideration. Another set of arguments believes the use of PMS in legal cases supports the view of women as irrational beings, subject to raging hormones that control their behavior, and who are not hormonally fit or responsible beings. This side also believes that while intense depression and anger may be caused by PMS, there are outlets for these feelings other than physical aggression toward oneself or others. By contrast, it is interesting that testosterone is not a defense entered by men for acts of violence; not even by the men Silber studied (see chapter 5).
>
> Rather than a clinical entity, PMS may be a cultural construct that allows women to "rage" once a month. Given that our culture historically sees women as a creature of her hormones and denies women legitimate expressions of anger, or frustration, a once-a-month "release" of anger, frustration, or other culturally-labeled negative emotions allows women to vent and reinforces the belief that women are victims of our physiology (Tavris 1992). It is interesting that the clinical diagnosis of PMS would place it under a psychiatric as opposed to a medical classification. Is PMS another cultural double standard, a newly found physiological-behavioral phenomenon that we do not fully understand, or an interaction of the two (Martin 1987)?

In the past two chapters, male and female adult sexual and reproductive anatomy and physiology have been presented. Their similarities are notable: they share a common hormonal system and functioning which varies by degree, amount, and patterning. Many of the structures are both homologous and analogous with each other. These similarities will be reinforced in the chapter on embryology and sexual differentiation in utero. In essence, on a biochemical basis, men and women are more similar than they are different.

Much of how and what we define, label, and respond to as sexual, male, female, masculine, or feminine is as much a function of cultural patterning as it is biology. Culturally, we learn to attribute positive and negative connotations to our bodies, behaviors, thoughts, and feelings relative to sexuality. As bio-cultural beings we are sexual creatures. Hopefully this presentation of anatomy and physiology has helped to illustrate:

a. our commonness as a species

b. our similarities as males and females

c. that our sexual behavior is linked to a biological foundation and

d. that much of our life cycle sexuality such as puberty is an interaction of biobehavioral components.

Reproductive Technology

In addition, reproductive technology developed within the past 30 years particularly since the mid 1970s, can impact tremendously on our physical evolution as hominids. Currently, artificial insemination by husband (AI-H), or donor (AI-D), in vitro fertilization (IVF), chromosomal filtration for gender selection, embryo transplants, amniocentesis, and chorionic villi sampling (CVS), are all available as alternative means of direct heterosexual contact for reproduction.

Sperm banks for **AI-D,-H** are found in most major cities in the U.S. Over 1,000,000 **artificial inseminations** are performed each year. A generation of artificial insemination babies has reached adulthood, some of whom are trying to locate their biological fathers. Minimally, artificial insemination means that men are no longer directly needed for impregnation; only their healthy ejaculate is. AI-D is used by some lesbians who want to be both biological and sociological mothers without having P-V intercourse. A very simple process is involved. All one needs is a fresh, healthy ejaculate sample, a syringe, and an ovulating female.[4] What are the potential consequences of AI-D,-H for partnering, parenting, and for men? This is not a balanced situation. Men still need women to carry the fetus and give birth to it. There are no egg donors as there are sperm donors. Surrogate mothers are not as accepted as are sperm banks and artificial insemination.

In vitro fertilization, "test tube babies," is now rather common in the U.S. and Great Britain for couples for whom the woman has irreparably damaged fallopian tubes. This procedure involves surgically extracting a mature egg from the woman's ovary, combining it with a fresh ejaculate sample from her husband, and then implanting the conceptus (the fertilized egg) into her uterus. Available since 1978, several thousand babies have been conceived and born, apparently physically and developmentally healthy. While success rates for a live birth remain relatively low—about 20%, and the procedure is expensive—several thousand dollars per trial, it is used and effective (*Parade* 1989; NBC Special 1994; Ragone 1994).

Chromosomal filtration of X and Y chromosomes to preselect a female or male fetus is gaining in success and popularity. Developed in the late 1970s–early 1980s by a reproductive embryologist in San Diego, this method reportedly has an 85% success rate for Y chromosome filtration. From a spun ejaculate sample, the lighter Y chromosomes filter to the top and the heavier X chromosomes settle to the bottom of the tube. Chromosomes and semen for the preferred gender are then filtrated out and artificially inseminated in the woman. Given that there persists a widespread cultural preference for sons, including our own society, what are the implications of this procedure? What will happen with the "natural" sex ratio balance, if there ever was one, that was relatively untampered with except by cultural manipulations such as female infanticide?

Embryo transplants from one uterus to another are also being done. In this situation a woman is artifically inseminated with another woman's husband's ejaculate. The man's wife usually has blocked fallopian tubes or problems with implantation. After fertilization occurs in the "donor" woman, the conceptus or fertilized egg is carefully evacuated from her uterus after several days and implanted in the wife's uterus. In one case, the receiving parents were killed in a plane crash and the state of the floating embryo was of legal and social concern. In this specific case, the frozen, floating embryo was destroyed. In 1989, there was a court case in the U.S. in which a divorcing couple contested ownership of their fertilized eggs. The court decided in favor of the wife. She could have them implanted. Questions of paternity and future child support remain unanswered at this time. We do not have legal, social, cross-cultural, or evolutionary models or precedent to incorporate easily these phenomenon into our sexual and reproductive behavior and belief systems. These technological options force us to rethink our attitudes about life, abortion, parenting, and "normalcy."

Amniocentesis and **chorionic villi sampling, CVS,** detect chromosomal normalities, abnormalities, and gender in embryos and fetuses. By either withdrawing amniotic fluid from the amnion during the end of the first trimester (amniocentesis), or sampling chorionic tissue early in the first trimester (CVS), much chromosomal data can be obtained and used as a decision to continue or terminate a pregnancy. In the former, a second trimester abortion would be performed; in the latter, a first trimester abortion, if termination was selected. In either situation there is a small chance of spontaneous abortion (miscarriage). One of the controversies surrounding these procedures is that abortions can be performed as a means of gender selection; a practice believed to be occurring in India presently (Miller 1988; Crossette 1989).

Fetal reduction is one of the most recent fetal choices. Intended for use in a large multiple fetus pregnancy or where a twin, or triplet is seriously chromosomally or developmentally impaired, fetal reduction is highly controversial. Fetal reduction involves the induced abortion of one fetus which has serious problems to help increase the chances of survival of the other fetuses (Kelly 1990).

The impact of these technologies could change our reproductive practices and future. There are clear implications for altered sex ratio balances, the number of adult men needed in a population, concepts of sexuality and sexual relations, definitions of gender and sex roles, as well as of parenting and families. These are not Orwellian (1950) or *Brave New World* (1946) fantasies, but realities of late twentieth century life. Sexual and reproductive choices and decision making now are qualitatively different than in previous generations and in other cultures, including our own.

Summary

Chapter 6: Modern Human Female Anatomy and Physiology

1. Female sexual and reproductive anatomy and physiology are an expression of the CC + H-P-G axis formula.

2. Female hormonal functioning is generally described as cyclic, in contrast to the male's tonicity.

3. The menstrual cycle is a function of a negative feedback interaction of the H-P-G axis.

4. Many of the primary and secondary female sex characteristics discussed are analogues and homologues of the male's.

5. Differences in male and female body fat and muscle mass are culturally interpreted. Female primary and secondary sex characteristics are often dramatically responded to culturally. Much cultural interest is shown toward female sexual and reproductive functioning. This can include controversial genital surgery such as circumcision, clitoridectomy, hysterectomy, and infibulation.

6. There are several models developed in western culture to explain the variety of female sexual response.

7. Various diseases and cultural management of female sexual and reproductive structures affect a woman's fertility.

8. The menstrual cycle is a biobehavioral phenomenon.

9. The menstrual cycle is culturally regulated and associated with taboos in many societies, including the U.S.

10. While menstrual synchrony is well documented cross-culturally, menstrual cramps and Premenstrual Syndrome (PMS) may be a culture-specific, western phenomenon.

11. Anatomically, physiologically, and hormonally, men and women are much more similar than they are different.

12. Numerous technologies such as AI-D,-H, in vitro fertilization and chromosonal filtration developed in western cultures over the past fifteen to twenty years have the potential to radically change human reproduction.

OVERVIEW

Chapter 7

Fertility, Conception, and Sexual Differentiation

This chapter:

1. Defines fertility, sterility, infertility, and conception.

2. Delineates criteria for male and female fertility and infertility. Discusses causes for infertility in the U.S.

3. Discusses cross-cultural and U.S. reactions to fertility and infertility.

4. Discusses biological, cultural, and technological aspects of conception.

5. Defines genetic, gonadal or hormonal, and phenotypic sex, gender identity, and gender role.

6. Discusses the sexual differentiation process in utero.

7. Discusses Turner's Syndrome, Klinefelter's Syndrome, and the Supermale Syndrome.

8. Introduces biological, prenatal theories of gender identity, and sexual orientation.

Chapter 7

Fertility, Conception, and Sexual Differentiation

In this chapter definitions and criteria for **fertility** and **infertility**, **conception**, and the **sexual differentiation** of the embryo are covered. As part of the discussion of the differentiation process, three of the more common, random chromosomal errors are presented. Since some of the current biomedical theories of the causes of sexual orientation, transsexualism (TSs), and transvestism (TVs) are based on embryologic arguments, these theories will be introduced in this chapter as well. More complete coverage of homosexuality, transsexualism, and transvestism will be covered in chapters 12 and 13.

Fertility and Infertility

Physiologically **fertility** is the ability to impregnate a female with one's own sperm if you are a male, and the ability to be impregnated, that is to ovulate, have patent or open unblocked fallopian tubes and carry a fetus to term if you are female. In general, in current U.S. culture, one is assumed to be fertile unless shown not to be through certain sets of criteria. People also are assumed to be most fertile in their twenties, with fertility for women declining noticeably after age thirty-five until they are no longer fertile post-menopausally. Men retain their fertility, even though there is a reduction in the amount of semen and sperm produced per ejaculate, until they die (Woolf quoted in Marino 1993). By gender, criteria for fertility become more specific.

Biomedically female fertility is determined by age, regular ovulation, patent fallopian tubes, and cervical mucous. In contrast to men, women's fertility begins to decline around age thirty-five until she ceases to be fertile about one year after **menopause**, the cessation of **menses**. In addition, as she ages, the chances of **Down's Syndrome**, a chromosomal abnormality, and some other pregnancy-birth related problems increase. Thus, a female faces a "biological clock" concerning fertility. A second criteria for female fertility is regular ovulation. A woman can only become pregnant when she ovulates; ovulation is therefore a necessary condition for conception. Ovulatory problems are one of the more frequent causes of female infertility (Speroff, Glass, and Kane 1978; Stewart et al. 1979; *Boston Women's Health Collective* 1992). A third criteria for female fertility is tubal patency, which means open, unblocked fallopian tubes that allow for the union of sperm and egg and the fertilized egg's passage to the uterus. Cervical mucous is another factor in female fertility. The texture, color, density, and amount of cervical mucous changes during ovulation to allow sperm to pass through the os, the opening in the cervix (see chapter 6 for a more detailed discussion of this process).

Male fertility is defined relative to **sperm count, sperm motility,** and **form**. As stated in chapter 5, men continuously produce millions of sperm daily from puberty until death. During the sexual arousal and ejaculatory process, somewhere between 200–400 million sperm are ejaculated each time. It is believed that the large number of sperm ejaculated help to move the other sperm along the way to the fallopian tubes. While it takes only one sperm to fertilize an egg, the pathway to fertilization is fraught with danger for the sperm. Only about 200 of the sperm actually survive to reach the fallopian tubes, a trip which takes several minutes after deposition in the vagina.

As such, sperm motility or movement is a second important variable in male fertility. Semen aids in sperm motility and the vast numbers of sperm help each other along the way. Active, fast moving sperm have a greater chance of reaching the fallopian tube and being received by the egg than less active ones. A third component of male fertility is **sperm form**. In gross anatomic terms, sperm are composed of a head, midsection, and tail. The head secretes an enzyme to dissolve the surface coating of the egg to make sperm envelopment by the egg possible. The midsection contains the chromosomal material and the tail aids in sperm motility. All three sections need to be present and functional for impregnation to occur.

As a couple, partner physiological compatibility is necessary for conception. Partner compatibility includes a harmonious pH balance between the woman's vagina and the man's semen, thinned watery cervical mucous, and active, numerous sperm which can pass through the os into the fallopian tubes. While fertility is assumed, its importance is not taken for granted. Fertility is important in all societies; it is probably one of the few universal concerns in human sexuality. This can be seen in art forms, myths, folklore, and people's value on fertility and kinship through time and space.

Beliefs about fertility and conception are widespread and culture specific. While modern human groups know that it requires penile vaginal contact to conceive, various fertility enhancers are found cross-culturally. To enhance conception, potions are consumed, rituals are performed, seduction techniques are encouraged, and spirits are appeased. For example, Mayan women in Mexico may consult a *curandera*, midwife, or traditional birth attendant (TBA) relative to fertility concerns (Faust 1988). Among the Brunei Malay, the *dukun*, a healer, may be consulted for advice as well as potions to ingest (Kimball and Craig 1988). Among the Tiwi discussed earlier, certain female totems are believed to be responsible for conception. They can be sought out or avoided depending on whether a woman wishes "to have life breathed into her body" or to avoid conception (Goodale 1971). Among the Sambia, a horticultural group in New Guinea, whose adolescent sexual practices are dramatic and will be discussed in chapter 10, fellatio, or oral sex performed on the husband, is believed to "prepare a wife's body for childbearing by 'strengthening' her" (Herdt 1993:306). Among the Sambia, semen is the vital life essence which not only makes and keeps men strong and healthy, but is necessary for female fertility and embryonic development. Their sexual beliefs are representative of what Barker-Benfield (1975) refers to as the "spermatic economy." Widespread, cross-culturally, and at various times in the U.S., the spermatic economy is a belief system that focuses on semen, i.e., ejaculate, as a precious vital life substance that exists in finite supply and can be "used up" in a man's lifetime if he is not careful where, how, and with whom he "spends" (ejaculates). Sambian sexual beliefs exist in a culture in which **endemic** or ongoing warfare over scarce resources, including women, is common. Resources for food, shelter, and water are in limited supply because of imposing natural boundaries (sometimes referred to as **impacted habitats**). Women who do most of the food procurement, processing, and distribution are frequently seen as the enemy since marriages are often political alliances amongst warring factions. Women can also be perceived by the men as sexually voracious and potential depletors of treasured ejaculate. The idea that men form the baby and women "grow it" is not that different than western thinking of several hundred years ago when uteri were perceived as the receptacles of the homonucleus (little baby) "given" by the man.

Much of the effort to ensure and protect fertility both within and outside the U.S. rests with the women. Until recently, the dominant fertility patterns among women were extended periods of lactation followed by pregnancy. Continuously uninterrupted menstrual cycles varied by one or two live births and short or nonexistent periods of lactation are largely a middle class, western, twentieth century phenomenon (Frayser 1985; Beyene 1989). Since fertility is critical to the continuation of any group, it is a topic which is taken seriously by most of the world's peoples. This includes means of enhancing conception, avoiding conception to be discussed in chapter 11, and means of dealing with infertility.

Infertility is seen as a tragedy societally and individually, regardless of the culture's individual concerns about overall population pressures. **Infertility**, or the inability to conceive and bear a child, is a cause of societal and cultural concern. When it occurs, it is almost a universal grounds for divorce and individual grief (Cohen and Eames 1982). There is much cross-cultural variation in response to infertility. However, a fairly widespread constant is that the woman assumes and is seen as being responsible for the fertility problem (Frayser 1985). For example, among the pastoral Nuer in Africa an infertile wife becomes a "husband" to another, assumably fertile woman. The female "husband" becomes the sociological father to her wife's offspring by a male. This practice allows the continuation of the infertile woman's patrilineage, the descent system where you trace your family through your male kin (Cohen and Eames 1982). As with many other cultures, the Sambia believe it is only the woman who can be infertile. When infertility occurs in their culture, the Sambian male takes another wife, but does not divorce his allegedly infertile wife (Herdt 1993).

In both western and nonwestern societies, infertility is managed by cultural means. Common solutions can include divorce and remarriage, polygyny, adopting a child, and fostering, the latter being primarily a nineteenth to twentieth century, western alternative. "Aunting" and "uncling" also occur. In these situations, the infertile couple involve themselves intensively with the children in their extended kin group. This may include financial, social, psychological, and ritual activities, similar to what occurs in unilineal descent groups in nonwestern societies. Single-parent adoptions are also increasing for both men and women in western societies. This extends to international adoptions for singles and couples. Some international babies are seen as easier to adopt—girls, for example, may be easier to adopt in societies that strongly value boys. Boy babies may be preferred because they are seen as "brighter, stronger, and healthier" than girls (Whelehan's counseling files). The economic and social flexibility that has occurred for some people in our culture in the latter part of this century makes this option more feasible. While overpopulation may be a global concern and an issue for certain groups, for much of the Third World and for some ethnic groups in the U.S. such as the Amish, Hutterites, and African-Americans, it may be perceived as a threat of genocide by the larger society. Regardless of generalized concerns about population pressures, for the infertile couple who want their own biological child, it is a very remote, abstract argument. Given the universal value of fertility, it is understandable the anguish an infertile couple experiences in not being able to conceive. In some societies infertility is cause for divorce, grief, and loss of status, particularly for the female. Many societies actively try to treat infertility indigenously either through biomedicine, potions, alterations in behavior, and consultation with specialists (Faust 1988; Marmor 1988; Kimball and Craig 1988; Becker 1990).

Since much of the biomedical work on infertility has occurred in western societies in the last decades of this century, the focus of this discussion will be on the west. In the U.S., fertility problems are increasing, with at least 18% of the couples who are trying to conceive and bear a child unable to do so (Hatcher 1986; Stewart et al. 1979; Ragone 1994). In the U.S. a couple is defined as infertile after they have been trying for a year to have a child without success (Stewart et al. 1979; Ragone 1994).

Physiologically, infertility may rest with the man, the woman, or the couple. While the statistics on causal attribution vary, roughly 35–40% of the time the problem is with the male; 35–40% of the time the problem is with the female; and the remaining percentage is either couple incompatibility, behavioral, or unknown (Stewart et al. 1979; Hatcher et al. 1986; Ragone 1994). Physiological causes for both male and female infertility relate to the established criteria for fertility.

Male infertility can be due to low sperm count, motility problems, or deformed sperm. STDs, abusive-addictive drug usage, and congenital problems can cause infertility in males. Relative to numbers, a subfertile or infertile male is one whose sperm count is below 20–40 million sperm per ejaculate (Kelly 1988; Stewart 1979). This is the most common cause of infertility in men.

Sperm motility is another factor in infertility. As stated, sperm need to move quickly and continuously in order to reach a fallopian tube and be able to fertilize an egg. Slow moving, sluggish sperm probably will not survive the trip or be taken in by the egg. Problems with sperm motility are the second most common cause of male infertility in the U.S. Finally, a man may produce misshaped sperm, or sperm missing one or more of its necessary parts. A semen analysis, which notes sperm count, size, shape, and motility is a key diagnostic tool in a male fertility work-up.

Female infertility can be caused by endogenous hormonal imbalances, illnesses, or stress, as well as by STDs, endometriosis, pelvic inflammatory disease (PID), and drug abuse-addiction. The most common form of female infertility is due to problems with ovulation. The second most common problem is some form of tubal blockage, followed by a combination of the two. Diagnostic tests for female fertility problems are usually more complicated, extensive, invasive, and costly than for males. These tests include hormonal assays, measurements of tubal patency, and studies of cervical mucous.

Interactive male-female fertility problems may be behavioral or physiological. Behavioral problems include either too frequent ejaculatory-penile vaginal intercourse that depletes the sperm supply,[1] too infrequent ejaculatory intercourse, or ejaculatory intercourse at times when the woman is not ovulating. Physiological problems include pH imbalances between the woman's vagina and the man's semen, incompatibility between the cervical mucous and sperm, often referred to clinically as "hostile" cervical mucous,[2] and occasionally an allergic reaction by the woman to her partner's sperm (Stewart et al. 1979).

In "spermatic economies" (Barker-Benfield 1975) where semen (ejaculate) is seen as a precious life fluid that exists in finite quantities, too seldom P-V intercourse may be a cause of decreased fertility, interpreted as "infertility." Among some groups such as the Sambia, Mae Enga, and other horticultural groups in Melanesia/New Guinea, P-V intercourse occurs relatively infrequently, resulting in a low birth rate. As stated, women are also seen in these societies as sexually voracious, powerful, and dangerous—eager to "swallow" a man's precious and limited life essence (Herdt 1982, 1993; Gregersen 1983; Williams 1986).

Currently, there are a wide range of treatments having variable degrees of success available to infertile couples in the U.S. Male infertility problems may be treated by isolating and concentrating his viable sperm and then artificially inseminating his partner with them (AI-H), by artificial insemination donor (AI-D), or combining donor-husband sperm in artificial insemination. Generally, vitamin or drug therapies do not alleviate the condition. If the vas deferens is blocked, or a varicocele, a varicose vein of the scrotal sac exists, surgery may be helpful.

> **ASIDE:** A recent article indicates that AI-H may be more successful if the man makes love with his partner using a condom to catch the ejaculate, rather than masturbating into a specimen jar in a doctor's bathroom. The greater eroticism of partner lovemaking is believed to cause a more forceful ejaculation of younger, fresher, healthier sperm (McCarthy 1990; Medical Aspects of Human Sexuality 1991:16).

For females, treatments may be hormonal, surgical, or both depending on the situation. Ovulation problems often are treated hormonally. Tubal blockage problems generally are treated surgically to remove the source of the obstruction. These treatments are revolutionary, dramatic, and can be controversial. Some of the more controversial treatments such as in vitro fertilization (IVF), embryo transplants, embryo-sperm (gamete) implantation in the fallopian tube were discussed relative to their socio-cultural implications in the previous chapter. Even AI-D is controversial, since some adult AI-D babies have searched for their biological fathers (Francoeur ed. 1989).[3]

> **ASIDE**: Legal and social questions arise in our culture as to whether AI-D donor files should be open to AI-D children, and as to whose rights take precedent—the donor's right to anonymity or the child's right to know biological paternity. It is noted that AI-D donors are medically and genetically screened prior to being accepted as participants, that phenotype and sociocultural matching occurs between the donor and child's family, and that the donor's medical and social history data are available to the AI-D child's family.

Couple treatments range the behavioral-physiological spectrum. For those infertility problems caused by intercourse-related behaviors, education about ovulation, the timing of fertilization, and sperm supply can help to alleviate the situation. While this may appear to be a relatively "simple" solution, sensitivity to the couple's psychoemotional state is important. Making love by the calendar in order to conceive a child can produce anxiety, tension, spectatoring, i.e., observing how well you are doing, and can be less than a spontaneous, passionate, sensuous experience for both people.

Couple infertility due to sperm-cervical mucous or pH incompatibility may be treated with drugs, with AI-H as a by-pass mechanism, or with the use of condoms for awhile to see if the problem may self correct (Stewart et al. 1979).

U.S. lay and folk remedies abound. They include increasing the frequency of P-V intercourse and ejaculation which can actually decrease the sperm count, using different positions in intercourse, ingesting vitamins or aphrodisiacs, which enhance neither fertility nor virility. There is anecdotal reporting of two generations duration of "infertile" couples who conceive a biological child after adopting a baby.

Given that some people are currently delaying parenting in this country and have several sex partners in the interim, there is a rise in STDs that can cause long-term problems in both men and women. The rise in fertility problems may appear at an earlier age and may continue to increase. The financial expense, psychological and emotional costs are great for those affected by infertility. It is interesting that even with all the sophisticated technology to treat problems and knowledge about infertility that we have in this country, the responsibility for a fertility problem is still largely seen as the woman's. In a study of middle class, professional couples in the San Francisco Bay Area it was found that regardless of the physiological "cause" of the problem, the woman was expected to somehow "fix it." If the physiological problem was not the male, he offered support to his partner, but did not assume responsibility for its resolution (Becker 1990). This is not that far from the generalized western and nonwestern response of seeing the woman as responsible for fertility. About half of the infertile couples in the U.S. can be successfully treated so that conception and a live birth can occur (*Boston Women's Health Collective* 1992:500).

Conception

As the preceding discussion indicates, fertility is a necessary condition for **conception**. Biomedically, conception is the union of the sperm and egg, dependent upon regular spermatogenesis and ovulation. Conception is not the same as **viability**, or the ability to create and bear offspring. Viability necessitates implantation of the fertilized egg into the endometrium and the development and birth of a full term fetus. Most of the fertilized eggs do not result in a full term fetus (Allgeier and Allgeier 1991). In terms of reproductive success, there is a great deal of wastage. Relative to an individual who may or may not wish to impregnate or be pregnant, conception and viability may be akin to playing roulette.

Conception is regulated and interpreted through culture. In most cultures, there are explanations given as to why and when intercourse results in conception (Gregersen 1983; Frayser 1985). It is a myth in western societies that nonwestern groups do not know that heterosexual genital contact is necessary for conception. That members of these societies do not openly discuss this, particularly with western researchers, is not surprising. Specifics of sexual behavior and conception, particularly across gender lines (most researchers until recently have been and are male), are not topics of everyday conversation. Explanations for conception are embedded in people's views of sexuality, reproduction, and male-female relations. For example, the Tiwi, whom we discussed previously, believe that the essence of sexuality is female. While male totems, animistic spirit beings, are important in their patrilineal kinship system, where descent is traced through males, one's spiritual totems are inherited matrilineally, through females. A woman conceives through a given act of intercourse when her spirit totem breathes life into her body (Goodale 1971). Given that much of the embryonic process is still unknown from a western technological perspective, (e.g., Muecke 1979; Wilson 1979), and that new knowledge about sexuality continually unfolds, a measure of humility is needed in understanding these explanations. It was only a generation ago, when your authors were children, that children were often told that the "stork brought babies," or that they were picked from the "cabbage patch." Conception occurred by a "seed being implanted in a woman's 'tummy'," leading a number of girls who swallowed watermelon seeds to wonder if they would have a baby. These were common U.S. folk beliefs about conception and birth.

Due to AI-D, -H; IVF, embryo and fallopian tube transplants, and forthcoming technological advances in conception, heterosexual genital contact is no longer necessary for fertilization to occur. The full impact of these western developments on conception is still to be seen. Some potential consequences of these developments have been presented in chapters 5 and 6.

From a western biobehavioral perspective, there are several aspects of prenatal sexual differentiation and postnatal phenotypic expression that culturally define sexual physiological "normalcy." While there are a variety of ways to biologically define one's sex as given in chapter 2, four criteria need to be met. Prenatally, genetic or chromosomal sex and appropriate differentiation in utero need to occur. Postnatally, appropriate gender identity or the knowledge that you are male or female, and gender role development and puberty, i.e., sexual adulthood, need to occur.

Genetic or **chromosomal** sex is determined at conception. Genetic or chromosomal sex is the arrangement of either XX pairing for a girl, or XY pairing for a boy. While a range of chromosomal X and Y combinations is possible and may occur, only XX or XY are genetically normal. Based on genetic sex, sexual differentiation or gonadal sex develops in the fetus (Muecke 1979; Wilson 1979). Gonadal sex gives rise to **phenotypic sex**, or, the external and internal physical characteristics that allow a culture to label a child a boy or girl. These characteristics include but are not limited to such structures as the penis, testicles, vas deferens, or prostate in the boy, and the clitoris, ovaries, fallopian tubes, or vagina in the girl. The labeling of a child as a boy or girl is based on visual inspection of the genitalia and is referred to as **gender identity** (Money and Ehrhardt 1972). It is believed that children know their gender identity by the time they are 18–24 months old (Money and Ehrhardt 1972). Based on gender identity, **gender role** develops. Gender role, sometimes referred to as **script** or **scripting** (Gagnon 1979) is the internalization and acting out of culturally defined male or female behavior, affect, and attitudes. The ideal, at least in U.S. culture, is to have genetic, gonadal, phenotypic, gender identity, and gender role synchronized so that one looks, acts, thinks, and feels like a culturally defined boy or girl, man or woman.

The process of sexual differentiation in utero and attainment of gender identity and gender role has received a great deal of attention in western theory from the middle ages through Freud to the present, (e.g., Bullough 1976; Freud 1929 [1975]; Gagnon 1979). Sexual differentiation, the most physiological aspect in this continuum, has also received a great deal of scrutiny (Jost 1961; Sherfey 1972; Wilson 1979; Muecke 1979). These theories range from postulations that as humans, we are all embryologically female in our composition (Sherfey 1972 based on Jost 1961), to a very complex interpretation of the hormonal-anatomical differentiation process (Wilson 1979).

A simplified interpretation of Wilson's (1979) and Muecke's (1979) work on in utero sexual differentiation follows.

Genetic or chromosomal sex is determined at conception by the pairing of either XX or XY chromosomes for a girl or boy, respectively. As part of embryonic development, regardless of genetic sex, the following schema occurs:

1.   The embryo is sexually undifferentiated for the first six weeks of life. Phenotypic sex cannot be determined by visual observation.

2. Both male and female embryos contain both the **Mullerian ducts**, which will develop some female sexual and reproductive structures, and **Wolffian ducts**, which will develop some male sexual and reproductive structures.

3. The Wolffian ducts develop part of the urinary tract system in both males and females, specifically the ureters, the collecting tubules of the kidneys, and part of the bladder.

4. The presence of Wolffian ducts is a necessary condition for Mullerian duct development in the female (Muecke 1979). In the male, the following process occurs. At about six weeks of embryonic development, the male begins to sexually differentiate. Several hormones are released by the embryo to expedite this process. They are testosterone, the **H-Y antigen**, both of which facilitate male anatomic development, and **Mullerian Inhibiting Substance** (MIS) which closes the Mullerian ducts and causes them to atrophy (Wilson 1979). Based on the hormone release and the XY chromosomal arrangement, the penis, testes, and scrotum develop. The Wolffian ducts develop into the rete testes, a rudimentary structure, the epididymis and vas deferens. During the course of fetal development and with the help of testosterone, the other accessory organs appear (e.g., prostate, seminal vesicles). The testicles descend into the scrotal sac during the third trimester. If all goes well, approximately nine calendar months after conception an infant is born, is given the gender identity of boy, and his formal gender role socialization begins.

The girl's differentiation process begins later, around 10–12 weeks of fetal development.[4] While estrogen may be involved in later prenatal development, its role, if any, in the differentiation process is not as clear as is testosterone in the male's (Hyde 1994; Wilson 1979). The Wolffian ducts spontaneously close in the female. A Wolffian Inhibiting Substance, analogous to MIS, is not required. The genital tubercle, homologous and analogous with the males, develops into the clitoris. The **Mullerian ducts** develop into the uterus, fallopian tubes, broad ligament, and upper third of the vagina with the other accessory organs (e.g, labia minora) following. The presence of an X chromosome in both males and females; the need for MIS and testosterone release in the male and lack of comparable hormone release in the female; the live birth of an X or XO "female" and not of a Y or YO male, all lead some researchers, (e.g., Sherfey 1972) to take a strong stand that human embryos are innately female.

As stated, only XX or XY arrangements are considered to be normative genetic sex in this culture. There are, however, a number of other X and Y combinations that can occur. Three of the most common are Turner's Syndrome, Klinefelter's Syndrome, and the Supermale Syndrome. **Turner's Syndrome**, represented by an X or XO combination, occurs in a range of 1/4000 to 1/2500 live births (Mange and Mange 1980; Stine 1977). **Klinefelter's Syndrome**, represented chromosomally as XXY, occurs in about 1/500 live births. The **Supermale Syndrome**, depicted as XYY, occurs in about 1/700 live births (Mange and Mange 1980). See Table 7.1.

Probably about only 2% of the Turner's Syndrome individuals live to be born (Mange and Mange 1980). Turner's Syndrome individuals often have a number of severe physiological problems and frequently die in their twenties. They are sterile, have incomplete or rudimentary ovaries, uterus and fallopian tubes, a blind vagina that may be corrected surgically, immature postpubertal external genitalia, are often short, and have webbed neck or fingers (Money and Ehrhardt 1972). Webbing is a fold of skin between the neck and shoulder or digits. Turner's Syndrome individuals frequently are mentally retarded (Mange and Mange 1980; Stine 1977).

While having some female phenotypic sex characteristics, they do not meet the criteria of normal genetic and physiological development presented earlier in this chapter. Turner's Syndrome individuals chromosomally represent the single X chromosome that some writers have used to postulate the innate femaleness of the human embryo. While it is true that males need testosterone, the H-Y antigen,[5] and MIS to develop as phenotypically normal males, single X chromosome individuals are not physically normal females either. This position is believed by your authors to be a feminist bias, since an X or XO female is not a physiologically normal female. This politicization of the embryo, i.e., that males are "incomplete" since they need hormonal release during differentiation to become a phenotypic male and without it would develop female characteristics, may be a backlash reaction by some feminist researchers. The backlash could be a reaction against Freudian interpretations of the clitoris as a "half-formed" penis and the vaginal orgasm myth

(1929), as well as a general western tendency to present male culture as total culture in which women are defined in terms of their relationships to men. Perhaps bias needs to be recognized more openly whenever it can be found to avoid politicizing the differentiation process.

Klinefelter's Syndrome (XXY) individuals have an essentially male phenotype. They are sterile and tend to have underdeveloped primary and secondary sex characteristics. Atrophied testicles[6] produce low levels of testosterone, frequently resulting in gynecomastia, or breast enlargement, low libido, problems with errectile ability, and more fat than muscle mass per overall body composition. In essence, both male and female secondary sex characteristics appear. Some XXY individual's psychologic development has been labeled schizophrenic, which may be more perceived than real (Mange and Mange 1980; Money and Ehrhardt 1972).

The Supermale Syndrome (XYY) receives attention because of the alleged aggressive and physically violent tendencies of these individuals. According to some theorists (Allgeier and Allgeier 1991), the extra Y chromosome produces men who are not only taller and more muscular than XY males, but who also have a greater propensity for acting out violently. Many of these studies are methodologically flawed (Allgeier and Allgeier 1991). About 50% of the XYY males are sterile.

All three of these syndromes: Turner's, Klinefelter's, and the Supermale are believed to occur randomly. Since both Turner's and Klinefelter's Syndrome individuals are sterile, they are self-limiting. They do not reproduce. Fertile XYY males do not appear to be more likely to produce XYY sons than XY males. Causes of these chromosomal variations are unknown (Mange and Mange 1980).

Our society relies heavily on genetic, hormonal, and phenotypic sex characteristics to assign gender identity and gender role. Some nonwestern societies offer an interesting contrast to ours regarding how they define gender identity and gender role. Among the Sambia in New Guinea and in those cultures that recognize inter-sex or third sex individuals (e.g., the *nadle* among the Navajo, *berdache* among Plains Indians and other groups), maleness and femaleness are not defined entirely by phenotype (Herdt 1982; Williams 1986).

The essence of being male or female is important as a defining characteristic and is culturally valuable. It can include substances such as semen or menstrual blood as well as spiritual, aesthetic, kinesic, or occupational attributes. Therefore, cross-culturally a man

Table 7.1
Summary of Anomalies of Prenatal Differentiation

Trisomic X	47, XXX	ovaries	female	normal female	fertile	some mentally retarded
Klinefelter's Syndrome	47, XXY	testes	male	normal male	sterile	low testosterone, small testes and penis, tall, gangly, at risk for Genito-Urinary tract anomalies
"Supermale" Syndrome	47, XYY	testes	male	normal male	50% sterile	tall, prison study, impulsive problems with emotional intimacy
Turner's Syndrome	45, X	streaks of ovarian cells	female	uterus and fallopian tubes	sterile	no menstruation or breast development due to estrogen deficiency, short, webbed neck, treat with estrogen, Money study reports hyperfeminine
Androgen overdose (AO)	46, XX	ovaries	ambiguous male	normal female	fertile	synthetic progestin (40's, 50's), internal Wolffian structure not developed, problems labeled male, at puberty will grow breasts will menstruate (if partial fusion labia), Study 10 A.O. and 15 treated AGS: tomboys, athletic, more energy expenditure, not more aggressive than matched
Adreno genital syndrome (AGS)	46, XX	ovaries	ambiguous male	normal female	fertile	23 AGS raised as virilized (untreated) lesbian fantasy (N=10), erectile clitoris, 32% do not want kids, secretions of adrenal cortices gain control, ovarian hormones, erectile clitoris treated with synthetic cortisone allows ovaries to feminize, untreated will masculinize

Condition	Karyotype	Gonads	External genitalia	Internal structures	Fertility	Notes
Complete androgen insensitivity (AIS)	46, XX	cryptorchid testes (undescended)	female short vagina	no uterus, tubes, no prostate	sterile	breast development at puberty, no menstruation, female psychosexual orientation, female gonadectomy preferred intervention, study N=10 feminine compared to AGS
Partial androgen insensitivity	46, XY	cryptorchid testes	ambiguous male	no uterus, tubes, no prostate	sterile	better to label female, profile of AIS-labeled male mirror image of overandrogenized girls
Dominican Republic Syndrome	46, XY	cryptorchid testes	ambiguous female	vas deferens, epididymis, seminal vesicles no prostate	fertile; unable to inseminate; urethral opening in perineum	puberty, voice deepens, more musculature, penis grows, testes descended and enlarged, male role assumed

(Adopted from Money and Ehrhardt 1972)

can be phenotypically and genetically male, but be labeled female or something else (i.e., "not man" or "near man") by his affect, demeanor, or special talents as in the case of the *berdache*. Similarly, a phenotypic and genotypic female may succeed in being a warrior woman (Williams 1986). With the Sambia, oral intercourse among adolescent males is seen as a way of preserving and recirculating semen—a vital, life sustaining fluid believed to exist in finite quantities. Male-male fellatio during adolescence builds up male energy (*jergunda*) in order to carry the adult male through his heterosexual relations as a husband and semen-nurturer to his unborn children, as previously discussed. P-V intercourse, by definition then, is a potential drain of this energy and must be carefully controlled (Herdt 1982). It is important to note that these roles and definitions are seen as positive and generally carry positions of status and respect with them.

Transsexualism

Gender dysphoria is a clinical term that refers to people who are uncomfortable identifying with or behaving according to their culturally defined gender (Money and Wiedeking 1980). Gender dysphoria may take several forms. Among these are transvestite people who may cross-dress intermittently and/or for extended periods of time. It includes, but is not limited to, those who feel compelled to cross-dress and those who are aroused by cross-dressing. **Transsexual** people believe they are trapped in the wrong body and psychically identify with the other sex. Recently, another term has arisen to describe gender variance.[7] **Transgenderist** is a term that has come from the community of people who self-identify as transsexuals, transvestites, and those outside and between these kinds of gender variant options. Transgenderism supplants the dichotomy of transsexual and transvestite with a concept of continuity. It also reflects a growing acceptance of non-surgical options and the recognition that there is a great deal of diversity in gender variant identities not represented by the clinical and community categories of transvestite and transsexual (Bolin 1994:401). While the cause of a transgendered identity is unknown, various theories have been proposed to explain its occurrence.

Etiology: Prenatal and Biochemical Explanations

Since biomedical explanations for transsexualism explore prenatal and biochemical causes for its existence and expression, they are discussed here. Anthropological and cross-cultural explanations and examples for transgenderism including transsexualism and transvestism are discussed in chapter 12.

Etiology or seeking the causes of transsexualism is a prominent and prolific area of research activity. This question is addressed not just in the clinical areas as might be suspected, but is also discussed by anthropologists and sociologists as well. While paying respect to the diversity of clinical approaches, it is perhaps not unfounded to suggest that these perspectives tend to focus on the individual factors affecting the development of a gender identity at variance with morphological sex. Such a perspective tends to follow a disease model of gender variance.

In contrast, as will be discussed more fully in chapter 12, anthropologists have tended to look at larger domains stressing systems of interaction including socialization, gender relations, kinship, warfare, political and economic variables related to gender inequality (see especially Wikan 1977; Munroe, Whiting, and Hally 1969; Burton and Whiting, 1961 among many). Rather than viewing the individual as gender dysphoric as described in the clinical literature, the anthropological model seeks to understand how gender variation operates within the cultural system and how it links up with other cultural institutions including the belief system around gender. Therefore, etiology is considered by anthropologists in terms of social constructs as is gender.

Transsexual etiology is of interest because it throws light on the broader subject of how gender identity is established not only in transgendered populations whose identity is in conflict with their morphological sex, but also for the population at large for whom gender identity is not problematic. In addition, it provides insight into understanding the cultural construction of gender in our society, and how gender is integrated within systems of gender relations such as inequality between the sexes.

Transsexualism is of interest to embryology because it has been framed in the clinical literature as part of a nature/nurture or essentialist/constructionist debate. Money (1986) among others has questioned this polarized argument as simplistic and unrealistic, emphasizing the interaction of biology and culture. It is the clinical studies that we shall draw on here in our discussion of fetal development as these illustrate current debates in these fields.

Money and Ehrhardt (1972) and Stoller (1968) regard socialization variables as taking precedence over prenatal sex hormones in the formation of cross-gender identity, although these researchers acknowledge that there may be some unknown biological (hormonal-metabolic) factor in the prenatal environment that may play a role. Stoller's model points to maternal overprotection and paternal distance—either emotional or geographical—in transsexual etiology. According to Stoller, the child fails to identify with the father and becomes effeminate. Green (1974a and 1974b) supports the view that transsexuals share effeminate childhoods and then are subsequently channeled into transsexualism as their options for "normal" gender identity development have become closed off.

The parent of the study of transsexualism, Dr. Harry Benjamin (1966), has favored biogenic variables even in those cases where socialization may have clearly been a factor. Others taking this position include Starka, Sipova, and Hynie (1975) whose findings of lowered levels of testosterone in 17 male-to-female transsexuals, three transvestites, and four homosexuals has *not* been replicated and must be regarded as "atypical." The phenomenon of transsexualism is in fact noted for its normal hormonal profile in individuals whose identity is variant.

Others such as Eicher (1981) have suggested that the H-Y antigen may be atypical in transsexual people. In preliminary research Eicher found in his research population of 40 male-to-female and 31 female-to-male transsexuals, that 84% were at variance with the norm. According to Ohno (1979), "H-Y antigen is a cell surface component present in all male tissues and absent in genetic females," although this association is controversial as a predictor of sex. But Pfafflin (1981), who studied transsexuals with a control group of non-transsexuals, found that the non-transsexuals were atypical for H-Y antigen in about 50% of the cases, thereby questioning Ohno's thesis that H-Y antigen is a predictor of sex to begin with, much less having an influence on gender dysphoric individuals.[8]

Several studies have looked at how transsexual persons process hormones. Seyler, et al. (1978) found female-to-male transsexuals' hormonal responses to DES (diethylstilbestrol) were intermediate between the "normal" female and male pattern. Migeon, et al. (1969) administered estrogens to male-to-female transsexuals and control groups of non-transsexual males and females and concluded that the transsexual hormonal response to estrogen was atypical to both control groups. While some prenatal hormonal disturbance suggests itself, Seyler, et al. also offer the possibility of a psychological factor since it is well documented that psychological states may be as much cause as effect of hormones. Devor (1989) notes that cases in which females have been exposed to androgens in utero have not increased the incidence of transsexualism in those populations.

This presents a strong argument against the hormonal basis of gender identity variance.

The etiology of the western transsexual identity remains as undetermined. While there is strong support among some researchers for atypical socialization variables (Stoller 1968; Green 1974a and b; Money and Ehrhardt 1972), other research indicates the potential for a fetal hormonal atypicality (Benjamin 1966; Eicher 1981; Seyler 1978). At this point, the jury is not in. However, Devor points us in a direction for reconceptualizing the issue within a bio-cultural framework that may eventually lead to greater understanding of gender dysphoria. She presents an interactionist approach to the development of gender identity in which the environment may be seen to potentiate prenatal influences either by inhibiting their development or by enhancing it. According to Devor (1989:22):

> External environmental experiences set into motion a momentum which may be in continuation of pre-natal influences, or in contradiction to them. In either case, social factors may be capable of overriding most, if not all, prenatal influences. Social influences may actually reset the direction which future development of a hormonal system will take. They may act to suppress or enhance biological predispositions. If social forces continue to exert pressure over long periods of time, a chronic situation can develop which may crystallize into relatively stable physical configurations that reflect the direction of social pressures. In this way, hormonal abnormalities might be seen to be the result of chronic social abnormalities.... [O]ne might interpret the gross hormonal differences between socially normal men and women as being a result, rather than a cause of the chronic social pressures which males and females undergo in the process of becoming socially normal men and women.

She concludes that the brain and its interacting endocrine system "learn" behaviors just as humans acquire behavior through cognitive processes. Thus: "Not only is the human mind in dynamic interaction with its environment, ... so too is the human body changing, learning and growing through its experience within its environment" (Devor 1989:22). The transgender identity is discussed further in regard to cross-cultural evidence in chapter 12.

Homosexuality

In some ways the numerous attempts to define the causes of homosexuality, or the sexual orientation for same gender and emotional love partners are even more controversial than trying to biologically explain the causes of transsexualism. Having to explain these identities and orientations clearly illustrates the assumption and bias that we are born heterosexual and hence sexually and emotionally attracted to people of the other gender. While the continuum of sexual orientations and accompanying life-styles will be discussed in chapter 13, the biological explanations for homosexuality are introduced here.

Given that we spend little time in this culture explaining the causes and desire to be heterosexual, the amount of energy invested in explaining homosexuality is a clear indication of our cultural discomfort with the topic, the narrowness of our range of culturally acceptable sexual orientations and behaviors, and our continuing value on sex for reproduction.[8] At the same time, there is an increasing research interest which supports a prenatal disposition to one's sexual orientation, regardless of preference (Goldstein 1990; LeVay 1991; *Time* 1993).

Theories that explain the causes of homosexuality include psychoanalytic (Freud 1922 [1975], 1959), learning environmental (e.g, Storms 1980; Green 1978), and biological (e.g., Goodman 1983; Mayer-Bahlburg 1977 and 1979; Gladue 1984; LeVay 1991). Many of these theories assume some level of "unnaturalness" about homosexuality, and tend to be more explanatory of male (gay), than female (lesbian), homosexuality. With the exception of Mayer-Bahlburg (1979), they are basically nontestable, and are strongly focused on explaining how and why. The biological theories, by definition, tend to downplay sociocultural variables which influence behavior. The theories also tend to be post-hoc explanations, particularly the biological ones, and they don't clearly differentiate between orientation and behavior, and tend to ignore the cross-cultural record which illustrates our sexual flexibility, i.e., the Sambia.

Biological theories tend to focus on prenatal determinates and/ or postnatal hormonal patterns. In general, it is accepted that relative to genetic, gonadal, and phenotypic sex, homosexuals are gender concordant for these traits. Generally, it is increasingly more accepted that their gender role is compatible with their gender and biological identities (Kelly 1988; Crooks and Baur 1987).

Prenatal theories focus on hormonal release patterns during the differentiation process, particularly for males. The release of prenatal hormones plugged into receptors in the brain are such that a predisposition toward homosexuality may be established; but not enough to disrupt the development of phenotypic sex characteristics (Goodman 1983; Gladue 1984; Goldstein 1990). Postnatally, there is investigation not only as to the absolute levels of testosterone and estrogen in gays and lesbians, but also as to the timing and sequencing of these hormones and LH release (Gladue 1984; Mayer-Bahlburg 1977 and 1979). In general, both the pre- and postnatal hormonal evidence is inconsistent and inconclusive. Most frequently, hormonal studies on homosexuals indicate that their hormone levels are within the range of those who are heterosexual or straight. It would be very difficult to posit a cause-effect relationship between hormone fluctuation or variation, sexual orientation and sexual behavior (Mayer-Bahlburg 1977 and 1979; Gladue 1984).

The weaknesses of many biologically based theories include a homophobic bias, their researcher-admitted inconsistency, the posthoc nature of the explanations, their androcentric or male oriented bias, and nonrecognition of homosexuality as an intra and interspecies wide behavior. They often ignore the cross-cultural record where evidence supports sexual flexibility or adaptability (Gregerson 1983; Herdt 1982; Williams 1986; Bullough 1976). We know very little about the cause of sexual orientation in general. In the age of AIDS and the threat to our civil liberties regarding abortion choice, certain forms of birth control, and lifestyle options, the controversial nature of these theories on homosexuality may be used to feed the belief that homosexuality is deviant, or unnatural. Biological theories may be used to support a disease model. In some ways, the alleged sexual revolution of the 1960s did not fulfill its aim of sexual choice and acceptance. Attitudes toward homosexuality are one example of this. It is extremely important that we maintain a relativistic stance and regard homosexuality and bisexuality as evidence of our human variablity.

Summary

Chapter 7: Fertility, Conception, and Sexual Differentiation

1. Fertility, infertility, and conception are biobehavioral phenomenon.

2. There are a number of technological procedures in western cultures to deal with infertility problems.

3. There are a number of theories used to explain sexual differentiation in utero. Some theories used to explain transsexualism and homosexuality rely on interpretations of the differentiation process.

4. Genetic or chromosomal sex, hormonal or gonadal sex, phenotypic sex, gender identity, and gender role are cultural terms used to explain the prenatal differentiation process and postnatal development of identity and roles in the U.S.

5. Turner's Syndrome, Klinefelter's Syndrome, and the Supermale Syndrome are variations of XX or XY chromosomal arrangements.

6. Gender dysphoria is a sense of discomfort with one's gender identity.

7. Of all the known sexual orientations, heterosexuality is the only one our culture believes does not need causal explanation. Increasingly, it is believed that predeterminants for one's sexual orientation may occur prenatally.

OVERVIEW

Chapter 8

Pregnancy and Childbirth as a Bio-Cultural Experience

This chapter:

1. Examines pregnancy and childbirth as a bio-cultural phenomenon.

2. Views pregnancy and childbirth as a physiologically normal, healthy process in which complications may occur.

3. Examines childbirth as the means to culturally create and extend kinship.

4. Examines male participation in the female experience of pregnancy and childbirth.

5. Examines cultural responses to pregnancy, childbirth, and the postpartum period.

6. Explains the stages of labor.

7. Explains the non-interventionist-interventionist birth continuum and places our culture's birth practices along the continuum.

8. Discusses postpartum depression biologically and culturally.

Chapter 8

Pregnancy and Childbirth as a Bio-Cultural Experience

In this chapter, pregnancy and childbirth are examined as a bio-cultural phenomenon, composed of an integration of physical, socio-cultural and psycho-emotional variables. This chapter focuses on physiologically normal pregnancy and birth as a part of the hominid life cycle. Thus, it emphasizes sexual-reproductive functioning as it is culturally managed. It de-emphasizes problems in pregnancy and birth and de-emphasizes recent technological innovations which intervene in human sexual reproduction. This chapter looks at pregnancy relative to trimester fetal development and the importance of overall health and life-style in the course of normal pregnancy and birth.

Pregnancy and birth are key life cycle status changes. With the birth of a child, a kin group is not only begun (**the family of orientation**), but it is extended (**the family of procreation**). The birth of a child forms a union between extended kin groups that can include **blood** (**consanguineal**) and **marital** (**affinal**) ties. It frequently transforms the social status of the biological and social parents to adulthood (Raphael 1988; Mead and Newton 1967; Newton 1981). The individuals involved such as the mother, child, possibly the father, and the group are affected by the pregnancy and birth process. In this chapter, the cultural management of pregnancy and childbirth are discussed from western and non-western perspectives.

The anatomical and hormonal changes in a woman's body are matched by cultural concern over her pregnancy and the birth process. Many societies recognize pregnancy as a unique state, placing the woman, and by extension, the fetus under special rules

159

of behavior extending to diet, exercise, normal routine, social and sexual interaction, and cultural/institutional participation.

The Fetus

The humanness of the fetus is culturally defined. In current western thought the point at which the fetus becomes human is controversial, and is the topic of intense debate relative to abortion and certain forms of birth control. In colonial days, a fetus was not defined as alive until the onset of quickening or fetal movement, usually detected in the fourth month during the second trimester of pregnancy (Bullough 1976).

Late twentieth century biomedical explanations trace fetal development by trimesters. Major organ and system development occurs during the first trimester. In these first three months when neural, brain, muscular, and organ development is forming and sexual differentiation begins, the fetus is seen as extremely vulnerable to external factors such as drugs, pollutants, toxins, or X-rays. For this reason, early confirmation of pregnancy and modification of diet, drug use, sleep, and exercise are seen as very important to healthy fetal development in western women. The second trimester, months 4–6, involves elaboration of skeletal, muscular, and system development. It frequently is an end to the nausea experienced by some western women and breast tenderness caused by high levels of progesterone secreted by the mother. The third trimester, months 6–9, is generally a period of major weight gain in the fetus and ongoing maturation of systems in preparation for postnatal life. Hearing, however, is the only system totally mature at birth. The testicles descend into the scrotal sac during the third trimester.

The Pregnant Female

The parturient or pregnant woman's body during these three trimesters undergoes radical changes. There are surges and elevated levels of estrogen and progesterones in her body that may initially contribute to morning sickness or nausea. Morning sickness is not a universal phenomenon. It is not reported as often in nonwestern societies as it is in the U.S. This may be due to the higher complex carbohydrate diet in nonwestern societies than in the west. Current U.S. biomedical recommendations to the pregnant woman experiencing morning sickness are to eat

Figure 8.1 Fetal development
at 9 weeks. The fetus (right)
is connected to the placenta
(left) by the umbilical cord.

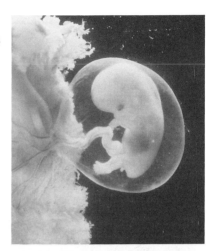

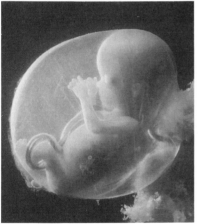

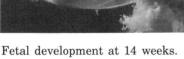

Fetal development at 14 weeks.   Fetal development at 20 weeks.

whatever she can swallow and keep in her stomach, regardless of
its nutritional component or value. Some food ingested is better
than none or food which is rejected (Erick 1994).

The woman's breasts increase in size by several pounds as the
milk ducts, stimulated by pituitary hormones such as LTH, prepare for lactation. Her uterus increases in size to accommodate the
fetus, **amnion**, or bag of waters, **placenta**, and **umbilical cord**.
Weight gain, slight separation of the pubis, slight lowering of the
red blood cell count, and slightly elevated blood pressure are all
normal physiological developments in pregnancy.

Labor and Birth

The birth process itself is a normal physiological event, in which problems may develop for either the fetus, mother, or both. It is estimated that in an otherwise healthy pregnant woman, about 92–95% of the births are normal (*Boston Women's Health Collective* 1976, 1984, 1992; Arms 1975).

This approach is a radical departure from the accepted twentieth century U.S. view that pregnancy is a physiologically dangerous, pathological state where the pregnant woman is placed in the sick role (Pritchard, MacDonald, and Gant 1985; Mitford 1992; Davis-Floyd 1989/1990, 1992; Williams 1989).

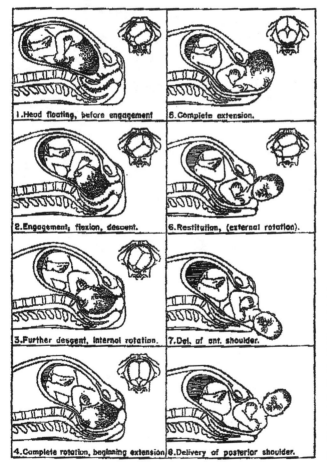

1.Head floating, before engagement.
2.Engagement, flexion, descent.
3.Further descent, internal rotation.
4.Complete rotation, beginning extension.
5.Complete extension.
6.Restitution, (external rotation).
7.Del. of ant. shoulder.
8.Delivery of posterior shoulder.

Figure 8.2
Major stages in the birth process. The fetus is shown in the uterus as labor contractions begin and progresses through the stages of birth.

While the exact causes for the onset of labor are unknown, it is probably a function of the interaction of fetal maturity and oxytocin, a labor-stimulating pituitary hormone. Labor is generally depicted as a three or four stage process depending on the medical text used.

Prior to or early in the first stage of labor, "**show**" or the mucous plug is usually expelled. Show is a bloody mucoid substance that covers the os (cervix) during pregnancy to protect the sterile environment of the uterine cavity during fetal development. The os fully dilates or opens during the first stage of labor. This generally takes several hours. A fully dilated os is 10 centimeters or five fingers in breadth. In U.S. culture, digital or finger vaginal exams are regularly performed during this stage to assess the state of dilation. The cervix effaces during the first stage of labor as well. **Effacement**, measured from 0 to 100%, is the gradual softening of cervical tissue. A "softer" cervix allows for greater ease in the passage of the fetus through the os. The bag of waters or amniotic sac may break during this or subsequent stages of labor, or the baby may be born with the amnion intact, called the caul. Transition, the second stage of labor, occurs when the cervix is fully dilated. Transition is the passage of the baby's head through the dilated os and may take several minutes to several hours. The third stage of labor is the actual birth of the baby through the introitus. The perineum stretches to accommodate the baby.

Stretching the perineum is culturally managed and an example of the bio-cultural nature of sex and reproduction. In most societies outside the U.S., the perineum is stretched through massage, application of warm compresses, or an upright birth position which allows the baby to pass through the introitus at an angle which fits the mother's body (Jordan 1993, 1983; *Medical Anthropology* 1981; Michaelson 1988). In the U.S., in over 90% of the births, the perineum is stretched surgically with an **episiotomy**, an incision in the perineum. This procedure is controversial. Proponents argue in favor of episiotomies as a means to reduce perineal tearing. Opponents to episiotomies state that tearing rarely occurs if upright birthing positions and massage are used (Arms 1975; *Boston Women's Health Collective* 1984, 1992; Jordan 1993, 1983; Michaelson 1988). In response to this opposition, "innovations" such as birthing chairs, originally developed for birthing women in Europe in the 1500s, have been introduced into maternity suites in some "progressive" U.S. hospitals as the newest, most female-centered aspect of high-tech childbirth!

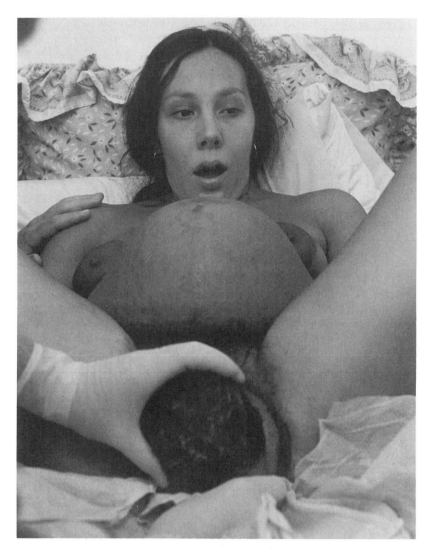

Figure 8.3 Baby emerging; natural birth, © Suzanne Arms.

The fourth stage of labor consists of the expulsion of the placenta. During this period, the umbilical cord is cut. In most societies the baby is cleaned to remove blood and the **vernix,** the waxy, protective coating, which covers the baby in utero. In most societies, the baby and the mother have some contact at this time. This is the first stage of infant-mother bonding. Again, this is a controversial belief in the U.S., where babies and mothers in hospitals

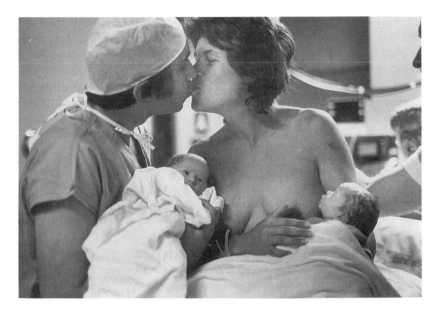

Figure 8.4 Mom and dad with twins following natural birth. Stanford, California, © Suzanne Arms.

frequently are separated shortly after birth for up to several hours, ostensibly to give the mother rest (Mitford 1992; *Boston Women's Health Collective* 1992).

After the expulsion of the placenta, the uterus starts a six-week process of involution or return to its nonpregnant size and shape. Involution is also culturally managed by massage in some nonwestern societies (Fuller and Jordan 1981; Jordan 1993, 1983), and through the use of drugs in the U.S. (Jordan 1993, 1983; Michaelson 1988). In many societies, women are encouraged to nurse immediately after birth in order to stimulate milk production and to help the uterus to involute. In general, our society still sees pregnancy as a bio-medical condition that needs to be medically managed. Other societies see it as a normal state in which problems may or may not occur[1] (*Medical Anthropology* 1981; Jordan 1993, 1983; Michaelson 1988; Pritchard, MacDonald, and Gant 1985).

While recognizing and attending to the physiological aspects of labor and birth, many nonwestern societies see birth as a significant social event. Birth unites women and reinforces their bio-social sameness (*Medical Anthropology* 1981; Frayser 1985). Birth attendants (to be discussed) are known, respected women in the cultures in which they practice. They assist, support and are with the parturient woman before, during and after childbirth (*Medical Anthropology* 1981; Mitford 1992; *Boston's Women's Health Collective* 1992). Birth also creates the family of procreation for the parents and family of orientation for the child. It involves a status change for the parents, called **matrescence** and **patrescence** by Raphael (1988).

Birthing Models

There are two general models developed as the cultural response to pregnancy: the **interventionist** and **non-interventionist**. These models exist on a continuum, since all cultures intervene in the pregnancies and births of their members. Those societies toward the non-interventionist end tend to view pregnancy and birth as a natural phenomenon and emphasize the socio-psychological dimensions over the physiological. In these societies, the well being of the mother and fetus are primary concerns with the expectation that the woman needs support. The physical birth process, while long and "laborious," is believed to generally occur on its own. Intervention in the actual labor process occurs when problems arise. This model is characteristic of societies outside the U.S. and the U.S. prior to the late nineteenth century (*Medical Anthropology* 1981).

The interventionist model is characteristic of current U.S. society, even with the range of birthing alternatives available such as birthing centers and rooms, family-centered childbirth, and rooming-in, where newborn and mother spend most if not all their hospital stay together. About 95% of U.S. births occur in hospitals. Birth is primarily viewed as a biomedical phenomenon with a "maximin" mindset (Davis-Floyd 1988). This view perceives childbirth as inherently dangerous and prepares for the crisis situation as a general rule. Thus, there is much medical and technological intervention in normal as well as complicated births (Davis-Floyd 1992). This includes routine use of internal and external fetal monitors to chart the fetal heartbeat, episiotomies, use of IV's and drugs in the vast majority of births (Michaelson 1988; Sargent and Stark 1989; Jordan 1993, 1983; Williams 1989). It also includes a rising Caesarean rate that averages about 20% of the hospital births in this country (Sargent and Stark 1989).

There are a variety of ways pregnancy through the **postpartum** or post-birth period are addressed culturally. These are discussed relative to birth attendants, the *couvade*, the confinement and birthing period, and early postpartum care.

Birth attendants are those people who minimally take care of the woman during her labor, the birth of the baby, and immediately after. Who these people are varies widely. Outside the U.S., they usually are female, they usually are known to the pregnant woman; and frequently they are midwives, women who are trained in pre- and postnatal care and indigenous birthing practices. These women attend to the socio-psychological and physical needs of the woman and her baby before, during, and after birth (Raphael 1988; *Medical Anthropology* 1981; Faust 1988).

The following passage is a description of a birth in a Mayan community in Latin America. It provides an interesting contrast to our biomedical approach.

Notes from a Field Log:
Doña Bernarda at work

I was up writing when there was a knock on my window—"Wake, my comadre." "Do we go?" "Yes." I hurriedly grabbed my camera, raincoat, and flashlight. Off we went into the dark, wet cornfields, a quick little step, over mud, ruts, and puddles behind the women who had come for Doña Bernarda. She was hard to keep up with— across the road and into San Cristobal Huixchochitlan. The house had two rooms. One was huge, with a hearth, straw, petates, and a cradle hung from the ceiling. The other room had a large altar with a huge saint taken out on July 25—the fiesta of San Cristobal. On a petate on the floor was Gloria Inez, age sixteen, giving birth to her first child. The room was lit by one light bulb high in the ceiling, and there was a warm glow of subtle browns. We had arrived around ten o'clock. Doña Bernarda examined her, massaged her; she was lying down. The hours that followed held a series of events and people. First, four of five women helped her— two aunts held her, with Doña Bernarda at her feet and me holding her hand. She was made to squat, first supported by an aunt, later by an older man, and finally by her husband. I took pictures for awhile, but, as a woman in the group, became part of the process—holding her hand, supporting her from the side when she squatted, covering Doña Bernarda with her rebozo when she felt cold and tired. She never stopped, her hand under the black Indian skirt (chinquete), her words and sounds constantly reassuring the tired, frightened girl, "Everything is all right," "The baby is coming well." Doña Bernarda pronounced the baby as "viene bien," but as the hours wore on, the aunts prayed and touched her back and front with two coins while praying to the Virgin to bring on the baby. The coins touching the body are an offering to San Ramon, patron saint of childbirth. All the while she was held, talked to, wrapped, changed, wrapped—a steady flow of soft, Otomi words—Doña Bernarda telling her how to push, grunting, encouraging her. Doña Bernarda sent for some yerba buena, and one of the women disappeared with a candle and came back with a handful of fresh picked herbs. They soon came with hot tea, but Inez had begun to vomit into the rags and rebozos which appeared from all over. Later she was given the yolk of an egg, then big gulps of hot tea. Doña Bernarda said she needed exercise; the older man lifted her up and shook her; they made her walk across the room twice. Doña Bernarda took some gray ball (tequesquite) from a paper offered by one of the women and broke the water. The tequesquite is a tough, salty rock which, rubbed on the "puente," ruptures the membranes. The scissors and iodine lay nearby on the altar. They pressed on her abdomen; the pains came irregularly and she tired. "Put holy water on her head so she doesn't sleep," said Doña Bernarda. She continued to squat—her husband behind supporting her, an aunt at each side, and Doña Bernarda on a piece of plastic covered with a cloth, a gray-white coat backwards and her large, brown rebozo over her shoulders. Women milled in and out, people came and were greeted; with each pain Inez covered her face—"Maria Santisima." Doña Bernarda said it was slow since it was her first—that she kept alcohol in case they fainted to revive them. The baby finally came around 12:30; Doña Bernarda pulled with strong, rhythmic strokes, and from under the black skirt emerged a clean, white baby with lots of black hair. She quickly wiped its nose, mouth and face; it cried heartily. The baby's cord was tied and cut, then it was wrapped completely in rags, and all attention was concentrated on the mother to remove the placenta. It came, but Doña Bernarda showed me a large lump protruding from one side of the vagina— "she was afraid and that's why that happened." Would Dra. Laurita come? I said "yes"—we agreed to bring her as soon as she arrived in the morning. Doña Bernarda washed the mother with rags and hot water, putting a hot compress on the lump. Then she turned her attention to the baby. Should she bathe it? she asked me. "What you always do, Senora," I replied. She laughed, and a basin was brought—she washed the baby vigorously with lots of soap and water, and especially its head. Its clothes were brought—a shiny, satin coat, green satin cap, and large flannel blanket. She made a bandage for the umbilical stump, painted it with iodine, and then wrapped white cloth and an Otomi belt around the middle. Then she finished dressing the baby, finally putting a white blanket over it. She then took clean water on her finger and reached way into the baby's throat and mouth, pulling out phlegm. Then she put a pinch of salt on her finger and put it in the baby's mouth—"She likes it." This is to bless the baby in case it dies before baptism. The atmosphere broke; the women began to smile. Someone made a bed next to the mother on the petate and they put the baby next to her under a large blanket. They took away

the rags, blood, placenta and plastic. Doña Bernarda washed with
soap and hot water. "What shall we do?" she asked. "Sleep," I
answered, and all laughed. "Here?" "I could sleep standing up," I
answered; and all laughed. She then discussed money with them
and said that Dra. Laurita would have to charge also. She said
she would return in the morning around eight. Both men and one
of the aunts accompanied us back to the house.

    Michele Godziehen-Shedlin is a Senior Staff Assistant at the
Center for Population and Family Health, College of Physicians and
Surgeons, Columbia University (Godziehen-Shedlin 1981: 13–15).

Since the mid-nineteenth century in U.S. culture, the entire
birth process has been increasingly medicalized, removed from the
home and familiar settings, and has become a male specialty.[1]
Hospitalization for childbirth usually means the woman is in an
unfamiliar, socially sterile, medical environment, attended by a
series of strangers. Often she is left alone for extended periods of
time, a practice rarely occurring in nonwestern societies where
social support is seen as important in the birthing process. In the
U.S., the physiological aspects of pregnancy are emphasized over
the socio-psychological. Women are delivered in the U.S.; they do
not give birth. In nonwestern societies, the locus of control rests
with the pregnant woman, in the U.S., with her medical birth
attendants (Martin 1987; Davis-Floyd 1992). In nonwestern societies, birth attendants are integrated into the fabric of the woman's
life; in the U.S. they are distinct, discrete entities who act on her.

    Examples of birth attendant-parturient woman relationships
from Greece, Latin America, and Egypt illustrate the importance of
this role. Shared characteristics of these birth attendants are that
they are female, they are older, experienced, respected people in
their cultures, and they may be involved in nonobstetric health
care roles as well. These birth attendants are known to the pregnant woman and are part of her social support system. They provide her with care, advice, and guidance from pregnancy through
the postpartum period.

    In Greece, this woman is called the *doula*. She is particularly
helpful in establishing a breastfeeding pattern and offering help to
the postpartum woman. She gives advice, helps with the daily
routine, and provides socio-emotional support in teaching the new
mother how to breastfeed. She temporarily becomes part of the
extended kin household (Raphael 1988).

    In Latin America, midwives are often called *doña*, a title of
respect (Faust 1988; Godziehen-Shedlin 1981; Sukkary 1981). These
women take seriously the overall health of their patients or clients,

and frequently care for their gynecologic, as well as obstetric needs. These midwives monitor diet, social activity, and if government-trained, may record maternal blood pressure, fetal heartrate, and other physical signs during pregnancy (Faust 1988). In much of Indian Latin America, pre- and postpartum massage and binding, where the women's abdomen is tightly wrapped, are part of prenatal and perinatal care. This necessitates the midwife's presence and her support of the pregnant woman (Fuller and Jordan 1981).

Egyptian midwives or *dayas*, also provide known, continuous pre- and postnatal care. They stay with the new mother for seven days after birth, taking care of both her and the baby. She, as with the *doula*, becomes part of the extended kin household, more like a family member than an outsider (Sukkary 1981).

Fathers have various degrees of involvement in the pre-postnatal process. In some societies, the sociological role of the prebirth father is highly ritualized through the *couvade*. The *couvade* is a culturally created biobehavioral phenomenon in which the father simulates the pregnancy and birth of the woman who is bearing his child. He may "experience" morning sickness and labor, and undergo comparable forms of food, activity, sexual taboos, and modifications of his daily routine that the mother of his child incurs. He may take to his hammock and experience simulated labor contractions during his partner's birthing process. The *couvade* is a cultural means of acknowledging and enjoining men to participate actively in the pregnancy and birth phenomenon (Kottak 1991; Raphael 1988).

For most of the twentieth century in the U.S., fathers were not allowed to participate actively in pregnancy and birth. They were forbidden to be with women during labor and birth. They were seen as economic contributors, but not as interested, involved participants during the pregnancy. From the mid-late 1960s, childbirth advocates have actively encouraged greater participation by fathers, extended kin and friends. This has resulted in the involvement of a greater number of fathers and others in their women's pregnancy, birth and postpartum care (Raphael 1988). Is the father-involved childbirth movement in the U.S. a version of the *couvade*?

Confinement refers to the period spanning the pre- through postpartum (birth) phase. Most societies, including the U.S., modify the pregnant, birthing, and postpartum woman's behavior in various ways. Dietary and sexual habits are often changed. Pregnant women experience restrictions of their daily routine relative to work, sleep, social activities here and in other societies (Frayser 1985). For example, we caution women to limit their intake of all forms of drugs during pregnancy, especially caffeine, nicotine, alcohol, and recreational drugs. We encourage a "balanced diet" rich in grains, calcium, fruit, and vegetables. In the Maya Yucatan, women are encouraged to eat chicken and soups (Jordan 1993, 1983). Our postpartum sex taboos usually end within 6 weeks; among the Bushmen they last several years (Frayser 1985).

In societies outside the U.S., most women give birth at home or in familiar surroundings, attended by a midwife who is known to the woman. In the U.S., about 95% of the women give birth in a hospital, often separated from family and friends and attended to by a number of strangers. The actual birth of a child in the U.S. falls to the interventionist end of the continuum. Most women want and receive drugs for pain, have an episiotomy, and undergo a "prep," a surgical preparation procedure that routinely involves shaving the pubic hair, and receive an enema, in order to "clean out" the intestines. Most women are hooked up to IVs and fetal monitors, both external and internal (Davis-Floyd 1992). Rarely does a woman wear her own clothes or jewelry; jewelry is removed and she wears a hospital gown (Michaelson 1988).

Outside the U.S., pain may be relieved with herbal remedies or talked through. Childbirth pain is seen as normal and tolerable, part of a process to get the baby born (Jordan 1993, 1983; Newton 1981; Godziehen-Shedlin 1981). Traditionally, episiotomies were unknown. The perineum stretches through the upright birthing position, nonrupture of the bag of waters, massage, and hot compresses (Jordan 1993, 1983). "Preps" are unknown and the woman's own clothing is usually worn. Babies and mothers are kept together after birth; babies usually are nursed immediately and whenever they cry. It is interesting that the counterpart of high-technology childbirth in the U.S., "family-centered childbirth," advocates procedures and behaviors that are common, widespread practices in cultures outside the U.S. These include recent "advances" such as birthing rooms and birthing chairs, having women move around during labor, or having a childbirth coach[2] present to help the woman. They involve as well the controversies over preps, episiotomies, and shorter hospital stays. Postpartum infant-mother contact, a given in societies outside the U.S., is beginning to be re-established here through the practice of rooming-in, where for most of a 24 hour period, the neonate and mother share a room.

Postpartum

The postpartum period, a biosocial event, extends from the birth of the baby until the woman resumes her full prepregnancy roles and new status in the society as a mother and adult. This may take several weeks as in the U.S., or longer in other societies such as the Bushmen where postpartum sexual taboos last 2–3 years (Frayser 1985; Murdock et al. 1964). Biologically, the woman's body returns to a nonpregnant state: the uterus involutes, and the menstrual cycle resumes, irregularly at first depending on whether she nurses. Nursing is common outside the U.S. and engaged in sporadically and for shorter periods of time in the U.S. Nursing may inhibit ovulation when it lasts for greater than 18 months, when it occurs regularly and without interruption, and when it is correlated with relatively low body fat in the lactating female (Frayser 1985; Ember and Ember 1990). Breastfeeding then serves as a means of birth control under these conditions. For most U.S. women, nursing is **not** a reliable means of birth control for several reasons. First, most U.S. women do not breastfeed long enough for the hormonal suppression of ovulation to occur on a regular basis. Secondly, most U.S. women do not nurse regularly enough and they give "supplemental" feedings-bottles of juice, formula or solids, and thus interrupt the rhythm necessary that is established by frequent, regular nursing. Thirdly, most U.S. women's body fat is too high to suppress the H-P-G axis regulation of ovulation. Again, nursing as a means of birth control is NOT recommended for most U.S. women. Nursing itself is nutrionally complete for younger infants, and helps to protect them from disease by supplying them with their mother's antibodies. HIV infected breastmilk can be passed from the lactating woman to the nursing infant. There is international debate and controversy regarding whether women with either unknown HIV status or who are HIV infected should breast feed (see chapter 14 also) (Altman 1998).

Postpartum depression, well documented in the U.S., and less so elsewhere, is both physiological and cultural. The elevated levels of estrogen and progesterone during pregnancy drop dramatically after birth. In addition to this endogenous hormone withdrawal, many women in the U.S. do not have extended kin and nonkin social networks and models for child rearing and social support. They are expected to read about child rearing and turn to the experts, health care, and social service people, for help with parenting. Frequently, they do not know what they are doing, and are alone at home and isolated from other adults with one or more infants, toddlers, and other young children. The response to this situation may be "post-partum depression" (*Boston Women's Health Collective* 1992).

In nonwestern societies, an individual exists as part of the larger social group, generally the extended family. A postpartum woman is part of that group as a continuous aspect of her life. Her midwife, as discussed, may also join this group briefly in the first few days or weeks after birth. While the confinement period may extend through part of this time, it also offers benefits: rest from the daily routine, regular food and relaxation from overall social and familial expecta-tions. At the same time, mother-infant bonding occurs. The relative separation of woman and baby from hour-to-hour social obligation may also provide immunity from infection for the infant.

The pregnancy through postpartum continuum is an example of a physiologically widespread phenomenon that receives cultural attention wherever it occurs. The variability and forms of interpre-tation are cultural specific. They range from highly technological to highly psycho-sociological.

Summary

Chapter 8: Pregnancy and Childbirth as a Bio-Cultural Experience

1. Pregnancy and childbirth are bio-cultural phenomenon. Cultures intervene in the management of pregnancy and birth in a variety of ways.

2. While this chapter treats pregnancy and birth as a physi-ologically normal process, the dominant view in U.S. culture for over 150 years has been that they are dan-gerous processes and need medical management and intervention.

3. Birth creates and extends kin groups.

4. Pregnancy generally is discussed in terms of trimesters relative to fetal development and changes in the woman's body.

5. Cultures involve the father of the child in the woman's pregnancy and birth in a variety of ways.

6. Labor and birth are a four-stage process which is man-aged culturally in a wide range of interventions.

7. The postpartum period is culturally defined and involves biological and social dimensions.

OVERVIEW

Chapter 9

Early Childhood Sexuality

This chapter:

1. Introduces definitions of childhood and parenting.

2. Outlines the various functions of the family.

3. Discusses kin groups and family forms including the nuclear family and the extended family.

4. Presents the importance of residence patterns and their relationship to descent patterns.

5. Provides an overview of kinship and its various forms and structure.

6. Reviews the major theories of the incest taboo, including issues of universality and the interrelationship of kinship structures with the incest taboo.

7. Discusses incest in the United States.

8. Traces historically the changing perspectives of western children's sexuality.

9. Introduces major theories of childhood sexuality including the psychoanalytic views of Freud and Horney as well as those of pediatrician Dr. Spock.

10. Presents evidence of the development of children's early capacity for sexual response as well as self-genital and child-child sexual experimentation in the U.S.

11. Presents an overview of nonwestern childhood sexuality to emphasize its diversity and the cultural shaping of children's development.

Chapter 9

Early Childhood Sexuality

Definitions

The subject of early childhood sexuality is a highly charged one. To think of sexuality in association with children brings to mind images of child molestation, incest, and the violation of innocence. To fully understand childhood sexuality, we will focus first on the cultural interpretation of childhood as a specific period in the individual's lifecourse with a beginning and an end, marked and recognized by social and physiological changes. We will be continuing our bio-cultural focus in this chapter by looking at the physiological features of childhood as well as the meaning that cultures assign to the physical body, particularly the genitals.

An issue closely related to childhood is the subject of parenting. Western views of parenting are concerned with biological paternity or paternity certainty. An individual is usually aware of who his/her mother is. It is the father that may be in question. Even in a minority of cases such as adoption, this western quest for the alleged "real" parents is documented by people's desire to find their "biological parents" often at great cost and expense. We suggest, however, that family is socially constructed through the meaning we give to biology.

Paternity certainty as a cultural emphasis only occurs under certain conditions, specifically those related to male-centered residence or living patterns and descent through the male line. It is also important under certain kinds of subsistence strategies, specifically in horticultural societies that are under resource pressures and in agricultural societies. Finally, concern with paternity certainty is found in societies in which males hold privileged positions. It is hypothesized with a good deal of evidence that these conditions are correlated and historically related to one another (Martin and Voorhies 1975). Therefore, concern over biological paternity is not random and can be understood in relation to the food quest and adaptation.

In light of this, it is important to remember that one of the unique aspects of human evolution is cooperation. Cooperation relates to parenting. Individuals who are not necessarily biologically related have, in our evolutionary past and under various cultural conditions, made a valuable contribution to survival of the group through parenting and nurturing the young. It is part of our human capacity to bond.

It has been posited by physical anthropologists, that one of several factors accounting for female hominids' long post-menopausal lifespan is their contribution to survival of infants and children through caregiving. Human females are unique in their long post-reproductive lives. What information is available from our closest relatives confirms that reproduction and the lifespan converge in nonhuman primates. It is only in human females that longevity far exceeds the period of reproductive fertility. As grandmothers and socio-partners, we can imagine the enormous contribution of human females to the survival of the group in evolutionary terms. Hominid females, in the past and today act as role models, caregivers of the young, parent substitutes, providers of a vast reservoir of knowledge in regard to the socialization of the young and as educators of new parents. Like most human behavior, human parenting is a biologically adaptive behavior that interacts with the socio-cultural dimension. This is also true of kin groups in which biological relations are culturally shaped.

It is easy to forget given our truncated western nuclear families, that children, kin, and family are the core of culture. Children are born into a kin group in which descent is reckoned. For the majority of the world's cultures, descent is **unilineal** (Holmes and Schneider 1987:387). This refers to tracing descent through only one side of the family, either the mother or the father, unlike Americans who reckon descent through both the mother and father. Kin groups and descent are important for our discussion because these relate to the social positioning and cultural meaning of children in a society, issues of parenting and parenting roles, and why paternity certainty is important under some conditions and not others.

Kinship is important for ordering human social relations and creating groups and boundaries. Throughout the course of evolution and in most non-technologically complex societies, the family is the primary unit for production and consumption. Kin groups are formed through marriages and reproduction. Marriages should not be confused with mating which is defined in terms of premarital and extramarital sex. "Marriage is the socially recognized union of two or more individuals who are usually, but not always of the opposite sex" (Cohen and Eames 1982:124). These individuals will share economic, reproductive, and sexual obligations. The functions of marriage are numerous and relate to these obligations. Marriage provides a setting which facilitates infant survival as well as a stable setting for children's socialization. Group survival is enhanced by extending social relations and providing sexual outlets and economic advantages to the participants (Holmes and Schneider 1987:388). As mentioned, all groups have incest taboos which prohibit sex and marriage with various kin and hence structure kin groups.

There are two forms of marriage rules. **Exogamy** prohibits marriage within one's own group. According to alliance theory exogamy creates economic and political relationships between groups that might otherwise be in conflict. **Endogamy** is marriage that must take place within the group. This may include groups of relatives, the tribe, or even a caste or class. Endogamy does not refer to marriage within nuclear families, but may include marriages between cousins or other related individuals. Descent theory regards endogamy as a vehicle for families to contribute to cohesion and solidarity by keeping wealth, power, and prestige within the group (Cohen and Eames 1987:121–122).

Families are primarily of two types, the nuclear family and the extended family. Nuclear families consist of the two parents and their biological progeny. This form is the preferred type in the U. S. at this time, although other forms of marriage are prevalent such as: the blended family which represents children from previous marriages; single female or male headed households; extended families and childless couples. Extended families include **consanguineal** (blood) and **affinal** (marriage) relatives in addition to the nuclear ones. In non-technologically complex societies, the extended family usually includes relatives from either the male or female lineage and is the most common family form (Winick 1970:203).

Family forms represent adaptations to different livelihoods. For example, the nuclear family is adaptive in situations which require mobility such as in gathering and hunting societies and among complex urban industrial peoples. Nuclear families are associated with **independent households** in which the new couple may practice a form of residence pattern called **neolocality**, where the couple lives apart from their parents in their own home and/ or in a different area. Where the couple resides is very important for understanding marriage patterns, household composition, and descent. In many societies, the newly married couple practices a form of **joint household residence**, living with or very near one of the married partner's parents. In **patrilocal** residence forms, the couple lives with or near the husband's family, and in **matrilocal** residence, with or near the wife's family. Matrilocality and patrilocality are the residence rules which lead to the formation of joint households. Statistically, about 55% of the world's societies practice patrilocality, 25% of the societies practice matrilocality and neolocality is practiced in 5%. Fifteen percent practice some other alternative. These may include **avunculocality** in which the couple lives with the groom's mother's brother, or **bilocality** in which the couple may pick which residence rule they wish to follow. **Duolocality** involves the married couple living apart from one another so that only blood relatives constitute the household and is referred to as a **consanguinial** family or blood related family in contrast to **affinal** relatives which connotes relatives by marriage (Cohen and Eames 1982).

Winch and Blumberg (1972) have proposed that independent households are found in simple societies like gatherers and hunters and complex industrial societies, while the joint household occurs in intermediate societies with domesticated food economies like pastoralists and horticulturalists. In the case of intermediate societies, Winch and Blumberg (1972) argue that the domesticated food economy provides sufficient resources for joint households who can benefit from the additional household labor as well as the group cohesion. These are not societies in which farm technology is available. Conversely, the independent household in urban industrial societies may be evidence of the capacity of individual households to support themselves, as well as for the necessary mobility required in these kinds of economies. They are similar to gatherers and hunters who also require mobility (in Cohen and Eames 1982:137).

There are several types of marriage represented in the cross-cultural spectrum. These include single mateships which are composed of two spouses. According to Ford and Beach (1951), 16% of the world's societies require this form, while in 84% men are permitted to practice polygyny if they can arrange it. **Polygyny** refers to multiple wives while the generic term for multiple spouses is **polygamy**. However, in 49% of the polygynous societies single mateships are actually the rule, since a man must have a certain amount of economic wherewithal to acquire additional wives. As is illustrated by the Tiwi of Australia, once a man has demonstrated that he is a good prospect as a husband in terms of his status and economic condition, having several wives enhances his economic standing considerably (Hart and Pilling 1960). In 14% of the polygynous societies, the only acceptable additional wife is her sister. This is known as **sororal polygyny**.

A small percentage (less than one percent of societies) practice **polyandry** in which a woman may have more than one husband (Cohen and Eames 1982:126–127). This pattern is common to Tibet, Nepal, India, and Sri Lanka, but is also recorded elsewhere such as among the Marquesans of the Pacific (Levine 1988; Suggs 1966). **Fraternal polyandry** occurs in Nepal and India where brothers may marry one woman and live patrilocally. All the brothers are recognized as the wife's husbands, and all take on the parenting role of father to her children. The wife has equal sexual access to all the brothers. In fraternal polyandrous societies, sororal polygyny may also be practiced (Schultz and Lavenda 1990:300–301).

Associated polyandry occurs among the Sinhalese of Sri Lanka (Levine and Sangree 1980), and is reported in the Pacific and among indigenous peoples of North and South America (Schultz and Lavenda 1990:301). Associated polyandry refers to a marriage in which a woman may have multiple husbands, but these husbands are unrelated. The Nayar represent one of the most famous of the anthropological reports of this kind of polyandry. The Nayar woman engages in a ritual marriage to a man from a linked lineage. After three days of seclusion where sex might occur if the wife is old enough, the couple parts and go their separate ways with no further obligations or relations. After this ritual marriage the woman is free to marry and have sexual relations with men of her own choosing as long as they are of the same caste or higher, i.e., endogamy. The only restriction is that these husbands cannot be brothers. This system is referred to as one of **visiting husbands**. The men and women in these relationships have more than one spouse. The households are consanguineal ones in that a woman lives with her children, her sisters, and her sisters' children. The marriages are valuable for their alliance functions, not economic ones. Men take on financial responsibility for their sisters and sisters' children (Schultz and Lavenda 1990:301–302).

Like polygynous marriages, polyandrous ones reflect adaptations to subsistence conditions. In Tibetan fraternal polyandry, polyandrous marriages help maintain family landholding units against possible subdivision through individual inheritance and partition of lands. The ecological conditions are such that households must pursue both agriculture as well as animal husbandry in order to survive. In such a situation a division of labor among co-husbands is advantageous (Goldstein 1987:39–48).

Other forms of marriage exist but are not widespread. These include a form of group marriage practiced by the Oneida utopian Christian community in New York. All members regarded one another as spouses. As a result of local hostility, this system was discontinued in 1879 (Gregersen 1983:134). Other forms include **woman marriage** as among the Nuer in which two women are married. One woman takes on the role of a social male and arranges for "her" wife to become pregnant by "another" man. The woman-husband then becomes the social father of the child (Cohen and Eames 1982:128–129; Blackwood 1984:56–63). The **levirate** and **sororate** represent marriage systems related to the death of spouses. In the levirate the wife will marry her dead husband's brother. Seventy percent of a sample of 159 societies reported on by Murdock preferred this form, while 60% preferred the sororate in which a man marries his dead wife's sister. One explanation of these practices suggests that re-marriage between relatives is important for maintaining the stability of the familial group which may be threatened by the death of the parent. Alliance theory argues that these marriage forms maintain alliances originally established by the dead sibling (Cohen and Eames 1982:128–129, 133–134).

Marriage creates kinship relationships between individuals and groups. We shall begin our discussion of kinship by focusing on descent groups. Descent groups are characterized by a permanent set of relations that are not changed by residence or death. These groups are formed through various principles of descent and represent ways in which human groups organize themselves. **Descent** is defined as "the cultural principle based on culturally recognized parent-child connections that defines the social categories to which people belong" (Schultz and Lavenda 1990:261). Descent groups are composed of people who recognize shared ancestry. It is the primary way people are organized in non-industrial and pre-state societies. While nuclear families are defined by common residence and are therefore impermanent, the descent group is permanent (Kottak 1991:297).

Two principles of descent occur worldwide. In order to illustrate the forms of descent and kin relations, anthropologists use a kinship diagram.

Kinship diagrams are more convenient than verbal explanations and allow us to see immediately how different kinship statuses are linked. In order to make a kinship diagram precise and unambiguous, all relationships in the diagram are viewed from the perspective of one status, labeled EGO. Terms of reference rather than terms of address are used—that is, terms we would use in taking *about* a relative rather than talking *to* one. In English, for example, we would refer to our "mother" but might address her as "Mom." The symbols used in kinship diagrams are these:

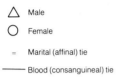

△ Male

○ Female

= Marital (affinal) tie

——— Blood (consanguineal) tie

Using these symbols, and English terminology, a kinship diagram of the nuclear family looks like this:

Figure 9.1 The kinship diagram.

Descent systems are either **unilineal**, tracing one's ancestors through one side of the family and not the other, or are **non-unilineal**. Non-unilineal descent is also referred to as **bilateral** or **cognatic** descent and is based on the principle of tracing descent through both parents equally (Schultz 1990:264). Approximately 40% of the world's societies are non-unilineal. These includes bilateral, ambilineal and bilineal descent discussed below. The American descent system is called the bilateral kindred.

A **bilateral kindred** is a group that forms around a particular individual. It includes all people who are linked to that individual through the kin of both sexes. These are the people Americans conventionally call relatives. Relatives form a group only because of their connection to the central person or persons, known in the anthropological kinship terminology as ego (Schultz and Lavenda 1990:265). Bilateral descent is shown in figure 9.2.

The principle of **ambilineal** descent forms groups known as the **ramage**. In ambilineal systems, descent is traced from a founding ancestor through either the male or the female line or both. **Bilinial** (or double) descent, is similar to ambilineal descent in that descent is traced through both the patrilineage and matrilineage, but each control different areas of activity and property. The Yako of Nigeria are a good example of this. Non-movable property such as houses, farms, and groves of trees are inherited through patrilineal descent. Movable property such as cattle and home furnishings are inherited through the matrilineage (Cohen and Eames 1982:154).

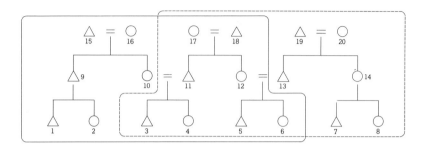

Figure 9.2 Bilateral descent. In this system the kindred is individual-centered, has no clearly stated boundary, and membership overlaps because individuals can belong to any number of kindreds. A simplified portrayal of a kindred is shown here. The solid line encloses the kindred of Individual 3 and includes his sister (4), his parents (10, 11), his uncle and aunt (9, 12), his cousins (1, 2, 5, 6), and his grandparents (15, 16, 17, 18). The dashed line encloses the kindred of Individual 6 and includes her brother (5), her parents (12, 13), her aunt and uncle (11, 14), her cousins (3, 4, 7, 8), and her grandparents (17, 18, 19, 20).

The unilineal principle of descent traces kin membership through either the mother's or the father's line exclusively as we mentioned. See figure below "Matrilineal and Patrilineal Descent" (Holmes and Schneider 1987:381).

Matrilineal descent traces membership through the female line only, while **patrilineal** descent traces group membership through the male line only. In these systems, an individual will trace ancestry through either their matrilineage or patrilineage. Unlike the clan, the lineage has demonstrated descent in that members can trace their kinship and exact relationships to one another and a founding ancestor. In contrast, stipulated descent occurs among people who claim to be related to a common ancestor, but are unable to document the exact relationship because the ancestor is either hypothetical or very remote. Stipulated descent is associated with clans. Figure 9.3 shows matrilineal and patrilineal descent.

Clans are groups whose membership is based on the principle of unilineal descent. Clans are defined as a descent group who are affiliated through the belief that they have a common ancestor, even if that ancestor cannot be directly traced. The ancestor may be a person or a mythical being. Consequently, clans have greater size and generational depth than the lineage. While the lineage is the smallest unit of unilineal descent and may be

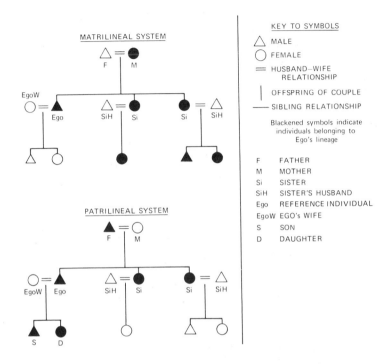

Figure 9.3   Matrilineal and patrilineal systems of descent.

a part or a segment of a clan, even larger groups may occur. When a society is divided into two large kin groups, this is referred to as a **moiety** system. "When clans are grouped into more than two large units, we call the units **phratries**" (Cohen and Eames 1982:151). Having discussed marriage and family forms, the organization of human groups based on marriage and children, and affinal and consanguineal relationships, we now turn to the subject of sexuality and kin.

Incest Taboos

Childhood sexuality, like adult sexual expression is managed by culture. One of the ways in which it is managed is through proscriptions against incest. The incest taboo is cited as an example of a cultural universal, although as we have seen earlier, there are exceptions to the taboo. The **incest taboo** is defined as a "[u]niversal prohibition against marrying or mating with a close relative" (Kottak 1991:47). The incest taboo refers to those family members that Americans call the immediate family. In virtually every society, sex is prohibited between an individual and her/his siblings, parents, and children. These people are termed "primary relatives." Anthropologists continue to debate the origin of the incest taboo, since it is such a widely shared institution. As a consequence of this debate a number of theories have emerged. We shall review several of the more prominent perspectives.

Psychoanalytic theories represent one kind of explanation about incest. Perhaps the most famous psychoanalytic view of the incest taboo is Sigmund Freud's theory of the Oedipus complex. The Oedipal complex has had an important impact on early- to mid-twentieth century anthropological reflection on the subject, particularly in the anthropological school of thought known as culture and personality. This subfield regarded personality as causal in shaping culture's more expressive aspects such as art, religion, and mythology. Not all anthropologists agreed with Freud's formulation. In fact, the anthropologist Bronislaw Malinowski in 1927 challenged the view that the incest taboo was a result of the existence of the Oedipus complex.

Freud's theory of the Oedipus complex was derived from his work with Western European clients. The Oedipus complex is represented as one stage in the psychosexual development of a child. The first phase a child experiences lasts from birth though one year of age and is the oral stage in which the infant's interests center on the mouth as a source of pleasure. Freud regarded pleasure seeking in the human as a given, as instinctual. The anal stage occurred at approximately two years of age. It is the period in which the child achieves control of his/her bladder and anal sphincter and finds pleasure in this sensation. However, toilet training can cause conflicts as the child's wishes may be controlled by external pressures such as toilet training practices. The Oedipal phase or *phallic* stage is characterized by love, hate, envy, and guilt, and occurs from about three to five years old. This is followed by a period of latency, from about six to twelve years of age, when the sexuality of the Oedipal phase is repressed. At puberty interest in sexuality is reasserted. Libido, or the desire for sex, which Freud considered panhuman and natural, underwent various expressions and repressions through these developmental stages. Freud felt these stages were embedded in our human biology and were universal. Freud's theory of development is linked to his view that sexuality is both a conscious and unconscious force throughout the life course. This was an astonishing view in light of his Victorian milieu.

The Victorian era is named for the reign of Queen Victoria who ruled from 1837–1901. The romantic period, which began around the 1800s, fed the Victorian era. The Romantic era celebrated nature and the unspoiled. These notions filtered into the Victorian period. At this time the concept of the noble savage flourished, derived from anthropological reports of "exotic natives" communing with nature in its unspoiled and pristine state. Middle-class English children, too, were seen as having a nobility derived from purity. They were viewed as yet unspoiled by civilization. As the industrial revolution gained momentum, children " . . . became the last symbols of purity in a world which was seen as increasingly ugly" (Sommerville 1990:198). Children were glorified during the Victorian era. In short, they were next to the angels in virtue. While adult Victorian's had to suppress their impulses in an era where sexuality and sexual symbolism were not publicly expressed, children represented innocence of desire. They were in a state of natural privilege associated with "childhood goodness" (Sommerville 1990:204). Sommerville (1990:209) has suggested that this view of children did not extend to the U.S. during this same time period. Middle-class American perspectives of children demanded competence and performance in contrast to British views of children who "symbolize[d] the innocence which a severely repressed society felt it had lost."

Sigmund Freud's work on childhood sexuality followed closely on the heels of the Victorian period. In 1905 Freud published his *Three Essays on the Theory of Sexuality*. Freud felt that human sexual energy, which he called libido, was present at birth and that through the course of children's development, this energy became focused in different body zones during the different stages of psychosexual maturation. This was a very different notion about the child than the Victorian one of innocence. For Freud, the infant was charged with an undifferentiated sexual energy; that is, he or she could find sexual pleasure in the entire body in the erotogenic zones. This is what Freud meant when he referred to the infant as "polymorphously perverse." Libido is an energy rewarded by the "aim" of pleasure. The erotogenic zones were areas of the body through which libido could be discharged. Freud emphasized these in his stages of childhood. The oral stage of development focused on the mouth area as the first zone for pleasure. This is the pleasure the child derives from sucking and nursing. The second zone emphasized the anus and was termed the anal stage of development.

This was the period where the child has learned to control her or his bowels and finds pleasure in the process of evacuation. The third or phallic stage actually refers to the phase when children, between three or four years of age, explore their genital areas and find self-stimulation to be pleasurable. This stage which accented the genitals also included the Oedipus complex which occurs around the ages of five or six. For boys, as will be discussed shortly, the Oedipus complex is resolved out of fear of the father and identification with him (Appignanesi 1979:76–88). The little girl conversely desires her father and resents her mother (Gleitman 1987:352–353).

The latency stage follows the Oedipal phase. This was the period from about six years old to puberty when children, according to Freud, lost interest in sex. This phase initiated the end of the four stages of infantile sexuality. It should be remembered that sexual interests were seen as repressed but not eliminated from the psyches of children (Appignanesi 1979:92).

The Oedipal phase is the cornerstone of Freud's theory of psychosexual development. Freud named this after Oedipus, the tragic hero in a mythical story of a man who married his mother unknowingly and, upon finding this out, blinded himself as punishment. According to Freud, the young boy covets his mother and wants to eliminate (kill) his father. In contrast, the young female desires her father and regards her mother as a competitor. Her complex is called the Electra complex. These wishes cause the child to feel fear and guilt toward the same sex parent. The Oedipus and Electra complexes are resolved through renunciation of the love object and identification with the same sex parent (Gleitman 1987:351). The gender bias in Freud's theory was clearly expressed when he wrote: "It does little harm to a woman if she remains in her feminine Oedipus attitude. . . . She will in that case choose her husband for his paternal characteristics and will be ready to recognize his authority" (Freud 1940a:99 in Sayers 1986:101).

To account for the Oedipus complex as universal, Freud turned to an evolutionary explanation. From Freud's perspective, the almost worldwide appearance of the incest taboo could be viewed as a mechanism to prohibit that which we desire. But from where did this taboo and desire arise? Freud addressed this in *Totem and Tabu* ([original 1927], 1961). At some early and unspecified point in time there existed a "primal horde" in which a father kept a harem of women, but expelled his male children. The expelled brothers colluded and murdered their father and ate him so they could have sexual access to their sisters and mothers. However, after their dastardly deed, they felt a great deal of remorse. Out of respect for their slain father's wishes, they renounced their mothers and sisters. This was the beginning of the incest taboo which prohibits sex and marriage between immediate blood relatives. Freud cited evidence cross-culturally of the ritual totemic meal which he interpreted as a symbolic re-enactment of this original crime. To Freud, the totem animal represents the father who is ritually eaten in commemoration of this event. According to Freud, the consequences of this primal scene have been transmitted to all of us through the collective unconscious, presumably somehow inherited (Freud 1950:34 in Bock 1988:32–36).

In response to Freud, Malinowski argued that the Oedipus complex was not a cultural universal, but was relative to the particular family structures found in a given culture. Since Freud's theory was based on Western Europeans with family constellations in which the father was the dominant figure, Malinowski tested this theory in a situation in which the family constellation was quite different, that of the Trobriand Islanders. He addressed this in his book *Sex and Repression in Savage Society* ([original 1927] 1961). He reasoned that different family structures would likely lead to different kinds of conscious and unconscious conflicts in individuals (Brown 1991:32). As we have seen, the Oedipus complex proposes that a young male child will desire his mother and want to get rid of his father. As the child matures this complex is then outgrown.

But the Trobriand Islander's had a very different family structure. They were a matrilineal society in which kinship was traced through the mother. In fact, the Trobrianders believed that the procreator of a child was a dead kinswoman of the mother. Although the mother was the primary authority figure, she had a warm relationship with her children. In contrast to the European system in which the father was the authority figure, the mother's brother among the Trobriand Islanders assumed the role of disciplinarian. He was also the person from whom the child would inherit. In contrast, the father (like the Trobriand mother) had a warm and affectionate relationship with his children.

In the Trobriand Islands, it was the mother's brother who earned the boys hostility; and it was his sister, not his mother, whom the boy desired. Trobriand children were subjected to a rigorous brother/sister incest avoidance rule (taboo) at puberty (Brown 1991; Bock 1988). Unfortunately, the comparable complex for women—the Electra complex—has been relatively unexplored. The Electra complex "received much less attention from Freud and almost none from Malinowski . . . " (Brown 1991:32). This is typical of the "unmarking" and silence around women that occurs in patriarchal societies and which is reflected in scientific theorizing. Despite this shortcoming,

> Malinowski's finding became "the cornerstone for the thesis propounded by relativists of all persuasions"—anthropological and nonanthropological, Freudian as well as anti-Freudian—that . . . the Oedipus complex . . . is a product of Western institutions and, more particularly, of the Western "patriarchal" family structure (Spiro 1982:1 in Brown 1991:33).

Malinowski's theory was a functionalist interpretation of the incest taboo that focused on its role in the maintenance of society. For Malinowski, the incest taboo was necessary to maintain order because without it confusion in positions and family statuses would occur.

Other anthropological thinking on the subject includes those theorists who suggest a biological explanation, proposing that some sort of genetic avoidance mechanism exists against inbreeding. Arguments are mustered on both sides of this debate. One side argues that deleterious genes have greater opportunity for expression in incestuous unions; the opposing side counters that genetic cleansing can in fact occur through this very process leaving the familial line healthier in the long run. We know evolution works upon genetic variation and that variation is a very effective mechanism for coping with environmental changes. Breeding outside of the immediate family provides the genetic variation upon which natural selection can act. Shepher (1983) has reviewed the literature on the consequences of incestuous inbreeding in human populations and has found that 42% of the offspring were nonviable (in Brown 1991:123–124). "If the figures Shepher cites are even approximately correct, mechanisms to avoid the cost of incest between close kin are quite expectable" (Brown 1991:123).

The primate data is also suggestive. There is evidence that both male monkeys and female apes emigrate from their natal groups thereby decreasing the potential for incestuous matings (Kottack 1991:317; Anne Pusey 1980 in Walter 1990:441). Those that oppose the biological explanations and promote more culturological perspectives cite the evidence of preferred marriage and incest regulations associated with parallel and cross-cousin marriage in unilineal societies. The distinction between these sets of first cousins is not one made by Americans. A **parallel cousin** is the child of ego's mother's sister or father's brother. A **cross-cousin** is the child of ego's mother's brother or father's sister (see figure on page 191) (Kottack 1991:315–316).

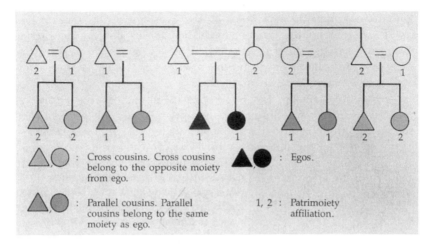

Figure 9.4 Parallel and cross-cousins and patrilineal moiety organization.

In societies which prefer cross-cousin marriage, parallel cousins are regarded as belonging to ego's descent group and are seen as similar to one's siblings. They may even be referred to by the terminology for brother and/or sister. Because cross-cousins are not regarded as consanguineal relatives, incest prohibitions apply only in the case of parallel cousins (Kottack 1991:213–214). Marriage with cross-cousins is the preferred form cross-culturally. About thirty percent of the world's societies have a system where cross-cousins are the preferred marriage partners (Cohen and Eames 1982:125). There are some societies, for example, among Middle Eastern groups, which prefer parallel cousin marriage.

At this point, the issue of exogamy is relevant. While the incest taboo is defined by avoidance of sex between primary family members, rules of exogamy refer directly to marriage (Spradley and McCurdy 1989:91). When the incest taboo is expanded to include groups of people outside the family of origin, this is called exogamy. Whole groups of people may be excluded as potential partners if they fall into certain relationships with "ego," the person from whom one is tracing relationships. The groups may be lineages, clans, moieties, or tribes. To clarify the issue, rules of exogamy refer to marriage, but the incest taboo may be expanded to cover a much wider range of prohibited individuals than just biological parents and siblings. Marriage prohibitions can include people who are not consanguineal relatives. Thus, exogamy may include incest prohibitions, but may also extend much further (Cohen and Eames 1982:66). "The incest taboo forces people to practice exogamy, to seek their mates outside their own groups" (Kottak 1991:315). This is often discussed anthropologically as the extension of the incest taboo.

Some anthropologists believe the incest taboo arose as a vehicle for establishing alliances outside the family. In fact, Edward Tylor's adage "marry out or die out" is one explanation for the origin of the taboo. According to Tylor, the "hatreds and fears" associated with closed families forced people to extend alliances to other families and therefore to build societies (Shapiro 1958:278). This explanation is typical of the functionalist approach in anthropology which maintains institutions exist to fulfill needs. Other notable anthropologists, such as Leslie White, George Peter Murdock, and Levi-Strauss have explained the incest taboo along these same lines with various refinements and additions, yet maintaining the theme of alliance building.

In 1922, Edward Westermarck offered an explanation for the incest taboo which argued that the taboo reflects an absence of sexual desire expressed in a "the horror of incest." Incest rules, therefore, exist for those who have gone "awry" or deviated (Brown 1991:119). Westermarck felt that the lack of erotic desire that people raised in proximity feel toward one another was an evolved sentiment ingrained in the psyche through biology. This theory is often summarized as "proximity breeds contempt."

The **Westermarck effect** has garnered substantial support, particularly as it relates to sibling incest. Spiro (1958) and Fox's (1962) analysis of Israeli kibbutz and Wolf's study of marriage in China (1966, 1968, 1970) provide provocative evidence for this theory. Spiro found that children who were raised together as cohorts on kibbutzim as part of social planning to reduce the role of the nuclear family, practiced sex and marriage avoidance of one another as adults (in Brown 1991:120). Fox interpreted this as support of the Westermarck effect. He concluded that in societies in which children are raised with close physical intimacy, they will not have sexual desires for one another; and the incest taboo will be more like an afterthought since siblings will not desire one another anyway. However, societies in which siblings are raised in the absence of physical intimacy, are more likely to have strict taboos since desire will need curbing. Fox's hypothesis integrates a Freudian thrust not incorporated by Westermarck (Brown 1991:120).

Arthur Wolf's investigation of marriage in China has intriguing implications for the Westermarck effect. Wolf studied two forms of marriage practiced in a Chinese village in Taiwan (1966, 1968, 1970). The minor form of marriage was one in which a girl was adopted into her future husband's family at an early age and raised as a member of the family. In the major form, marriage took place in adulthood without any previous familial association between the partners. Wolf (1970) suggested that wife adoption would lead to a sexual aversion in the couple who was reared together. In comparing major and minor forms of marriage, he found that in the minor form there were 30% fewer offspring, the divorce rate was 24.2% and extramarital relationships were found in 33.1% of the marriages. In contrast, major marriages had a 1.2% divorce rate and 11.3% rate of extramarital sex (Wolf 1970:503–515). This evidence strongly supports the contention that familiarity leads to disinterest and even aversion.

This is by no means an exhaustive review of anthropological theories of the origin of the incest taboo, but only highlights some of the more prominent ones. The incest taboo may be explained by a convergence of several theoretical positions,[1] for example it:

1. Establishes alliances and extends peaceful relations beyond the group.

2. Promotes genetic mixture.

3. Preserves family roles, guarding against socially destructive conflict (Kottak 1991:319).

We would add that the Westermarck effect of childhood proximity leading to lack of desire could be added to this list as a vehicle of psychological conditioning. Thus, the incest taboo is a consequence of what people would not want to do anyway under certain conditions. Walters argues that in societies where children are exposed to prohibitions that prevent the familiarity of the Westermarck effect, then desire for the forbidden could arise. In such a case, the incest taboo would operate as a mechanism to prevent incest (1990:440). We would like to point out that despite its universality the particular form, focus, meaning, and response is extremely variable cross-culturally. This diversity is especially apparent in the case of exogamy, in which the incest taboo is extended. In spite of the near universality of the incest taboo, incest does occur. We have discussed the special situations of exceptions for elites. We want to turn now to incest in the United States which is primarily non-consensual and involves victimization. This is because children and adolescents are not in a position vis-à-vis their parents and older siblings to give informed consent.

It is difficult to get accurate estimates of the incidence of incest. One report suggested that 250,000 children are victims of incest, with half of these cases involving fathers and stepfathers (Kelly 1990 citing Russel 1983). Depending on the report, statistics for incest range from four percent in Gebhard's (1965) study to recent figures of 15% in the work of Becker and Coleman (1988) (in King et al. 1991:380), and 27% in Hunt's (1974) survey (in Francoeur 1991:613). Estimates as high as 70% in alcoholic families are cited in Morgan's work (1982). It must be remembered that although a stepfather or stepmother is not a biological parent but a social parent, the severity in terms of trauma may be the same for the victim. The victim has had a trusted parental figure violate her or him.

The survivors of incest suffer a host of psychological problems and maladjustments throughout the lifecourse as a result of their experiences. These include sexual acting out, sexual dysfunctions as adults, low self-esteem, self-blame, and self-destructive behaviors. They may experience psychological problem not unlike the Vietnam veterans who suffered Post Traumatic Stress Syndrome.

Incest should not be confused with children's curiosity about their own genitalia and that of their siblings that occurs around two to three years old. In one study, 15% of the women and 10% of the men in a college population reported this behavior with 75% brother-sister exploration and 25% same gender behavior (Finkelhor 1980 in Francoeur 1991a:111). But even in such a benign sibling context, there is still the possibility for coercion and possible incestuous sexual abuse. Twenty-five percent of this population, mostly women, were uncomfortable because force was used (Finkelhor 1980 in Francoeur 1991a:111). Incest survivors are invariably the less powerful person in such interactions. Generally, the survivor is not in a consensual position because of differences in authority or because of actual or threatened physical force.

The double standard of the western gender system is reflected in the incest statistics in the U.S. According to Stark (1984), 85% of the incest victims are female with "only 20% of the sexual abuse of boys and 5% of the sexual abuse of girls perpetrated by adult females" (in Kelly 1990:356). According to Herman and Hirschman's (1981) study of American women, 20% of the female population has experienced some form of incest. Stepfathers are more likely than biological fathers to engage in incest. Statistically speaking, incest is initiated between 6 and 11 years of age for the girl, and may last two years or more (Stark 1984 in Kelly 1990:356).

Children who are survivors of incest are in a disadvantaged position in relationship to the abuser. One common response is accommodation since they may feel they have no other alternative. They may be conflicted by the love and trust they feel for the biological or social parent. Or they may not want to cause family problems by telling; very often they have been manipulated or threatened with consequences if they do tell. Thus, the accommodation strategy is one of serious denial. Researchers note that some survivors have no direct memory of the incest as adults. Only through psychotherapy do they become aware of the memories so long denied and hidden.

One aspect of father or stepfather incest is the purported acceptance of the situation by the mother. In such cases, the mother is reported to default in her role as spouse, and the daughter or stepdaughter takes the mother's role in relation to her father or stepfather. In such situations, the young female is doubly injured in terms of betrayal of trust, both by her father/stepfather and by her mother who tacitly allows the relationship. The mother, not usually overtly, may permit it to go on because it is part of the family denial system. However, this is a very controversial theory and some researchers such as Chandler (1982), Ward (1985) and Myer (1985) deny that the mother has any role at all (in Francoeur 1991:615 in Kelly 1990:356). Certainly, the degree of the mother's tacit involvement varies depending on the family dynamics.

The father's role as perpetrator is more clearly identified. In fact, the sexually abusive father or stepfather follows a pattern in which gender inequality is an important factor. The father is typically hyper-masculine in his disregard and respect for women and children. We may view him as oversubscribing to his sex role. Another contributing cultural factor, which may also play a part, includes a valuing of women for their youthfulness. Finkelhor (1984) has identified four cultural and psychological factors that converge to place children at risk for sexual abuse by fathers or stepfathers. These factors are embedded in a patriarchal system of masculinity and include the following ideologies and values: (1) in sexual relations men must dominate; (2) most emotional needs are regarded by men as sexual ones; (3) an inability of men to identify with the needs of children; (4) a belief in patriarchal privilege and perogative and the special rights of the father/stepfather within the family (see Phelan 1986:531–539).

Discussions of adult-child sexual contact, whether incestuous or non-incestuous, are highly charged with emotion in American society. Given our cultural mores and values, much adult-child sexual contact very probably will have negative consequences for the child. However, the universality of the negative consequences of adult-child sex for children in the U.S. is questioned by some researchers (Nelson 1989a, 1989b).

In an intensely debated article, Nelson posits that it is important to learn if *every* incestuous sexual experience is always negative for the child (Maltz 1989). Nelson offers several conditions that may mitigate the deleterious consequences for the child. These include: (1) an emotionally bonded nonsexual relationship between the individuals; (2) sexual contact which occurs between people relatively close in age; and (3) an egalitarian relationship between the parties (1989). Given these three variables, an incestuous relationship may *not* always result in long term emotional scarring and damage for the child.

Pat Whelehan has dealt with a small number of adult male and female incest "survivors" over the past three years. Some of these individuals did experience extreme emotional damage with long lasting negative consequences relative to their self-esteem and romantic relationships with others. Three of these individuals came through the experience intact. The differences? The three individuals who viewed their incest/adult-child sexual experience positively were males who were not coerced and who had a positive nonsexual relationship with their older sexual partners. They checked in with your author because they wondered if "something was wrong with them" because they felt "okay" about their experiences and who they are currently as sexual adults. The other survivors fit the model of incest "survivor." They are women who were coerced and had hostile or nonbonded sexual relations with males who were not only older, but in a position of authority over them. They experienced much emotional and psychological trauma from their earlier sexual contact (Whelehan counseling files).

While it is difficult to remain objective about this issue given our culture's view on adult-child and incestuous sexuality, it is important to do so in order to help those individuals who have been traumatized to regain a positive sense of self and sexuality. We need to know when such contact may or may not be harmful and then consider the individual's unique experience across this continuum.

Theories of Childhood Sexuality

Freud's theories of childhood sexuality present a conflict model of development. Each stage has its unique set of conflicts centered around the erotogenic zones. Neo-Freudians such as Karen Horney (1885–1952) and Erich Fromm (1900–1980) among others challenged the Freudian view that erotogenic zones and stages of development are the key to understanding human behavior. Rather than focusing on the zones, some neo-Freudians felt that the relationships people had with one another were primary. This represents a shift from biology, libidinal energy and erotogenic zones to social relations in understanding childhood development (Gleitman 1987:355–356).

Horney took issue with Freud on several specific points in his theory of childhood sexuality. She contested Freud's position that children did not recognize gender differences at birth by asserting that children knew "intuitively" the differences because each gender had sensations of being penetrated and penetrating. Horney brings a less male biased view to the psychoanalytic scenario of child development. She argues that both genders envy one another's genitalia. This is in contrast to the Freudian perspective which proposed that women have penis envy, but men have no comparable womb envy. Later in her career, she challenged the Freudian notion of the Oedipus complex. She regarded the Oedipus complex as situationally derived from a particular kind of family dynamics where emotional dependency of the child was combined with self-centered and unresponsive parents (Sayers 1986:40).

Dr. Spock's Baby and Child Care book has had a tremendous impact in the area of child development as well. His book has sold over thirty million copies and has been published in 39 languages (Spock and Rothenberg 1985). Since 1945, Spock's book has been revised to accommodate changes in sex roles and family structure in the United States. As he neared his 80th birthday at the time of the last edition, Spock added a co-author in anticipation of the need for a successor. Dr. Spock passed away March 15, 1998 at the age of 94. Spock's model of childhood development, like Freud's, was also based on stages of development, for example, he notes that "[b]oys become romantic toward their mothers, girls towards their fathers" (Spock 1985:435). For Spock, these feelings about wanting to marry the other gender parent happen between the ages of three and six years old. Such feelings are important for preparing the child for adult sexual attraction and relations. He states "[w]e realize now that there is a childish kind of sexual feeling at this period which is an essential part of normal development" (1985:447). Again like Freud, Spock perceived that this attraction brought up sentiments of rivalry, jealousy, and fear toward the same sex parent. For Spock, the resolution follows as a natural process. "Nature expects that children by 6 or 7 will become quite discouraged about the possibility of having the parent all to themselves" (Spock 1985:436). This marks the end of this phase of attachment which will be "repressed and outgrown" and which is succeeded by an interest in other activities such as athletics, education, and same sex peer involvement (Spock 1985:437). Spock is rather Freudian in that he regards these interests as caused by the sublimation of sex that takes place from ages six until about twelve years of age.

Spock advises parents not to give in to their children's feelings of rivalry by refraining from overt affection with one another. It is important for children to be confronted with the fact that they cannot ever marry the parent. Spock's account is obviously influenced by Freudian theory of the Oedipus complex (1985:437). However, Spock's re-presentation of Freud is much less sexually oriented around the pleasure principle. For example, Spock regards the interest in sex that occurs between two and one-half and three and one-half years of age as part of a much broader pattern associated with the "why" stage of curiosity (1985:451). He regards this as part of children's natural curiosity about why the genders are different and where babies come from.

Cultural Relativism and Childhood Sexual Behavior

The messages many American children get from their parents is that sex is bad and should be delayed until marriage and that love must precede sex. This is part of a cultural "ideal of childhood sexual innocence . . . that children and adolescents need legal protection from all sexual contact and, in some cases, from sexual information and contraception as well" (Konker 1992:147). In addition, this message is frequently gender biased. The double standard for male and female sexuality is patently apparent to the young child although it may not be to the parents (Darling and Hicks 1982; Francouer 1991:107). It is interesting to note that young boys cause their parents less concern when they express an early interest in sex and romance than do young girls who may be thought of as "precocious" (Kelly 1990:166). This is evidence of the double standard applied at a very early age so that little boy's sexuality is far more acceptable than little girl's among North Americans.

Evidence from a variety of sources verifies the childhood capacity for sexual response. In males, in utero penile erections have been reported as early as 17 weeks. Orgasm in boys occurs as young as five months old (Kinsey et al. 1948:177). Orgasm is reported in girls as young as seven months old (Calderone 1985; Bakwin 1974). The human physiological sexual response function is clearly in place early in infancy. This should not be confused with the meaning we give to that response in terms of adult sexuality. The child is merely exploring her/his body and responding to pleasurable sensations. Childhood capacity for sexual response and pleasure is part of our biological heritage. It is experienced very differently from adult sexuality.

Ford and Beach (1951) have classified American society as a sex negative culture. A more contemporary view might regard American society as sex ambiguous, rather than purely negative, since elements of the positive coexist with the negative. Sex negative aspects are evident in attitudes toward childhood masturbation. Yet the evidence shows that children:

> [w]hen left alone, . . . spontaneously explore their bodies, their genitals and experience their developing sexual nature. Sex play and exploration are major factors in a child's development. Even when discouraged or prohibited by adults . . . children manage to explore their bodies and their sex organs (Francoeur 1991:107).

Sexual curiosity is a normal part of childhood growth and development. It is expressed in play activities and seems to be a component of learning one's sexual identity and who one is as a member of a gender category vis-à-vis body awareness.

Goldman and Goldman (1982) in a comparative study of Australian, Swedish, English, and North Americans, found American children were less knowledgeable about sexuality than the others. This may reflect generalized trends in North American socialization practices. It may even become worse in the future if widespread fear about AIDS negatively impacts sexual practices, or if parents respond to fears about incest by not expressing affection to their children.

Western children experience several stages in their sexual development. From about two or three years old we see children expressing interest in their own genitalia and engaging in various kinds of sex play with other children. The doctor-nurse game is a common one whose sexual connotation is acknowledged among Americans. Masturbation is prevalent among three to six year olds (Konker 1991:149). According to Kinsey et al. (1948, 1953), by five years of age 10% of the boys and 13% of the girls will have experienced some kind of sex play.

Although Freud argued for a period of sexual latency between the ages of six and twelve, Borneman's (1983) research on childhood sexuality along with that of Goldman and Goldman (1982) has seriously challenged this view. These researchers report a capacity for sexual response from infancy throughout the lifecourse.

In preparation for the adolescent period, children develop affectionate relationships with special friends during childhood. In western societies, where there exists a well defined concept of romantic love and heterosexuality, children also express romantic interests. Ninety percent of the 9–11 year old children in one study reported having a boyfriend or girlfriend (Broderick 1972). Although this is the period cited by Freud as one of latency and lack of interest in sexual partners, children continue to be increasingly interested in sex as they approach puberty. American sexual norms shape an inverse relationship between public expression and sex. As children become increasingly interested in sex, their public expression of that interest in terms of self-exploration, play with others, and questioning is generally suppressed. While sexual interest increases, sexual expression decreases. The same may be said of same gender sexual exploration and romantic attachments. Generalized homophobia is presented at an early age and compounds broader sex negative trends which children in the United States are exposed to relatively early.

It must be pointed out that there is a great deal of variation in parenting perspectives regarding sexuality from the extremely punitive to the permissive in our pluralistic society. Children also learn sexual mores from peers and the media as well. The consequences of these childhood experiences can have a profound impact on sexuality later in life. Sex therapist Dagmar O'Connor fears that the concern over AIDS will give children the notion that "sex will kill" and this will enhance an existing western trajectory of fear of sexual feelings (Jackson 1990:184).

Noted sexologists John Money and Gertrude Williams challenge deeply embedded western values when they state "a childhood sexual experience, such as being the partner of a relative or an older person, need not necessarily affect the child adversely." They continue, "no matter how benign, any adult-child interaction that may be constructed as even remotely sexual qualifies, *a priori*, as traumatic and abusive." According to Konker (1992:148) the evidence is unconvincing that adult and child sexual contact is inherently deleterious. Money cites the recent social trends toward sexual conservatism as potentially harmful ("Attacking the Last Taboo" 1990:72). For example, funding has been curtailed for research on sexually troubled children as a consequence of such concerns (Jackson 1990:186). A fear exists that the response to child abuse may have gone too far. Adults may become uncomfortable expressing affection to children and children may consequently be denied affection and touch, which is so important in their development and capacity for bonding.

It is with this *caveat* that we explore childhood sexuality cross-culturally. Evidence from the more restrictive societies such as ours will be contrasted with more liberal approaches to childhood sex. The classification of societies as restrictive, semi-restrictive and permissive by Ford and Beach (1951) represents a continuum of sexual mores. This is not to suggest that restrictive and permissive societies are "pure" types without variation as we mentioned previously. Undoubtedly within this classification scheme, there are necessarily mixed and contrasting cultural elements in regard to sexual expression. However, Ford and Beach's approach is nonetheless a useful tool for comparison and contrast in spite of its limitations in recognizing overlapping diversity. Some of our examples will include incidences of adult-child sexual interaction. It must be remembered that relativism is essential as well as an awareness that such interactions cannot be viewed from a western and adult perspective. These examples must be regarded as contextually framed.

We will turn to Ford and Beach (1951:178–192) to begin our review of childhood sexuality. Ford and Beach's study of human sexuality employed the Human Relations Area Files files and relied on a sample of 190 societies, although the sample size fluctuated in relation to available information for particular questions and hypotheses. The United States along with fourteen other societies, among them the Ashanti, Dahomeans, Kwoma, Murngin, Manus, and Trukese, are labeled as restrictive societies in that children are denied any form of sexual expression at all. How children are sanctioned varies from reprimands around masturbation to more extreme measures. For example, among the Kwoma of New Guinea, a woman has the right to hit a boy's penis with a stick if she catches him with an erection (Ford and Beach 1951). Ford and Beach note that sex negative cultures such as these maintain a similar attitude about sex education by keeping sexual information away from children. As we shall see in our discussion of adolescence, a society that is restrictive in terms of childhood sexual exploration may not be so restrictive for adolescents, or may have a double standard for females and males. Regardless, punishment and discipline does not stop children from exploring their bodies. For example, Trukese children will play at having intercourse although they will be punished with a whipping if caught. Other groups in which children's sexual expression occurs despite negative sanctioning includes Haiti, Manus, and the Kwoma.

Semi-restrictive societies are defined by Ford and Beach as those in which there may be formal sanctions against sexual behavior, but these are not enforced or regarded with great concern.

The Alorese have a formal restriction against children's sexual expression, but unless it is blatant, the adults will overlook it in the case of older children.

Permissive societies are those in which there is a liberal attitude about children's sexuality. These societies are also permissive about issues of sexual education. Included among these are: the Copper Eskimo, Easter Islanders, Hopi, Ifugao, Marquesans, Samoans, Lesu, Tikopia, Wogeo, and Yapese. The Ponapeans, for example, give a full and detailed sexuality education to four and five year olds. In societies in which children have access to covert observations of adult sexuality because of the sleeping arrangements, this may serve as a form of sex education. We are not suggesting that adults engage in sex publicly and in full view of their children. One of the components of human sexuality, apart from certain ritual situations, is its privacy. However, where families share a room, as among the Pukapukans, children will take advantage of the opportunity to observe and to learn about sex, despite their parents' efforts at discretion.

Among the non-restrictive societies are those in which adult-child sexual interaction occurs. These are societies in which adults may stimulate the genitalia of children and include the Hopi, Siriono, the Kazak, and the Alorese (Ford and Beach 1951:188). It must be remembered that this kind of stimulation on the part of the adult cannot be understood from the perspective of western adult sexuality. Such adult to child behavior occurs for a variety of reasons. It may be done to calm the child down or give it pleasure, not unlike the way in which an infant or child stimulates her/himself. It may also be considered part of the maturation and development of a child's sexual functioning. There is a great deal of heterogeneity in such practices and the Lepcha of India provide an excellent example of a society that condones childhood sexual expression. Because the Lepcha believe sexual intercourse is important for stimulating maturation, eleven and twelve year old girls engage in full coitus. According to Ford and Beach: "Older men occasionally copulate with girls as young as eight years of age. Instead of being regarded as a criminal offense, such behavior is considered amusing by the Lepcha" (1951:191). Other cases of adult child sexual interaction includes finger defloration of female infants among some Indonesian societies, Australian aborigines, and Hindu Indian groups. The Tontonac of Mexico invite a priest to deflower the infant girl about a month after her birth, followed by the mother's defloration of her six years later. Among the Kubeo, an old man would deflower the eight year old girl by stretching the vagina until three fingers could be inserted (Gregersen 1983:263, 268). In fact, child-adult sexual contact is institutionalized as part of initiation rituals in at least twenty countries (Konker 1991:148).

Such cultures are also permissive toward auto-stimulation or self-pleasuring as well, i.e., masturbation. Societies which are tolerant of self-stimulation include the Pukapukans and the Nama Hottentot in which public self-masturbation by children is considered acceptable behavior. Other societies are indulgent of children's early efforts to imitate adult copulation. Trobriand children even engage in oral and manual genital practices with no parental objections. As children approach puberty different restrictions or proscriptions may be placed upon them. We shall discuss these later childhood experiences in greater depth in our discussion of adolescent sexuality in chapter 10.

The subject of cross-cultural childhood sexuality requires cultural relativism and sensitivity. It is all too easy to allow our own ethnocentric attitudes about sex to color our understanding of the cross-cultural record. Kelly (1990:223) has coined the term **erotocentricity** to define the process whereby we allow our own culture's sexual attitudes, values, and mores to bias our understanding of sexuality in other cultures. This will continue to be important as we discuss adolescent and adult issues in sexuality. Relativism is vital not just in respect to the cross-cultural record, but in regard to the variety of sexual experiences and expressions found in complex western societies.

Summary

Chapter 9: Early Childhood Sexuality

1. Childhood and parenting are discussed.

2. The functions of marriage and the family are presented.

3. The subject of kinship is introduced along with kin terms, including residence and descent.

4. Some of the major theories of incest are offered along with the consequences of incest.

5. Theories of childhood sexuality are reviewed including the psychoanalytic perspective and Dr. Spock's perspective on childhood sexuality.

6. Attention is given to childhood sexual expression in the United States and cross-culturally.

OVERVIEW

Chapter 10

Puberty and Adolescence

This chapter:

1. Defines and contrasts puberty and adolescence.

2. Discusses rites of passage as initiation ceremonies that facilitate the transition to adulthood.

3. Presents the three phases of rites of passage.

4. Uses the Sambia as a case study of rites of passage in which ritualized homosexuality occurs.

5. Outlines adolescent sexual behavior in nonwestern as well as western societies.

6. Addresses the issue of adolescent sterility.

7. Reviews the topic of sex education in the United States.

8. Compares and contrasts adolescence in the U.S. with nonwestern transitions to adulthood.

Chapter 10

Puberty and Adolescence

Puberty and Adolescence: A Bio-Cultural Phenomenon

Americans are known for regarding adolescence as a tumultuous time and distinct phase in the lifecourse (Mead 1961). In fact, it is this western based view of adolescence as a life crisis that inspired researchers such as Margaret Mead in the now classic *Coming of Age in Samoa* (1961, orig. 1928) to explore the cross-cultural record to search for the causes of teenage trauma.

It should come as no surprise to learn that the ethnographic spectrum challenges this view of adolescence. Adolescence is not only *not* universally regarded as a stressful period, but not all peoples regard it as a distinct phase in the lifecourse. For purposes of linguistic convenience, we shall use the term "adolescent" to refer to the teenage years, bearing in mind it is not a cultural universal. While people in the U.S. have a scheme that divides the lifecourse into the five phases of infancy, childhood, adolescence, youth, and adulthood, other peoples have far different views. According to Oswalt (1986:99): "[a]boriginal peoples usually recognized few formal age grades, and by the time children were eight years old, they were working for or with same sex adults." Although non-industrial peoples do not elaborate the adolescent stage as do we, most peoples of the world recognize that the period of sexual maturation as one in which readiness for adulthood occurs.

Before continuing, it is appropriate here for us to define our terms since adolescence is often confused with puberty. **Puberty** is generally defined biologically. It is the period in life when secondary sexual characteristics develop and a person is in the process of becoming capable of reproduction (Offir 1982:521; Kelly 1990:512).

Puberty usually takes from two to four years. In girls, the ages for the onset of puberty are nine to sixteen, and in boys, from twelve to sixteen. A detailed discussion of the physiology of puberty has been described in chapter 5: Modern Human Male Anatomy and Physiology, chapter 6: Modern Human Female Anatomy and Physiology, and chapter 7: Fertility, Conception, and Embryology. The physical changes characterizing puberty include development of pubic hair and body hair, deepening of the voice, breast development in the female, widening of hips in the female and shoulders in the male, maturation of genitalia and capacity for reproduction (Ford and Beach 1951:170–171). In females, menarche occurs at the same time as a widening of hips and about a year or two subsequent to the development of breasts. Girls in the U.S. on the average reach menarche around twelve years old, although the age at which this occurs is affected by diet and nutrition (Spock 1985:495). The cross-cultural evidence suggests that nonwestern females are generally not fertile following menarche (Ford and Beach 1951:172; Schlegel and Barry 1991). In addition, there is an increased interest in sex as a result of the physiological changes that happen with puberty (Byer and Shainberg 1991:375). However, it must be remembered that interest in sex is extremely variable in individuals (Kelly 1990:167). How one experiences the biological baseline of puberty is shaped by social structure, subsistence technologies, stratification, kinship and descent, cultural values such as conservative or liberal attitudes of parents, religious orientations, and a number of other factors (Schlegel and Barry 1991).

Menarche is one of the more obvious indicators of puberty. The onset of menstruation has followed a historical trend of occurring at an increasingly younger age. In western countries, menarche is now reached two years earlier than it was in 1900 and four years earlier than in 1840 (Moore 1980:138–139). In nonwestern countries, the pattern for menarche occurs at a younger age in urban areas than rural ones. Economic variables are also important. In higher economic strata menarche begins earlier than in lower classes of the same population (Eveleth and Tanner 1976 in Moore 1980:140). Better nutrition and overall body fat seem to partly account for this change "although additional cultural and genetic influences also contribute to this phenomenon . . . " (Eveleth and Tanner 1976 in Moore 1980:140).

While puberty is identified by a series of physical changes, **adolescence** is a cultural construct defined as a "period of emotional, social and physical transition from childhood to adulthood" (Kelly 1990:505). Adolescence is the time of life that spans puberty and ends with adulthood. It is not limited to the teenage years. Robert Francoeur (1991a:41) notes that as a period of psychosocial maturation it may even extend into the late twenties in some societies such as the United States and Ireland. If it is recognized as a distinct stage, how and when it begins as well as what it means differs a great deal throughout the ethnographic spectrum. In the West, adolescence as a distinct phase has varied historically as well.

Adolescence did not appear in the West until the nineteenth century. It was not until the 1800s that adolescence also came to be regarded as a time of conflict, especially among the upper classes. Prior to that time, youth was regarded as a period of preparation for adulthood that covered a long time span with gradually increasing responsibilities. There were various symbolic markers for the transformation of youth into adulthood. This was related to the economic situation in which working class children at 10 or 12 years old entered occupations as apprentices, in the case of boys, or perhaps as domestic servants for girls. These decisions were made by their parents (Sommerville 1990:211–212).

> Having entered upon their calling children were playing parts in the adult world and had a recognized status there, even though they were not yet considered adults. Rights and responsibilities came gradually, with a number of milestones on the way to maturity. In some respects they were still considered children for years afterward. But the adult world was not a foreign and unknown territory to them (Sommerville 1990:213).

Other options varied by class for girls and boys. Upper and middle class boys were also involved with training for their profession by the latter part of their teenage years. During this period, the age of marriage was around fifteen years old; and puberty did not happen until much later, around 18 or 20 years old (Francoeur 1990:114).

The Industrial Revolution changed the nature of work and consequently the apprenticeship as a vehicle for integration of youth into society was lost. As the population grew so did employment opportunities for the middle and upper classes. The aristocrats sought positions in the military and government which, in turn, had to be expanded to accommodate their need for employment. At the same time, education for the upper and middle classes also kept youngsters in the home under their parents' guidance. The results of these trends were the creation of adolescence as a separate phase in western youths' lives characterized by their separation from the world of work and adults (Sommerville 1990:216).

Along with this trend, the age of puberty also gradually dropped from the late teens and early twenties to what it is today (Aries 1962).

Turning to the cross-cultural data, we find that the most elemental ways that peoples categorize and define themselves are based on two criteria: age and gender. Adolescence is an example of an **age-grade** which is a grouping of individuals who share a certain age range. "Responsibilities, rights, and obligations of individuals change as they progress through different age-grades" (Cohen and Eames 1982:411). In many societies one's age grade is a significant part of one's life and place in society. This is particularly true in societies in which other ways of identifying and categorizing individuals such as on the basis of power, wealth, and status are absent. Age grading is related to **age sets**. An age set is "[a] nonkin association in which individuals of the same age group interact throughout their lives" (Oswalt 1986:432). One of the consequences of initiation ceremonies is to foster age sets among initiates.

Puberty Rituals: Initiation Ceremonies as Rites of Passage

In many societies there is "[c]eremonial recognition of a major change in social status, one that will permanently alter a person's relationship with members of the greater community" (Oswalt 1986:437). Such ceremonies are called rites of passage or a passage ceremony. It should be remembered that not every society marks the transition from childhood to adulthood with elaborate ceremonies. We are an example of a society that does not.

It has been proposed that the absence of rites of passage rituals is the cause of much trauma in our status transitions (Shapiro 1979:283). Rites of passage often include initiation ceremonies in which a child takes on the new status of adult. In 1909, Arnold Van Gennep devised the rites of passage model as a tripartite scheme and explained the functions of these ceremonies. According to Van Gennep (1960:3):

> Transitions from group to group and from one social situation to the next are looked on as implicit in the very fact of existence, so that a man's [and a woman's] life come to be made up of a succession of stages with similar ends and beginnings: birth, social puberty, marriage, fatherhood and motherhood, advancement to higher class, occupational specialization, and death. For every one of these events there are ceremonies whose essential purpose is to enable the individual to pass from one defined position to another which is equally well defined.

Rites of passage have three distinct phases: separation, transition, and incorporation (Van Gennep 1960:12). In separation an individual is removed from his/her previous world or place in society. Transition, or liminality, are "threshold" rites that ready an individual "for his or her reunion with society." This is the phase in which the initiates undergo training for their new position and learning new information important to their anticipated status. As part of this, they are likely to experience some sort of ordeal or testing. Incorporation rites integrate individuals back into their group. This includes a return to the community after the initiate has been socially and perhaps even spatially removed. It is the public recognition of the person's new status in society (Van Gennep 1960:21, 46, 67).

In the case of the pubescent, rites of passage function to ease the journey from the status of child to that of adult. According to Chapple and Coon (1942), changes of status are disturbing for personal and social relations within the group. Initiation ceremonies, like other rites of passage, help ease and facilitate the transition to adulthood. They do this for the novice who must experience an identity shift as s/he takes on a new position, as well as for the broader social group who must now accept the youth as an adult.

Initiation rites are often stressful and include rigorous tests, hazing, isolation from previous associates, and/or painful ordeals. These attributes provide symbolic referents for learning about what the new status as adults includes. Adolescent rites of passage provide an opportunity to practice and gain knowledge about adulthood. According to Chapple and Coon (1942:484–485), these ceremonies restore equilibrium to the individual as well as the community. In addition, the dramatic, painful, and stressful elements may help prepare the youth for the "stresses" of adulthood as well as function to enhance group solidarity (Oswalt 1986:106). The rites of transition are particularly important as a period in which the novice is liminal, "betwixt and between statuses" according to Victor Turner (1967). By occupying this liminal status, the pubescent individual will be in a unique position to unlearn their position as a child and take on their new responsibilities as adults.

There are gender differences in rites of passage that may be related to differential socialization. Chodorow (1974) has argued that female and male socialization are contrasted in terms of continuity and discontinuity. She believes that this is based on the universality of women as caregivers of children. Therefore, the majority of children experience a female during their early years, but each gender experiences this female differently. Male socialization is discontinuous, in that they must eventually experience a separation from the domestic worlds of their mothers, while females do not. Young females learn to identify with their mothers and other women as associational role models and need not learn a new identification as they mature. In contrast, the boy's association with women during childhood prevents him from making an easy masculine role identification (Chodorow 1974:54). The boy's role model is a cross-sex one in contrast to the associational one of little girls. The model is based on notions of the traditonal nuclear family in which father is the breadwinner and mother works in the home.

Chodorow's argument is related to theories about why initiation ceremonies are often so much more elaborate and at times more severe for males than females, although female ceremonies are more prevalent. There are a number of ideas on this subject. Burton and Whiting's (1961) cross-sex identity hypothesis suggests that young men in polygynous and patrilineal households are likely, for a variety of reasons, to assume a feminine identity. In order to switch their cross-sex identification with their mothers to that of men, a severe initiation ceremony is mandated. It is designed to impress upon the boys that the world of men is more important and valued than that of women. According to this theory, severe practices such as circumcision or genital mutilation are guaranteed to catch the young man's attention and to reverse the cross-sex identity. An important facet of these ceremonies is a mockery or "put down" of women which contributes to the masculine self-definition as *not* feminine. This argument has been criticized because the cross-sex identity is not demonstrated by the researchers but only speculated. Other less psychodynamically focused theories include Young's socio-cultural interpretation of initiation ceremonies.

Young's (1965) study of rites of passage found that male initiation ceremonies were designed to incorporate men into the entire community, while female initiations integrated women into domestic groups. Young postulated that this was related to the differential roles and tasks that each sex was assigned in society: women a domestic one and men a public role. For Young, male solidarity was an important feature overlooked by the psychoanalytic approaches of researchers such as Burton and Whiting (1961 in Bock 1988:117). Solidarity was also a salient feature for the female household as well. Young explained that female initiations were less elaborate because the domestic sphere is more private and less extensive as opposed to the public world of males (1965:106). The domestic arena is also continuous and familiar in that females are born into it and exposed to information about their role as they are growing up. This was suggested in Chodorow's (1974) approach as well.

Female initiation ceremonies will often ritually mark menarche. This event is regarded as a very important one in many cultures. How a society responds to menarche is related to the broader cultural context such as attitudes toward women in general, womens' positioning in society, forms of social organization, and beliefs about menstrual blood. For example, New Guinea highland groups are well known for their view of menstrual blood as polluting. Menarche rituals are as varied as there are cultures. In some groups, like the Gussii, clitoridectomy may be part of the ceremony. In others such as the Tlingit, a girl was confined for at least a year with a series of proscriptions around her behavior, e.g., she must not gaze at the sky or must scratch an itch only with a stone (Oswalt 1986:108–109). Among Andaman Islanders (Service 1978:60–61 in Moore et al. 1980:140):

> Once menarche is attained, a girl in this society is secluded in a hut for several days. Her behavior is closely regulated by prescriptions concerning bathing, posture, speaking, eating, and sleeping. The personal name used during childhood is replaced by one taken from a plant in bloom during the ceremony and is retained until the next rite of passage (marriage). A boy reaching puberty does not experience physical isolation but is singled out by the occasion of an all-night dance held in his honor. Scarification of his back and chest further emphasizes his coming change in status. Dietary restrictions are enforced for a period of a year or more at the end of which time the boy is given a new name.

Schlegel and Barry (1980: 696–715), in a study of 186 societies, found that societies that emphasized female initiation ceremonies were more likely to be gatherers and hunters. They suggested this was because reproduction is important for foraging groups who have a low population density. Such ceremonies emphasize the importance of the life-giving attributes of women. In contrast, initiation rites in small scale plant cultivators (like non-intensive horiticulturalists) emphasize equally both girls and boys. However, rigid separation of the sexes during the rites of passage is enforced. This accents the cultural importance of gender differentiation in such societies (Schlegel and Barry 1980:712). In fact, this characteristic is not uncommon in puberty rites in general. Despite variance in social organization, a general feature of female initiations is that they are centered on fertility while male's are focused on responsibility. In horticultural societies with both male and female initiation ceremonies, same sex bonding is an important function of the initiation ceremonies where homosocial relations (same gender) are an integral part of such cultures (Schlegel and Barry 1980:712).

Case Study of Rites of Passage: The Sambia

Gilbert Herdt's study, *The Sambia: Ritual and Gender in New Guinea* (1987), describes the rites of passage of Sambia male adolescents as they pursue adulthood. What makes the Sambia of particular interest is that their initiation into manhood involves an extended period of institutionalized homosexuality. All males among the Sambia will have experienced the roles of fellator and fellatee with other men during the course of their initiation and journey into manhood (see also Herdt 1981,1988).

The Sambia are a highland Papua New Guinea group characterized by warfare and male privilege. As is typical in groups like this, there is great disparity in the status of men and women, with men being privileged and dominant. At a prepubescent age boys go to live in all male clubhouses, where for the next seven years they will fellate the older males (teenagers and men in their early twenties) who share the clubhouse with them. It is only by swallowing the ejaculate of the older boys that a young boy can hope to grow into manhood. Manhood is defined by semen, which is regarded as a very powerful substance. Boys are believed to be without semen. So it is important that a boy consume as much as possible in order to have an ample supply of it. The obvious way to get this is by fellatio, or oral insemination. Semen has a power known as *jerungdu* (Herdt 1987:101).

Like sexuality in general, there are rules and practices regulating homosexual behavior. When a young man reaches about twenty-five years old, he is married. However, he must remain attentive to preserving his supply of sacred fluids, lest his spouse, who is regarded as potentially dangerous and polluting, sap him of his strength and use up his semen during intercourse. Societies like the Sambia are noted for female pollution avoidance rituals, which dramatize women's social inequality to men. Female pollution avoidance rituals may restrict menstruating women by confining them to a special structure, e.g., a menstrual house. These rituals are based on beliefs that females are impure. Avoidance of women and concepts of female impurity associated with the female menstrual cycle contribute to status inequalities and disparities between the sexes (Herdt 1987). Such rituals do not occur in societies in which women share power with men, but tend to be found in patrilineal societies and societies in which women have low prestige (Zelman 1977:714–733).

The purpose of initiation among the Sambia is to make men out of boys. Little boys inhabit the world of women and are dangerously contaminated by it. As a result of this, they are not quite masculine. Masculinity is not something that is seen as "naturally" occurring. *Jerungdu* must be acquired as the source of masculinity. In addition to the feminization of young boys that is a result of their association with their mothers, there is the problem that males cannot manufacture their own semen. To make matters worse they can lose *jerungdu* through ejaculation. The Sambia ritualized homosexual initiation resolves this dilemma of manhood. Fellatio is the means to acquiring an initial supply of semen. The proof of the power of the initiation among the Sambia is that young boys provide evidence that it works by becoming bigger, physically strong, and assertive. In the end, " . . . the idealized masculine behaviors of initiation have remade the boy into the image of a warrior" (Herdt 1987:104). Through the course of initiation the boys have learned the cultural values associated with masculinity and with it the secrets of manhood hidden from Sambian women. For example, heterosexual coitus is particularly dangerous for men since through it they can lose their power. The initiates must learn how to drink white tree sap to replenish their source of power, once they are past the stage of ingesting semen through fellatio (Herdt 1987:164).

The Sambia are a provocative example to contrast with our own western concepts of manhood and sexuality. During most of the initiation cycle the initiate is not permitted any heterosexual activity. In the early stages, the boys act as the fellators and ingestors of semen, which contains the power to make them grow into manhood. The initiates are prohibited from masturbation or anal sex as well. In other words, they have no sexual outlets other than wet dreams. But from about fifteen years of age through eighteen, the boys enter the third stage of initiation when they become bachelors and inserters rather than fellators. Their ability in the "inseminator" role proves:

> . . . that they are strong and have *jerungdu*, because their bodies are sexually mature and have semen to "feed" to younger boys. They feel more masculine than at any previous time in their lives. So the bachelors go through a phase of intense sexual activity, a period of vigorous homoerotic activity and contacts, having one relationship after another with boys. Their sexual behavior is primarily promiscuous, for the initiates are concerned mostly with

taking in semen, while the bachelors mainly desire sexual release
through domination of younger boys . . .
　　Eventually Sambia adolescent boys become more interested
in females (Herdt 1987:162).

The purpose of the homosexual behavior is to acquire semen so
that the youths may ultimately marry and achieve fatherhood.
Around the age of seventeen, the bachelors enter the fourth stage
of initiation in which they are permitted interaction with women.
From the beginning of the initiation until the fourth stage, they
have not been in contact with women. In their late teens and early
twenties, the initiates go through a fifth stage of bisexuality as
married men which is followed by a sixth stage of adulthood in
which heterosexuality is practiced (Herdt 1987:107). This stage is
associated with the birth of the man's first child. Thus, the birth of
a child is the marker for adulthood.

Adolescent Sexual Behavior: Nonwestern

The Sambia initiation ceremony provides an excellent example
for beginning our discussion of adolescent sexuality. As we have
seen, the Sambian male's first experiences involving the genitalia
are as fellators from about age 7 through age 14. At fifteen they
then move into a new stage where they can experience sexual outlets
for the first time (other than wet dreams). Prior to this they were
prevented from doing so. In this phase they are the bachelor recipi-
ents of fellatio. The inseminator role moves with them as they take
on a new status as newlyweds (they do not cohabit with their
wives), but continue homosexual behavior as inseminators. The
inseminator phase is a very pleasurable one for males. From about
the ages of 14–19 there is no bonding with females. This is a period
in which males experience "profound homoerotic pleasures" (Herdt
and Stoller 1989:33).
　　This case clearly points to the difference between sexual behavior and sexual orientation. The Sambia initiation ceremony is
an important vehicle for sex education facilitating the young man's
transition to a heterosexual lifestyle. The young male whose primary sexual experience is a homosexual one is gradually shifted
into the newlywed status which allows him to make the transition
in sexuality by going through a bisexual phase. In addition, "[t]he
customary first sexual intercourse between spouses is fellatio" usually taking place in the late teens or early twenties (Herdt 1987:164).
Women are given sexual instructions at their menarche ceremony.

It is interesting that cultural practices help ease the male into his change of lifestyle. The newlywed bride resembles the young initiates in appearance. The bride covers her breasts and wears a noseplug similar to that of the young males that are the fellatees. In addition, the wives perform fellatio on their husbands. "The bride's similarity to boys and the fellatio, thus help to provide an erotic bridge between the homosexual and heterosexual lifestyles" (Herdt 1987:165). Herdt's research on the Sambia challenges clinical theories that early homosexual experiences lead to later adult homosexuality (Schlegel and Barry 1991:109).

The Sambia are a fascinating example of nonwestern sexuality as it is experienced in adolescence and into young adulthood. Male-female relations are part of a well reported phenomenon of sex antagonism in this particular area of New Guinea. Women are regarded as potentially dangerous because of their ability to pollute and deplete men of their semen. Wives are acquired by politically motivated and arranged marriages. Because they come from outside the group (exogamy), they can never be completely trusted. While the arranged marriages may help create alliances, they are also potentially disruptive because at any time the bride's family might become enemies. Warfare plays an integral part of Sambia homosexual as well as heterosexual behaviors (Herdt and Stoller 1989:32). It must be kept in mind that our concepts of homosexual, bisexual, and heterosexual are in the context of our western experience. The Sambia do not have categories analogous to ours. Their heterosexual, homoerotic, and bisexual behavior is obviously vastly different from the western one and cannot be translated into our western clinical and homophobic perspective (Herdt and Stoller 1989:31–34).

Turning now to our more general topic of pubescent and adolescent sexuality, nonwestern and western, we need to point out that male and female adolescent sexuality may occur at different times. For example, among the Tiwi, adolescent women have sex with post-adolescent men, not men of their own ages (Hart, Pilling, and Goodale 1988). Age of marriage is an important variable to consider since sex may occur in the context of premarital, marital, and/or extramarital sexual behavior. In countries in which individuals marry late, for example Singapore, where women marry at 24.4 years old, the premarital sex period may extend into the twenties. Many of the studies of premarital sex in developing nations target students as the research population so the ages may span both teens as well as the early twenties. Conversely, marital sex may also include the pubescent age groups since in many societies people who marry in their teens may be regarded as adults. Marriage in most traditional societies ends adolescence as a cultural stage (Schlegel and Barry 1991:109). In 74% of forty societies, males marry at eighteen years or older and in 69% of forty-five societies females marry at seventeen or younger. The modal age of male marriage is eighteen to twenty-one years old and for females it is twelve to fifteen years old. Men tend to be older than their wives at first marriage, and hence experience premarital rules for a longer time. However, it is important to remember that males generally have greater access to a double standard that allows them more premarital sexual freedom (Frayser 1985:208).

It is important to remember for our discussion of pubescent and adolescent sexuality that the meaning premarital sex has for westerners is not necessarily convergent with that in other societies. In addition, sexual behaviors westerners include in a repertoire of premarital sex also varies. In some cases cross-culturally, data from the Human Relations Area Files biases the definition of premarital sex by focusing on heterosexual penile-vaginal intercourse. The degree to which premarital sex among youths is approved of, disapproved of, and even condemned will vary cross-culturally as well. Despite the diversity, all cultures have rules regarding sex relations with appropriate partners. A variety of factors interact and are related to rules relating to premarital sex. Martin and Voorhies (1975) have found that societal approval of premarital sex is related to the type of social organization, population density, subsistence and resource patterns, all of which are directly related to the status of women. Relevant also to this discussion is the role of adolescent sterility and cultural attitudes toward children conceived nonmaritally.

Ford and Beach's (1951) codification of cultures as restrictive and permissive based on a massive review of 190 different societies provides us with a useful approach in describing adolescent sexuality. This approach has been refined and continues to be an important source for understanding how and why adolescent sexuality is structured. Most notable is the work of Barry and Schlegel (1991) discussed later in this chapter. Again, this model of ideal types represents what in actuality is a continuum. Ford and Beach recorded fourteen very restrictive societies in which children are prevented from sexual expression and acquiring sexual knowledge. However, sex with the onset of puberty is allowed for girls in ten restrictive societies, and for boys in one; the Haitians. "For the most part these peoples seem particularly concerned with the prepubescent girl, believing that intercourse before menarche may be injurious to her" (Ford and Beach 1951:18). In most of the African societies studied by Ford and Beach, boys were prevented from having sex before their initiation ceremonies. In some societies rules against sex after puberty may remain restricted or may actually be intensified. To restrict premarital sex among young people, societies will 1) separate the sexes, 2) chaperone females, and/or 3) negatively sanction premarital sex (Ford and Beach 1951:182). Of these measures the first is the most successful, while the third has not proven to be a deterrent to the highly motivated youngster (Ford and Beach 1951:183–184). One of the means restrictive societies use to ensure control of youngsters' sexuality is by placing a value on virginity. Some may even have tests of this virginity through demonstrations of bloodied cloth or defloration ceremonies (Ford and Beach 1951:186).

In semirestrictive societies, there may be formal proscriptions directed at teenage premarital sex, but these are not regarded as serious offenses. Prohibitions against premarital sex for females specifically occur in twelve societies, for older children in two, while sanctions against both sexes are found in thirty-four societies (Ford and Beach 1951:187). There are three permissive societies that allow coitus for adolescent boys; the Crow, Siriono and Tongans; one society that allows premarital permissiveness for girl's only, the Thonga; and one that limits permissiveness to the commoner class, the Nauruans. There are forty-three permissive societies in which there are no gender specific restrictions. The only restrictions are those around incest, which would be expected (Ford and Beach 1951:190).

> By the time of puberty in most of these societies expressions of sexuality on the part of older children consist predominantly of the accepted adult form of heterosexual intercourse, the pattern which they will continue to follow throughout their sexually active years of life (Ford and Beach 1951:190).

While Ford and Beach documented the variation in adolescent sexuality, other anthropologists have been interested in explanation, asking questions such as how social structure influences premarital sex norms. Schlegel and Barry's *Adolescence: An Anthropological Inquiry* (1991) provides valuable insight in this regard. A summary of some of their findings and review of the research is presented below to show how sociocultural features pattern sexual behavior (Schlegel and Barry 1991:109–121).

1. For both sexes, adolescent permissiveness is related to the absence of a double standard.

2. Adult adultery for women and men is frequent in societies that are permissive for adolescent sexuality.

3. Premarital sexual permissiveness for females is associated with simpler subsistence technologies, absence of stratification, smaller communities, matrilineal descent, matrilocal residence, absence of belief in high Gods, absence of bride's wealth, high female economic contribution, little or no property exchange at marriage, and ascribed rather than achieved status, an evaluation of girls as equal or higher than boys.

Recently, concern for AIDS has resulted in an increase in research on the subject of sex, including sexual behavior among youths in countries such as Thailand, Philippines, Taiwan, Hong Kong, Sri Lanka, Japan, Malaysia, Micronesia, and Melanesia (Sittitrai 1990:173–190). These are known as Pattern III countries reflecting the transmission of AIDS from other areas through immigration and tourism (see chapter 14: HIV Infection and AIDS for a more detailed discussion). Unfortunately, our western heteroerotic bias has focused primarily on male populations and heterosexual behavior. In the majority of Pattern III countries studied, about 20% to 50% of the adolescents report having engaged in premarital sex. Yet, these same countries also place a value on female virginity (Sittitrai 1990:177). There are some studies available of adolescent sexuality that include homosexual and/or bisexual behaviors. Most of the studies of homosexuality cross-culturally do not focus on the specifics of homoerotic sex practices among youth. However, a Japanese study reported 7% of the male and 4% of the female research population experienced some male or female homosexual behavior such as kissing, petting, and/or mutual masturbation. These researchers did note that the age of the first same-sex experience was 15–17 years old with partners usually older (Sittitrai 1990:178). This kind of specific data is essential for understanding adolescent sexuality so important in regard to HIV transmission. A great deal of research remains to be done in this area. Schlegel and Barry (1991) have made a substantial contribution in this regard. Using the cross-cultural correlational approach, they compared homosexual behavior among pubescent boys and girls in 24 societies. They found "In virtually all cases, if homosexual relations are tolerated or permitted for one sex they are for the other as well." In addition, their research indicates that cross-culturally, homosexuality in adolescence tends to be transient and " . . . appears to be a substitute for heterosexual intercourse when intercourse is prohibited or access to girls is problematic" (Schlegel and Barry 1991:126).

Adolescent Sexuality: Western

In U.S. society, adolescence is commonly considered a period of anxiety. It has been linked with a capitalist ideology of competition and gender stratification as well as that peculiarly western view of personhood as one of autonomy and independence. Western adolescents may well wonder if they measure up. Both males and females at this time experience concerns over adequacy. Males are worried about their penis size and females are worried about the size of their breasts. This is tied to their newly developing sense of self as sexual beings. While there is no relationship between breasts or penis size and one's capacity for sexual functioning, these myths prevail.

We would like to point out that the data on western adolescent sexuality is biased toward the white middle class and does not typically include variation by class and ethnicity. In addition, it is important to know that we are citing general patterns and trends, not stereotyping or assuming that these features are true for all. Generally, adolescent males and females are socialized differently in regard to human sexuality and their respective roles in the process. Both sexes may encounter increased sexual interest as a result of the physiological changes accompanying puberty. Sexuality is an integral component in the dual gender system of socialization. Males learn that they are the initiators of sex and that this is a reward that they will, in all likelihood, have to work for in some way. Because they are the initiators they also face the risks of sexual rejection.

Females learn that they are the keepers of a desired resource, sex. They are not the direct sexual initiators, although frequently they are the initiators of courtship through non-verbal actions and displays (Perper 1985). Adolescent sexuality is related to broader conceptions of gender ideologies. Young females are taught that once a male gets out of control sexually there is no stopping him. So adolescent girls are warned not to tease a boy and get him sexually excited. In addition, sexual experiences for adolescent females are connected to intimacy needs, relationships, and love. Adolescent female sexuality is not centered on orgasm or the genitalia in contrast to adolescent boy's sexuality with its genital focus and demand for physical gratification (Hyde 1985:290). Simon and Gagnon (1973) refer to this as a **relational** ideology in comparison to a **recreational** ideology (in Delameter 1989:46). Sex is a developmental process in western society and over the course of time female sexuality develops a genital component while male sexuality may mature into "more complex, diffuse sensuous experience" much like that of the adolescent female (Hyde 1985:290). In this manner, as people age in western society, their sexuality becomes more alike than different.

Adolescent sexuality in the west includes masturbation as a component, as it does cross-culturally. A male's first ejaculation is usually experienced during masturbation. In five percent of the cases, it occurred during homosexual activity, and in 12 percent as a wet dream according to Kinsey et al. 1948 research. Generally, the rate of female masturbation is somewhat lower than males as is their sexual activity in general, both homosexual and heterosexual (Kinsey et al. 1953). Fifty-eight percent of adolescent boys and 39% of adolescent girls reported masturbating at least once in more recent research (Sorensen 1973). In a study by Gagnon et al. (1970) of high school students, 77% of the males and 17% of the females self-reported masturbating twice a week or more. It is difficult to acquire reliable information about masturbation rates.

Adolescent sexual experimentation also includes the custom of making out, enhanced by the car culture of the 1950s and 1960s. Making out usually refers to kissing, but it also may escalate into petting. Petting includes everything up to and short of vaginal intercourse, including oral and manual practices. Through petting, adolescents learn to negotiate cultural rules against vaginal intercourse by discovering alternatives that lead to orgasm. For example, in a study by Newcomer and Udry (1985), 25% of the males and 15% of females in a population who had no previous experience with heterosexual coitus engaged in oral-genital practices.

Adolescents also engage in homosexual and heterosexual intercourse. Kinsey's statistics, although dated (1948 and 1953), are revealing considering the time frame. Sixty percent of the males and 33% of the females reported at least one homosexual or lesbian experience by fifteen years old. This data again support the highly flexible nature of our sexuality and the distinction between behavior and orientation.

The following statistics provide some evidence of heterosexual behavior:

Table 10.1
Age of Sexual Intercourse

Age	Girls	Boys	Study
under 15	5–17%	19–38%	Gordon and Gilgun 1987
by 19	75%	no data	New Woman Sex Report 1986
by 19	50%	78%	Zelnik and Kanter 1980

A recent study of women aged 15–44 (n = 8000) by the Alan Guttmacher Institute found that female teen sexual activity, especially that of Caucasian middle and upper classes, is on the increase in the 1980s. They cite the following findings (in "Teen Sexual Activity Rises" 1991:25):

> The percentage of girls aged 15–19 who reported engaging in sexual activities increased from 47.1% in 1982 to 53.2% in 1988.
> The percentage of sexually active girls in the 15 to 17 year-old age bracket rose from 32.6% to 38.4% in the same period.
> In 1988, 58% of sexually active teenage girls reported having had two or more sex partners.
> In 1982, 48% of the sexually active girls age 15 to 19 reported that contraceptives were used in their first sexual intercourse. In 1988, 65% of the girls 15 to 19 reported that contraceptives, mostly condoms, were used in their first sexual intercourse.

The U.S. Center for Disease Control reports that 72 percent of American youths have had sexual intercourse by the time they graduate, and forty percent have had sexual intercourse by the ninth grade.

The evidence suggests that sex education (which includes AIDS education) must be having some impact because 58 percent of 15–19 year old males also used a condom during their first intercourse (Sonenstein and Pleck 1989 in Kelly 1990:171). It is possible to become pregnant during first intercourse and it is also possible to contract the HIV virus as well. We cannot assume that American adolescents are using contraceptives regularly. According to Wallis (1985), it is not until adolescents have been having sex for a year that they initiate contraception. Using contraception is difficult for adolescents since it is loaded with symbolic meaning about oneself as a sexual being. By using contraception the adolescent must acknowledge engaging in premeditated sex. This may lead to conflicting feelings regarding values and sense of self as a "moral" being. With western adolescence, then, comes a number of questions for the individual concerning sexuality and contraception.

Adolescent Sterility / Fertility

The decision to engage in sex is not without risks. AIDS and
other STDs are endemic in this culture as are the consequences of
an unwanted pregnancy. Teen pregnancy is a particular problem in
United States. Why is the U.S. adolescent pregnancy rate higher
than in other countries? What factors influence teen pregnancy
from a socio-cultural perspective? One possible contributing factor
is a phenomenon known as **adolescent sterility**. This is presented
in greater depth in our discussion of birth control in chapter 11:
Topics in Adult Sexuality: Human Sexual Response and Birth Control. Adolescent sterility (or subfertility) refers to a period between
the age of menarche and reproductive maturity, about twenty-three
years old. Despite the onset of puberty, intercourse is less likely to
result in pregnancy than in a reproductively mature woman (Ford
and Beach 1951:172–173). In a number of societies, adolescent females are permitted premarital sex, yet pregnancies are unlikely
to result (Ford and Beach 1951). These are the societies that Ford
and Beach (1951:190) list as permissive, that is, there are no sexual
restrictions on adolescent sex activity short of incest regulations.
The list is too long to reproduce here, but examples include the
Ainu, Aymara, Trobrianders, and Yapese.

Adolescent sterility allows teenagers and youths to experience
their sexuality as well as learn and grow as individuals. It provides
for an extended period of practice and discovery about oneself as a
sexual being without the burden and the additional concern and
responsibility of parenthood. Sexual experimentation in these societies allows for a far easier transition to marital sex than in societies which prohibit sexual experimentation during adolescence,
and then expect that the couple will be able to reverse their attitudes about sex after marriage (Ford and Beach 1951: 195). Adolescent subfertility in combination with permissive cultural attitudes
also allows for an important period of learning without the status
changes that accompany childbirth. In most societies, childbirth
provides men and women special status, sometimes it may confer
adulthood while in others adulthood may occur prior to parenthood. Regardless, having a child is viewed as an important status
marker.

As we suggest in chapter 11 in our discussion of birth control,
adolescent sterility does not seem operative in U.S. populations. Based
on the critical fat hypothesis (Frisch 1978), it is believed that ovulation occurs only when certain levels of fat have been reached. This
is directly related to the requirements for successful reproduction

and lactation. Lancaster (1985:18) attributes the loss of adolescent sterility in adolescent girls in contemporary societies to:

> [s]edentism combined with high levels of caloric intake [that] lead to early deposition of body fat in young girls and "fool" the body into early biological maturation long before cognitive and social maturity are reached (Lancaster 1985:18).

Without adolescent sterility as a damper on fertility in combination with adolescent sexuality without contraception, the result is a very high adolescent pregnancy rate in the United States. For every 1,000 females, 110 aged fifteen through nineteen become pregnant (Henshaw et al. 1989). This differentially impacts adolescent African American females where adolescent pregnancy is twice as high as Caucasian females in this age group.

What is interesting is that in comparison with 37 other countries, the U.S. pregnancy rate is " . . . higher than that of almost any other developed country. While American adolescents are no more sexually active than young people in other industrialized countries, they are much more likely to become pregnant" (Byer and Shainberg 1991:386). Unfortunately, these statistics compare Caucasian American adolescent females, not African Americans or other minorities. Nevertheless, these differences may be accounted for in part because we do not have comprehensive sex education programming, readily available birth control, and a sex positive attitude in this culture in contrast to other industrialized societies (Wattleton 1990).

What is the price of adolescent pregnancy? Girls who are under 18 years of age who give birth are "half as likely to graduate from high school as those 20 years old or older" ("The Tragedy of Parental Involvement Laws" 1990:5). In addition, a recent Minnesota study found that 80% of the seventeen-year-old mothers in their research population never graduated from high school. Other problems prevail. There is a high mortality rate among children of teenage mothers due in part to low birth weight and associated illnesses. In addition, teenage mother heads of household are seven times more likely to be poor and need public assistance ("The Tragedy of Parental Involvement Laws" 1990:5).

Sex Education in the United States

Directly related to the problem of teenage pregnancy is the issue of sex education in the United States. Sex education in the U.S. can be contrasted with traditional non-technologically complex societies. As we discussed previously, families whose sleeping quarters are shared provide an opportunity for children to become aware of the sexual activities of their parents and/or siblings. Sex in these situations is undertaken with discretion, and we know of no cultures in which children and youths are allowed to watch openly. For example, among the Mangaians, a sex-positive Polynesian culture in which premarital sex is the norm, parents do not discuss sex with their young children. But because Mangaians live in one-room houses, the opportunity for discrete observation does occur. However, the specifics and details of sex and reproduction are learned outside the home, not unlike in the U.S. (Marshall 1971:109).

Without rites of passage to help them make the transition to adult status and also to provide avenues for learning about sexuality, where do western adolescents learn about sexuality? Research suggests that it isn't from parents, but rather from peers. This is not to suggest that parents have no influence on their children's sexual behavior. In fact, according to Fisher (1987:484):

> A few studies have examined the issue of family relationships and their influence on sexual/contraceptive behavior, leading to the general conclusion that premarital sexual activity is less likely when the families relations are good. Lewis (1973) reported . . . several family relationship variables were related to sexual behavior in females. Jessor and Jessor (1975) reported that the mother-child relationship was related to sexual activity, and Joregensen, King and Torrey (1980) found that a satisfying relationship with the father correlated significantly with regularity of contraceptive use and regularity of effective contraceptive use. Fox (1981) found that the quality of the mother-daughter relationship was a strong predicator of whether the daughter was sexually active or not, and Darling and Hicks (1982) concluded that adolescents are more likely to have intercourse if they perceive themselves to have poor communication with their parents.

The mass media should not be underestimated as a source of sex education. In fact, media importance takes precedence over the peer group according to Darling and Hicks (1982). Other sources of information include both the public and private school systems in the U.S., which offer sex education curriculums. These, however, are not uniform in terms of content and coverage.

Concerns over sex education gained prominence in the 1960s as a result of SIECUS (The Sex Information and Education Council of the United States) and AASECT (the American Association of Sex Educators, Counselors and Therapists). Research indicates that the question of sex education is not "if" such programs should occur but rather "what kind of approach," since surveys indicate that 75% to 85% of adult Americans favor sex education (Kelly 1990:337; Gallup poll). Bear in mind that "sex education programs in schools are still the exception rather than the rule" (Kelly 1990:337).

Sex education programs vary a great deal as to their success rates. While a number of stances may be taken, we have an excellent example of what doesn't work: the Reaganesque "Just Say No" approach of the 1980s. This was the result of a 1981 effort to reduce teenage pregnancies without advocating birth control through the Adolescent Family Life Act, a congressional act that funded programs promoting premarital abstinence. One study even found that "participants [in one AFLA project] engaged in more sexual activity than controls" (Troiano 1990:101).

Since scare tactics have proven ineffectual, sex education curriculums whose goals are to reduce adolescent pregnancy through the use of contraception can be very successful. Such programs must take a multidimensional and comprehensive approach (Troiano 1990:101). By that, we mean that not only should the mechanics of reproduction be addressed, but the psychological and social aspects as well. Concern over the threat of AIDS has recently given a new impetus to sex education. Other societal trends are also reflected in new approaches to sex education development. For example, sexuality is in a historical niche where it is now regarded as an important and very natural component of one's life. This view is also related to trends in which sex and procreation were separated resulting in a greater emphasis on sex for pleasure. These patterns are in their incipient stages and just beginning to be felt in sex education which is still suffering from conservative paradigms of fear and abstinence.

The evidence in regard to abstinence models is intriguing. Apparently sex education programs do not impact the likelihood of sexual activity one way or another, but rather may actually increase the likelihood of contraception and hence affect pregnancy and STD transmission including HIV infection (Kirby 1985; Zelnik and Kim 1982 in Kelly 1991:337–340). This is expressed in the contrast between European sex education programs and those in the United States. According to Robert Francoeur (1991:125), European programs take for granted that adolescents are having sex and their approach consequently focuses on the issues of how to combat STDs and pregnancy. "Americans are mainly concerned with keeping teenagers from being sexually active and enjoying it" (Francoeur 1991:125).

Comparison and Contrast: Preparation for
and Transition to Adulthood

Adolescence is a culturally constituted phase associated with puberty. As we have discussed, whether a culture even acknowledges a period of adolescence differs, as does the length of time allocated to such a stage. Therefore, adulthood and the age at which we reach it also differs considerably. Despite the variability of how and when children reach adulthood, cultures provide mechanisms for the change of status. As we have seen, this occurs through rites of passage.

As we introduced earlier, Mead's study of Samoan girls' adolescence challenged our own western conceptions of adolescence as a period of strife due to the pubescent surge of hormones. Mead's study refuted this view in a controversial analysis that revealed a harmonious adolescence for Samoan girls. Mead was as interested in American adolescence as she was in Samoan adolescence and has provided some clues about the western adolescent experience at the time. Some of her ideas still ring true since the publication of her book, *Coming of Age in Samoa* in 1928, although parts of her conclusions are dated. For example, her frustration hypothesis with its obvious Freudian dimension is a questionable explanation. Mead's interpretation of why adolescence was such a torturous time for American youths rested on the idea that urges for sex were frustrated and suppressed by norms against teenage sex. This ignores other dimensions of adolescence as a period of growth. In the U.S. these norms were associated with the age of marriage which is ideally delayed until after graduation from high school. Mead felt that teenage sexual norms that allowed making out and petting just flamed a libidinal inferno that must ultimately be repressed. Teens expressed their frustration with this through rebellion and revolt (Davis nd:3). From this view, American culture emerged as a repressive one and Samoan culture was regarded as permissive because these sexual urges were not frustrated.

Like Mead, Spock (1985) has also accepted rebelliousness as a given for adolescents. He states: "It isn't often realized that the rebelliousness of adolescents is mainly an expression of rebelliousness with parents, particularly the rivalry of son with father and girl with mother . . . " (Spock 1985:503). In contrast to Mead, Spock regards rebelliousness as a powerful and positive force leading to the establishment of autonomy in the individual and ultimately to creative change in society. While the question of U.S. adolescent strife is interesting, there is danger in perspectives such as Mead's and Spock's when applied to a complex society such as ours. Dona Davis' (n.d.:4) comment here applies equally as well to Spock, although it addresses Mead.

> This . . . comparison of types . . . not only stereotyped sexual behaviors in non-Western societies, it stereotyped adolescent sexual behavior in our society. Researchers ignored the social complexities of sexual styles among adolescents as well as the many and various ways in which young people negotiate the rules of their culture to achieve sexual satisfaction.

While there is a widespread belief that adolescence is a period of turmoil for American teens, we must be careful not to overgeneralize. It is important to bear in mind the importance of ethnicity and socio-economic-status as factors effecting how adolescence is experienced and expressed. This is not to say that we cannot describe some of the patterns among American adolescents, but that we must remember these represent trends of usually the white middle class, and by so doing gloss over the variety of expressions of U.S. adolescence by different ethnic groups.

Many societies provide rites of passage for adolescents to help facilitate the transformation from the status of child to that of adult. As we have seen, the transition to a new position in society may be marked by new sets of rights and obligations as well as relations with people, including kin and non-kin. It may include new expectations and changes in the individual's identity as well. In preparation for this transformation, rights of passage participants undergo a journey through three characteristic phases: separation, transition or liminality, and integration. The phases of transition facilitate new learning and the development of identity components necessary for the new status. Ritual activities demarcate and actually facilitate the transformation of the individual's identity. They impress upon the novice the importance of the new status and what it means to be an adult man or woman (or perhaps some other option) in that society. By being separated socially and/or even geographically from their families, novitiates are given an opportunity to develop themselves as future adults. In addition, their families and others who have previously related to them as children may now regard them in their new status as they are reintegrated into society as adults. In such a way it is clear to the neo-adult what their position and place in society will be.

Let us contrast this with the experience of the U.S. teenager. Again, this description does not refer to the various indigenous and ethnic peoples in the United States that may have very rich rites of passage. For example, many Native American peoples maintain their traditional rites of passage for young females and males including the vision quest, and the Jewish bar/bat mitzvahs also provide critical recognition of life passage changes. As the western child undergoes puberty with its accompanying physical signs, are there any rituals or rites of passage that publicly recognize these changes on a cultural level? While individual parents may celebrate their daughters' first menstruation when that occurs, very often it is treated with secrecy and embarrassment. With the growth of male body hair and deepening of the voice, a father may acknowledge this with "You're a man now, son!" But what does that mean? The meaning is not spelled out nor are the markers of adulthood evident. Where are the ritual referents to know when adolescence is over and adulthood starts? How does an adolescent know when this is going to happen?

In the United States, generally the transition to adulthood is a diffuse one unmarked ritually. We live in a ritually poor society. While the symbolic aspects of adulthood are few, there are several societal markers that give an individual the legal status of adult as opposed to that of minor. This is known as reaching the age of majority and includes issues such as: the age at which one is considered a consenting adult, the age at which an individual may be married, and the age one may be tried in court as an adult. These vary state by state. Other events that may contribute to adult status include economic independence, marriage, and the birth of a child. But none is sufficient in and of itself as a clearly defined event that identifies the individual as an adult. In short, adulthood for many Americans occurs in an unintegrated way, in contrast to U.S. ethnic groups and nonwestern societies in which distinct symbolic referents for adulthood are expressed ritually or ceremonially. The western adolescent finds her/himself no longer a child, but certainly not an adult. They are betwixt and between youth and adulthood in a society that provides very little in the way of well-defined status markers. In fact, they often receive conflicting messages from society about their positioning in the stages from childhood to adulthood.

While initiation ceremonies are well known for their ritual ordeals, the novices know that at the end of these tests that they will be unequivocally declared adults. In contrast, U.S. adolescents generally have no tests or tasks that once accomplished will identify them unequivocally as women and men. Without rituals of transition to guide the adolescent on a journey into adulthood, social adaptation and transition to adulthood may breed areas of conflict and tension (cf Shapiro 1979:283). One of these is in the area of sexuality. In fact, Miller and Simon (1980:153) have described adolescent sexuality as "behavior in search of meaning."

The American teenager is kept in a liminal status of no longer a child but not yet an adult despite their biological maturity. In our society, adulthood is associated with sexual rights and until that time teenagers do not "own" their own bodies—they do not have the freedom to experience their sexuality until they are adults. This is buttressed as well by the legal system. Thus, U.S. teenagers may be fully functioning sexual beings, but they are not regarded as having rights to that sexuality. In this regard they are still children. Their sexual desires are not regarded as legitimate as adult ones.

This situation stems from a variety of sources, including a western sex negative or sex ambiguous worldview, a dogma that confounds sex with romantic love, definitions of adulthood, and conflicting attitudes about contraception and abortion. While adolescent sexuality is a complex issue, one dominant perspective regards teenagers as too immature psychologically to handle the sexual experience, although we have seen this is not the case in many societies that are not sexually restrictive. We must however, pay attention to the cultural context of the adolescent experience in America. The western view of sex is that it is not to be taken casually (Clement 1990:58). However, there is a double standard that allows males more leeway in this regard than females. With this in mind, Reiss (1967) has described the American standard as one of "permissiveness with affection." This standard is one which stems from the equation of love and marriage, only it is expanded to also include premarital couples. As a result sex is problematic for many American adolescents. We would like to note in conclusion, that research in the area of ethnicity and class is needed to create a fuller picture of the transition to American adulthood and the sexual issues encountered. We believe that the cultural factors in the construction of adolescence must also be included. Unfortunately with few exceptions, the majority of research in this area is limited to the white middle class (see Schmidt 1977).

Summary

Chapter 10: Puberty and Adolescence

1. Puberty is a physiological phenomenon, while adolescence is a cultural one that may or may not be coterminous with puberty.

2. Rites of passage are introduced as rituals that facilitate the transition from childhood to adulthood.

3. Rites of passage have three phases and distinct functions in non-industrial societies.

4. Theories of rites of passage are addressed. Female ceremonies are more common, but male ceremonies are more elaborate and severe.

5. The Sambia are discussed as an example of a rite of passage in which homosexuality is institutionalized as a phase.

6. Nonwestern sexual behavior among adolescents is presented in non-industrial as well as third world countries.

7. Western, particularly U.S. teenage sexuality, is reviewed.

8. The role of adolescent sterility and its relationship to sexual practices is described.

9. Sex education in the United States is analyzed.

10. The transition to adulthood in North America is contrasted with the transition in societies that have rites of passage.

OVERVIEW

Chapter 11

Topics in Adult Sexuality: Human Sexual Response and Birth Control

This chapter:

1. Discusses human sexuality in the cultural context.

2. Introduces the topic of nonwestern sexuality by examination of the Tantric model of human sexuality.

3. Presents western theories of sexuality.

4. Outlines major issues in sexual dysfunction and its cultural implications.

5. Places birth control practices in a bio-cultural context.

6. Examines western approaches to birth control.

7. Examines nonwestern approaches to birth control.

Chapter 11

Topics in Adult Sexuality: Human Sexual Response and Birth Control

Human Sexual Response (HSR)

How we experience and express our human sexuality is a physiological response that occurs within the larger context of culture. Human sexuality like other behaviors varies cross-culturally and is influenced by a variety of factors. These include: adaptation to the environment (how people survive-make a living); social organization including family and kin groups, the structure and complexity of the culture, political and economic organization including rank, class, and power differentials as well as gender roles and relations, and the ideological value system. All play a part in shaping sexuality. The meanings that sex is given is integrated in a particular society. How people express and experience their sexuality is linked to the wider ideological system, such as beliefs about reproduction, menstruation, and pollution. Finally our sexuality is ultimately part of much wider ". . . worldwide economic, social, political and cultural systems" (Ross and Rapp 1983:57). Sexuality is an integral part of wider cultural patterns and clearly articulates with these in many ways throughout the ethnographic spectrum. " . . . [I]t is embedded in a complex web of shared ideas, moral rules, jural regulations, obvious associations and obscure symbols" (Davenport 1976:117). Despite a wide array of diverse sexual practices, there are two limitations on the expression of sexuality. These are limits imposed by biology and the internal logic and consistency of a culture (Davenport 1976).

By presenting human sexuality in its broadest possible scope we hope to avoid our own western erotocentricity. We want to avoid the assumption that our own values about sex are the best or right way, and that everyone else's are either exotic, strange, bad or wrong; in short "other." Our intent is to see the exotic in ourselves and to come to understand that we westerners are not the only ones with theories of human sexual response, nor are we unique in thinking about sex and giving it meaning. We will find that our western ethnocentrism is like blinders, shaping what we consider normal and abnormal, good and bad. Ethnocentrism prohibits us from accepting variation for what it is, the expression of bio-cultural diversity.

The Cross-Cultural Record and Human Sexuality

The ethnographic record is filled with sexual behaviors both familiar and foreign to us. Cross-cultural evidence highlights not only the heterogeneity of sexuality but its equally dramatic variation in meaning, metaphor, and symbol. The meaning of sexuality or how cultures construct sexuality, is crucial for understanding the subject of sexology. Vance and Pollis (1990:2) note that a perspective in the tradition of cultural constructionism has:

> ... suggested that sexuality was not a biological given determined by organs and acts but a profoundly social product in which bodily sensations were linked to sexual acts, identities and meanings in ways that were fluid and changeable over place and time.

Unfortunately anthropological research is far richer on the subject of heterosexual rather than homosexual, and bisexual behaviors, although research by Gilbert Herdt, Walter Williams, and Evelyn Blackwood along with others has done much to rectify the situation. Sexual practices reported for heterosexuals, regardless of marital status include vaginal, anal, and interfemoral intercourse (the penis is placed between the partner's thighs), cunnilingus (oral stimulation of the vulva), fellatio (oral stimulation of the penis), masturbation, and mutual masturbation (Davis and Whitten 1987:73). Homosexual behaviors cross-culturally include oral and anal sexual practices, and mutual masturbation (Ford and Beach 1951; Herdt 1988; Gregersen 1983, 1994).

What is regarded as erotic in one culture may not be considered so in another. Kissing, for example, is not a universally recognized erotic behavior, e.g., the Japanese and Chinese did not have it in their sexual repertoire until western contact influenced them (Ford and Beach 1951). The meaning that humans give to their sexuality varies even in terms of what is regarded as sexual. For some people breasts are not erotic but are a place where babies feed.[1]

The Mangaians, a Cook Island group, provide an example of how sexuality varies cross-culturally. The Mangaians live in a culture where sex is valued positively. For example, prior to marriage, youths are encouraged to have as many partners as possible. As you recall this is not unlike Mead's report of Samoan adolescent girls. In their early teens and twenties, the Mangaian youth may average sex as many as 18 or 20 times a week (Marshall 1971). Sexual expression is highly elaborated in this culture through terminology and a focus on pleasure. There is an emphasis on female satisfaction and multiple orgasms given her by her partner. Romantic love is not necessarily linked to sex, which in contrast to the dominant western view, is regarded as *preceding* affection and love (Marshall 1971).

From this description it should be apparent how culture determines the meaning of sexuality. Mangaians have a very strong and positive tradition of sexology transmitted through socialization and oral traditions of sex education. They are not unique among the world's cultures in their development of a positive and elaborate approach to sex. For example, India has a very old and complex tradition of sexology exemplified in the Kaama Suutra (*The Precepts of Pleasure by Vaatsyaayana*) circa A.D. 200–400 (Gregersen 1983:32). The Kaama Suutra and others like it are in a genre westerners refer to as sex manuals. Yet, we cannot conceive of these in the same way as our own western sex manuals. For the westerner science has replaced the sacred in giving us a "god's truth" about human sexuality.

In contrast to western sexology manuals, "the Kaama Suutra . . . was considered a revelation of the gods" (Gregersen 1983:32). This text described 529 possible positions for sex, categorized women by depth of the vagina and men by penis size, as well as gave helpful hints such as how to enlarge the penis (Gregersen 1983:55, 70, 94). The rather extreme number of positions was arrived at by assigning minor differences in detail to new positions (Gregersen 1983:55; 1994:62). The Kaama Suutra was not a taboo or deviant text, but was a widely accepted one. Hindu ideology provides an approach that contrasts sharply with the western one. Sex is regarded, for the most part, positively in Hindu religion as opposed to negatively in the Judeo-Christian tradition (Gregersen 1983:215; Francoeur 1992:1–8).

Hinduism is a complex and heterogeneous religion with many sects and different trajectories as well as diverse views on sexual practices evolving over 3,500 years. Several sects specifically regard sexuality as a means by which one reaches spiritual union with the higher powers of the universe. We shall focus on one of these sects, Tantra. Tantric traditions occurred prior to Hinduism and Buddhism and were subsequently adopted into these religions (Francoeur 1991:5). Tantra schools that occur in Buddhism have their own unique flavor (Devi 1977; Gregersen 1983, 1994; Francoeur 1992:1–8).

While Tantrism flourished in India in the eleventh and twelfth centuries, its conflict with general Hindu asceticism meant that Tantrism stayed a marginalized philosophy within broader Hindu traditions. However, Hindu ascetism is different from the Christian in that it coexists with attitudes valuing sex in moderation and the "religious power of sexual union" (Francoeur 1992:2). Additionally, Tantrism went against other Hindu beliefs including eating meat, having sex when a woman was menstruating, and the violation of caste prescriptions in certain of their sex rituals (Devi 1977:12–13).

The Tantric view of sexuality perceives sex as a religious experience. Sex is regarded as a vehicle to transcendence and is a meaningful component in the religious system. *Maithuna*, or sexual union, is the way for people to try to approximate what it would be like to merge with the sacred. This contrasts sharply with the western mind-body dualism inherent in Christianity in which the ultimate sacred experience is chastity; and sex is relegated to the world of the profane. Ideally, in its most extreme form, sex is tolerated in Christian heterosexual marital intercourse if it is in the interests of procreation. The Hindu tradition also paradoxically stresses the virtue of celibacy but for different reasons. In this case it is because of concerns over ritual pollution and the belief that semen is a source of male health and must be controlled. This ascetic aspect of Hinduism is part of some Tantric approaches although it is generally antithetical to the philosophy. It must be remembered that Hinduism also includes sexual love and pleasure as one of four goals on a journey to spiritual knowledge (Francoeur 1992:2; Garrison 1983).

The Tantric sect follows the teaching of books called Tantras. The earliest of these is considered part of the Hindu sacred text, the Vedas. In the Tantric view, sex is far more than mere biological satiation and includes the process whereby males and females can achieve a sense of common humanity or "communitas" (cf. Turner 1969). Through sex men and women have the possibility of transcending their gender and recognizing those qualities of the other gender in themselves. Through this process, the individual has access to her/his spirit being. In this regard:

> A doctrine of primal androgyny also pervades the Tantric approach: a male seeks out a female, and the female a male, because they do not know that the opposite sex is lodged within their own being. To realize this fully is important for the future life (Gregersen 1983:222).

The power of the universe is expressed through static inertia regarded as female and dynamic inertia considered as male. This spiritual power is created through the union of these opposite cosmic forces into a primal energy that ultimately can lead to spiritual perfection (Garrison 1983:7). This "unifying principle of being [is called] *prana*" (Devi 1977:14).

In comparison to western concepts of intercourse which usually consist of approximately two phases, foreplay and sex, the Tantric texts view sex as occurring in multiple phases: beginning with having one's thoughts dwell on it, keeping company with the opposite sex, flirting, intimate conversation, desire for coitus, firm determination, and physical copulation. For Tantrics sexual energy is a force that engenders religious ecstasy. "Tantric worship is through the flesh, with body and mind" (Devi 1977:16). This is nowhere as apparent as in the *Panchattattva* sex ritual, used in Bengali. This sex ritual exemplifies the path to oneness that is achieved through the mystical elements of coitus. An entire procedure, describing environment, breath, body attitude, mental attitude, time, and technique defines this sacred experience of coitus.

> **ASIDE**: Participants in the Rite of the Five Essences, a Tantric love ritual, for instance, use the five forbidden M's: wine or *madya,* a meat or *mansa,* fish or *matsya,* parched grain or *mudra,* and sexual union or *maithuna* in a kind of holy communion. This love ritual begins with an enhancement of the environment, with flowers, incense, music, and candlelight. The couple bathe and massage each other with fragrant oil. After alternate-nostril breathing and a period of meditation designed to hasten the ascent of the vital energies of the *kundalini* (see below), the couple chant a Mantra and envisage themselves as an embodiment of *Shiva* and *Shakti,* the supreme couple. With the woman on his right, the man kisses and caresses her whole body, from feet to head and back to the toes. The woman then arouses the male with caresses and kisses all over his body. Finally, after the woman moves to the left of the male, the couples move through a series of coital positions until each experiences the "transcendental power of love."

Usually, the male refrains from orgasm in order to retain the vital life energies of his semen (Francoeur 1991c:6–7).

This religious view of sexuality may be compared to the general tenants of Christianity. St. Augustine of Hippo is credited with influencing western Christianity with a negative view of sexuality. St. Augustine interpreted the story of the Garden of Eden as the site of original sin. Because Adam and Eve defied authority through the discovery of sex, that sin was passed on to all humankind. While sex was viewed as necessary for procreation, and hence not immoral, sexual desire and arousal were thought sinful. The story of Adam and Eve provides a framework for regarding human nature and a rationale for regulating sexual behavior, gender relations, and marriage. Celibacy came to be the exalted state as opposed to the Tantric view of sex as sacred. This is because the original sin of sex meant a fall from grace and was "antagonistic to spiritual liberation." "Christianity, for the most part, has not been able to integrate sexuality into a holistic philosophy or see sexual relations, pleasure, and passion as avenues for spiritual meaning and growth" (Francoeur 1992:7).

Western Theories of Sexology

Twentieth century western theories of human sexual response (HSR) have undergone modification in tandem with societal changes, particularly in the areas of gender roles. In this section we shall focus on several of the major sexologists of our time and their influence on the study of human sexuality. The history of western sexology testifies to shifting definitions of human sexual response over the course of time. Our discussion will emphasize twentieth century thinking, although these perspectives were obviously influenced by earlier ideas on the subject.

Generally speaking, sexology was dominated by the Freudian view of human sexuality up to and through the challenges offered by Kinsey and later researchers. The major changes in the twentieth century thinking have been on female sexuality. The Freudian view highlighted women's passivity and distinguished between two kinds of female orgasms, clitoral and vaginal. Men's sexuality was regarded as naturally important, active, assertive and dominant in contrast to women's. Subsequent research by Kinsey, Masters and Johnson, and Whipple and Perry questioned this view by concentrating instead on women's capacity for sexual arousal and pleasure. What emerged was a reformed sexology that centered on the similiarities in women and men's sexuality rathar than the differences. Women and men came to be considered as more alike in their respective sexual responses than different. Women were no longer regarded by western sexology as sexually passive and unresponsive. This theme of similarity continues to dominate the modern western view of human sexual response today.

The sixties marked an era of massive social and sexual unrest.

> Put briefly, men changed their sexual behavior very little in the decades from the fifties to the eighties. They "fooled around," got married, and often fooled around some more, much as their fathers and perhaps their grandfathers had before them. Women, however, have gone from a pattern of virginity before marriage and monogamy thereafter to a pattern that much more resembles men's . . . (Ehrenreich et al. 1986:2).

The sexual revolution was officially on its way in 1966/67 with media acknowledgment of it as a "social phenomenon," although scholars argue whether there was indeed a revolution (see chapter 12). According to Ehrenreich et al. (1986), the sexual revolution preceded the women's movement by two or three years. Both movements converged in the late '60s around women's desire for sexual freedom and for equality. This called for a change in gender relations and power inequity. It included the separation of sex from marriage and reproduction enhanced by the widespread availability of the pill beginning in 1960. Reports from the 1960s indicate that middle class women, both black and white, were engaging in premarital sex (Ehrenreich et al. 1986:40).

The meaning of sex was also under revision as well. Masters and Johnson (1966) demonstrated that women's sexuality was more extensive than previously thought, so that sex came to include more than just coitus. The discovery of the role of the clitoris in female sexual response was certainly one factor in this new view. Definition of sexuality now included additional behaviors such as cunnilingus which was only whispered about in the fifties. Ehrenreich et al. (1986) describe the de-medicalization of sex that occurred along with this expanded approach toward female sexuality. As Freudian theory, which dominated the medical perspective began to lose its foothold, numerous "pop" books on the subject of sex appeared. Sex experts and their popular books sprung up everywhere including J's *The Sensuous Woman* (1969), Comfort's *The Joy of Sex* (1972), Friday's *My Secret Garden* (1974) and Hite's *The Hite Report* (1976) among others.

These books represented more humanistic trends that reported on and offered sexual advice as the changing meanings of female sexuality were felt in the bedroom (Ehrenreich et al. 1986). These popular accounts were not scientific and did not include principles of sound research methodologies, but, in all fairness, that was not their purpose. They represented the "democratization" of sex, a popular genre that co-existed with newly emerging medical perspectives of the 1960s such as that of Masters and Johnson (1966).

Freud

Freud's contribution to the study of human sexuality affected western sexological thinking up through the 1950s. As we have discussed, Freud focused on phases of childhood and adolescence that were organized around his theory of psychosexual development and the Oedipus complex. Sexuality is the cornerstone of psychoanalysis as Freud conceptualized it. Freud's position was that male sexuality was the norm while female sexuality was passive and problematic. He identified two kinds of female orgasms. The first kind was the clitoral one that was related to early experiences with masturbation. It was considered an immature kind of orgasm. As a woman developed, the focus of her orgasm shifted from the clitoris to the vagina. The only mature orgasm was the vaginal one which was associated with female reproduction. According to Freud, women who experienced sexual pleasure in ways other than penile penetration were immature in their sexual development, and fixed in an earlier phase of development (Byer and Shainberg 1991:186).

Kinsey

Alfred Kinsey and his colleagues published the Kinsey Reports on male sexuality in 1948 and female sexuality in 1953. Their work had a profound impact on both the scientific community of sexologists as well as American's conception of sexuality. While their contributions are many, we shall highlight a few of the most significant findings. Kinsey and his colleagues concluded that male and female sexuality were not different at all (Ehrenreich et al. 1986). Their research in the 1950s was undertaken at the time of the postwar baby boom noted for the return to traditional gender roles. Their determination that male and female orgasms were alike was in direct contradiction to the Freudian position discussed previously.

The Kinsey approach to human sexuality was a highly medicalized and scientific discourse in which the affective component of sex was unacknowledged. Betty Friedan, an early feminist and author of *The Feminine Mystique* (1963), criticized the Kinsey

reports for just this reason stating that sexuality was presented
"... as a status-seeking game in which the goal was the greatest
number of 'outlets,' [or] orgasms" (Friedan 1963:263 in Ehrenreich
et al. 1986:43). Ehrenreich et al. (1986) suggests that Kinsey's
dedication to a scientific sexology that could be quantified had a
profound influence on American's conceptions of sexuality. It shifted
the focus of sex to the number of orgasms, which allowed Kinsey
and his colleagues to consider a wide variety of other behaviors
besides married heterosexual coitus, e.g., homosexuality, masturbation, premarital and extramarital affairs at a time prior to the
changes subsequently encountered in "sixties" sex.

Masters and Johnson

   William H. Masters and Virginia E. Johnson succeeded Kinsey
in the development of a scientific sexology. Their interest went
beyond Kinsey's to bridge the time gap since his work and to provide data on physiological responses during sexual stimulation
(Masters and Johnson 1966). Masters and Johnson's conclusions
were garnered from their massive clinical research effort which
resulted in the publication of the now famous *Human Sexual
Response* (1966).
   Masters and Johnson are as well known for their research
methods as they are for their findings. The participants in their
study included 382 women from 18 to 78 years old and 312 males
from 21 to 89 years old, both married and unmarried (Masters and
Johnson 1966:12–13). Although their contribution to sexology was
significant, their research had limitations. It was over-represented
by subjects with formal education and was biased by class, economic privilege, and race. Of the nonwhite population only 11 family units were African-Americans (Masters and Johnson 1966:12–15).
Finally, homosexual data, while collected, was not presented in
their book, *Human Sexual Response* (1966).
   The research setting of the laboratory allowed for direct observation and measurement of physiological changes and responses
during a variety of sexual activities including manual and mechanical manipulation and intercourse in different positions (Masters and Johnson 1966:21). Their clinical and high tech methods
included artificial coital equipment, e.g., artificial penises equipped
to measure female sexual response, as well as various kinds of
monitors to measure genital and other physiological responses. The
clinical procedure was enhanced by in-depth interviews.
   Masters and Johnson documented four phases in the human
sexual response cycle (HSR): excitement, plateau, orgasm, and the

resolution phase (Masters and Johnson 1966:4). Variations during this four-scheme model differed by gender. Male response cycles varied among individuals primarily along the dimension of duration of the response, while females differed by duration and intensity.

The **excitement** phase for both males and females is characterized by engorgement of the blood vessels (vasocongestion) in the pelvic area and increased muscle tension. Lubrication of the vagina occurs along with the swelling of erectile tissues including the clitoris, labia, vaginal opening, and penis during this phase. The **plateau** stage precedes orgasm and its duration may vary. For males, full erection and engorgement of the testes are reached. For females, the vagina continues to expand and the uterus continues to elevate—both continuing from the excitement phase. During plateau, the **orgasmic platform** is reached. This is a process in which the vaginal opening is reduced in size because of engorgement of the surrounding erectile tissues (King et al. 1991:65–66; Byer and Shainberg 1991:182).

Orgasm marks the third stage. The female orgasm is characterized by three to twelve contractions at .8 second intervals (Hyde 1985:278; Francoeur 1991:182). Female orgasm lasts for 13–51 seconds while male orgasm is from 10–30 seconds (Bohlen et al. 1982 in Francoeur 1991a:182). Male orgasm is usually but not inevitably associated with ejaculation, although before puberty, in old age, and with some erectile dysfunctions, it is possible to experience orgasm without ejaculation. Orgasm and ejaculation are physiologically distinct. Reports of multiple orgasm in men are still controversial and undoubtedly clouded by definitional interpretation. Self-reports of multiple orgasm in Kinsey et al. (1948) are defined as occurring during one sexual encounter. In the Kinsey data, 15–20% of the teenage boys reported multiple orgasms with only 3% of the subjects claiming this ability after 60 years old. A number of sexologists accept the possibility of multiple orgasms, although Masters, Johnson, and Kolodny (1988:94) maintain that once ejaculation has occurred it is not possible for a man to have multiple orgasms (also Francoeur 1991a:184, 188, 190).

The final phase of human sexual response is that of **resolution**. This is the process where blood is released from the engorged areas, muscle tension is relaxed, and the body returns to its previous state. For the female, this process may last up to half an hour if orgasm occurred, or an hour if only the plateau stage was reached (Masters and Johnson 1966).

Masters and Johnson provided the evidence that discredited the Freudian view of two discrete orgasms for women. According to

Freud, a woman was doomed to vaginal frigidity if she could not make the transfer from the childish clitoral orgasm to the mature vaginal one. They documented that the clitoris was central in female orgasms whether from indirect stimulation during coitus or direct stimulation. According to Masters and Johnson, there is no purely vaginal orgasm. Ehrenreich (1986) points out that this was not really news since earlier research demonstrating this had been available "for decades," although the psychiatric doctrine had been tenacious in sticking to the Freudian view.

In addition, Masters and Johnson verified women's capacity for multiple orgasm. They challenged the myth that women's orgasms were like men's, single and requiring a refractory period of up to several hours before they could re-experience orgasm. As early as 1953, Kinsey et al. had reported that 14% of the female population were capable of multiple orgasms. Masters and Johnson found that women's multiple orgasms were no different than their single ones.

Masters and Johnson's detractors criticized their research for the same reason as Kinsey's: they had " '. . . reduce[d] human sexuality to physical responses,' though, of course, only physical responses are accessible to quantitative measurement" (Ehrenreich 1986:66). This model of human sexuality is androcentrically biased with its focus on penetration and orgasm as well as one dimensional in scope (Francoeur 1993:72).

Kaplan

Helen Singer Kaplan has written extensively on the subject of human sexuality (e.g., 1974, 1979, 1983, 1989). One of her many contributions was the modification of the four stages of human sexual response described by Masters and Johnson. Kaplan reconceptualized these as two stages: arousal and orgasm by lumping the excitement and plateau phases into one. She added a precursor stage to arousal that she called "desire" (Kaplan 1979). Her work on sexual dysfunction, *Human Sexual Inadequacy* (1974), is of particular interest since it represented a synthesis of the behaviorist approach of Masters and Johnson to sexual dysfunction with psychoanalysis. While Masters and Johnson's therapeutic model focused primarily on the symptoms, Kaplan's integrated concern for interpersonal interaction along with an emphasis on the unconscious as well (Ficher and Eisenstein 1984:143; Kaplan 1979).

Kaplan has also proposed that male sexuality and a female sexuality are distinctive. Subscribing to theories of the biological basis of sex differences, Kaplan believes that these different sexualities are due to testosterone, accounting for what she regards as a much stronger sexual drive in males than for females. In contrast, the female sex drive is more shaped by lived experience. Kaplan regards humans as monogamous pair bonders, although she leaves open the option for serial pairbonding (in Klein 1981:73–75, 77, 92).

Singer and Singer

In 1972, Singer and Singer offered another model of female sexuality that included three kinds of orgasms: the vulval orgasm, described by Masters and Johnson, the uterine orgasm, and the blended orgasm. The blended orgasm combined the vulval and the uterine types. Singer and Singer arrived at this model from evidence that it was possible for women to experience orgasm without "vulval contractions" or in Masters and Johnson terminology—contractions of the orgasmic platform (Singer and Singer 1972:256). Their research seriously challenged Masters and Johnson's clinical research that documented the vulval orgasm as the zenith of orgasm. Singer and Singer concurred with Masters and Johnson that the **vulval orgasm** is a result of either direct or indirect clitoral stimulation. They, however, described another type of orgasm they labeled the **uterine orgasm**.

> The "uterine orgasm" does not involve any contractions of the orgasmic platform . . . this kind of orgasm occurs in coitus alone, and it largely depends upon the pleasurable effects of uterine displacement. Subjectively the orgasm is felt to be deep, i.e., dependent on repeated penis-cervix contact (Singer and Singer 1972: 259–260).

This orgasm is characterized by interrupted breathing at which time orgasm and the expulsion of breath occur simultaneously (Singer and Singer 1972:260). Women who do not have their uterus may also experience this kind of orgasm (Byer and Shainberg 1991:187).

The **blended orgasm** combined characteristics of both the vulval and the uterine. It incorporated contractions of the orgasmic platform, but is experienced as deeper than a vulval orgasm and more akin to the uterine in that breathing is interrupted (Singer and Singer 1972:260).

Whipple and Perry—The "G Spot" Controversy

The **Grafenberg or G spot** is located approximately one or two inches inside the opening of the vagina on the anterior wall about halfway between the outer labia and the cervix (Francoeur 1991:148, 153). It is named after the 1950 report of Dr. Ernst Grafenberg, a German doctor, who first located and described this small area that is about the size of a dime (Ladas, Whipple, and Perry 1982:33). The G spot is an area of sensitive tissues that when stimulated may lead to orgasm. In 1982, Ladas, Whipple, and Perry's research and subsequent book, *The G Spot and Other Recent Discoveries About Human Sexuality* (1982), began a controversy. Ladas, Whipple, and Perry (1982) argue that the G spot is universal among females. However, more recent research indicates that the G spot may not be universal but present in only about 10% or less of the population (Alzate and Londono 1984; King et al. 1991:34). The G spot is of importance since Ladas et al. contended it is associated with its own unique kind of orgasm.

Ladas, Whipple, and Perry also suggested that there are three kinds of female orgasms. One is the **tenting** type of orgasm related to clitoral stimulation. This is similar to that described by Masters and Johnson in which there occurs an "orgasmic platform" in which the vaginal entrance constricts. Tenting refers to the effect when " . . . the inner portion of the vagina often balloons as a result of the lifting up of the uterus inside the abdomen" (1982:144–145). The **A-frame** orgasm is quite different than this variety and is described as a consequence of stimulation of the Grafenberg Spot. This orgasm has no orgasmic platform, rather "the vaginal musculature relaxes and the entrance opens" (Ladas, Whipple, and Perry 1982:144). In addition, the tenting is absent so that ". . . the uterus seems to be pushed down and the upper portion of the vagina compresses" (Ladas, Whipple, and Perry 1982:145). A third kind is referred to as the **blended** orgasm (Ladas, Whipple, and Perry 1982:151). This orgasm focuses on the pubococcygeal muscle, as in the tenting type of orgasm and the A-frame type where the trigger is the G spot, but the focus of muscle response is the uterus (Lada, Whipple, and Perry 1982:150).

The G spot research is important for another reason. Perry and Whipple (1981) have proposed that women can ejaculate a substance very similar to seminal fluid during orgasm. They hypothesized that between ten and twenty percent of women have this capacity. Perry and Whipple have proposed that the G spot was the source of that ejaculate; however, other researchers have contested this (e.g., Goldberg et al. 1983). Analysis of the ejaculate by Belzer et al. (1984) found the same enzyme produced by the male prostate, but noted that its chemical structure varied. Goldberg et al. (1983) found no difference between female ejaculate and urine.

The G spot and its role in female ejaculate must be researched further before conclusions may be drawn.

ASIDE: Because this discussion has focused on the clinical sexological aspects of human orgasm, the "nuts and bolts" of orgasm have been emphasized by researchers at the expense of the playful, liminal and wondrous. College students from one of your author's Human Sexuality courses offer descriptions of orgasm experiences as well as their ideas of what constitutes good sex.

DESCRIPTIONS OF ORGASM EXPERIENCES

It's great! It's the one time when every nerve in your body seems alive.

It was not until two years ago that I experienced orgasm on a regular basis. The first one happened while I was fully clothed and "dry humping." I was so excited that I started laughing hysterically. The guy thought he did something weird. It is an intense sensation that I can feel in my eyelashes!

It felt real good. It's one of the best feelings in the world. Sometimes it makes me get up and scream or if I can maybe knock down a wall or something.

It feels as though someone is tickling me with a feather. I can feel myself about to explode at times. You feel it coming on, the mind gets blank and the stomach feels light and then, let's just say

DESCRIPTIONS OF CONDITIONS FOR GOOD SEX (PARTNER, MOOD, ATMOSPHERE)

Partner that is responsive and likes to have fun with it. Not always serious.

Atmosphere can be anywhere because sex is fun everywhere. I prefer being alone with my boyfriend at home or in the park. Anywhere we feel close and intimate or just horny!! The craziest place was on the top floor of a parking garage!

Being sexually attractive, mood I'm in or the atmosphere as long as strangers are not around. If friends are around, it doesn't matter.

Television on with lights dimmed. Start by cuddling and kissing.

I must be alone in the bedroom with my girlfriend. I usually have to have the radio on. We usually have the lights out but a night-light on.

good. After sex I feel really close with my girlfriend.

I usually need my partner to stimulate me with his fingers or tongue in order for me to orgasm. I have never had a multiple orgasm and I am trying to achieve orgasm just from sex.

While you're having sex it's as if you never want to stop, as it keeps feeling better until finally . . . Complete satisfaction. Once the orgasm has been reached there is a feeling as if I am at peace with myself.

Hard thing to remember after it has happened. At first it builds up and then it just feels like you let loose— not physically, but more emotionally. I usually scream very loud, it seems uncontrollable. I cannot just orgasm, need partner to try to make me and give effort.

A very tense, contractive explosion! My body freezes and then melts in a matter of seconds.

Feels like your body is tensing and feels like you're losing sensation of things around you.

Tensing and relaxing—Great stuff.

My condition for good sex is a partner I am highly attracted to and feel comfortable with— yet not too comfortable to be totally used to. I like hours of foreplay leading up to sex and I like sex to be in the heat of summer or on a really cold day.

Spontaneous mood with a partner who really enjoys having sex with you. Some kind of rainstorm happening outside with loud slow music would definitely add to the atmosphere.

If it felt good, it usually was good—I like it loud and wild!

The right frame of mind!

Love. Great sex, in my opinion is practiced and not reached the first time, so I feel I must love my partner and appreciate their view and ideas of what is good sex.

No real particular conditions. Afternoon is always fun. Must both be in the mood. Romantic exciting mood is good too. Spontaneous.

Problems in Sexual Response

Sexual dysfunctions are physiological symptoms that are shaped by culture. This is not to say that the causes of sexual dysfunction are only biological. Sexual dysfunctions may have their origin in the unique psychology of the individual or may be responses to cultural tensions and stresses around the issues of sexual performance and gender relations. Men and women will occasionally experience problems in sexual functioning and orgasm. However, culture may play an important role in the patterning of sexual dysfunctions as the following case study illustrates.

Donald Marshall's study, "Sexual Behavior on Mangaia" is frequently cited as an example of a society in which sexuality is not repressed but rather celebrated. The activities of these Cook Island Polynesians have gained reknown. Premarital and extramarital affairs are reported as the norm. Sex is given much attention and is positively affirmed so that

> "... the average Mangaian youth has fully as detailed a knowledge—perhaps more—of the gross anatomy of the penis and the vagina as does a European physician" (Marshall 1971:110). Less than one out of a hundred girls, and even fewer boys—if, indeed, there are any exceptions in either sex—have not had substantial sexual experience prior to marriage (Marshall 1971:117).

Sex is not associated with love or marriage. It is believed that orgasm must be learned by women with the help of "the good man," and the expectation is that a woman will have two or three orgasms to the male's one (Marshall 1971:122).

Yet, all is not sexual paradise on Mangaia. The culture is pervaded with a "patterned ambivalence" (Marshall 1971:110). A rigorous division of the sexes occurs between four and five years old. In contrast to the emphasis on the sex organs (a variety of terms exist for describing male and female genitalia), is an extremely rigorous code of modesty. This emphasis on modesty is in sharp distinction to the sensuous dance and sexually explicit and ribald folktales of the Mangaians. In addition, according to Marshall (1971:111) "... [I]ntricate incest prohibitions are contrasted with the restriction of most social contacts to those that take place only between kin ... " Evidence suggests that this patterned ambivalence is tied to **tira**, a form of male impotency, illustrating bio-cultural dimensions of sexual dysfunction. This is a highly feared condition. Unfortunately, Marshall's research is primarily on male sexuality so that comparable information of female sexual dysfunctioning is not given.

Tira has four stages in its development. Stage one, **tiraora,** "lively penis," is characterized by high sexual interest and vigor. Stage two, **tiramoe,** "sleeping penis," refers to impotency, the inability to have an erection and lack of sexual desire. If untreated, young and old men alike will enter stage three of **tiramate,** "dead penis." In this stage erection and ejaculation are both impossible. **Tiran aro,** "lost" or "hidden penis," marks the fourth stage. In this stage the penis has "retracted into the body" and death will result if treatment does not occur (Marshall 1971:156). Treatment includes heated herbs and smoke therapy for the genitals. *Tira* is said to be common and is regarded as very dangerous. While Marshall could not collect data as to its actual frequency, birth data were indicative: 29% of men over 18 (n = 281), and 23% of women over 18 (n = 164) had no progeny. Out of a sample of people over 50 years of age (n = 129), 15% of the men (n = 78) and 20% of the women (n = 51) have not had children. Marshall's explanation is a bio-cultural one, which includes dietary factors along with performance pressure for men. Patterned ambivalence around sexuality may be a powerful factor in fostering *tira.*

Ambivalence toward sexuality may be found in both sex negative as well as sex affirming societies, since both restrictiveness and permissiveness are terms that include a complexity of variables. These occur on a continuum of attitudes and values within and between societies. For example, Mangaia is a sex affirming society in the sense that sex is not restricted premaritally and is generally regarded positively. Yet as previously mentioned, this is contrasted with gender segregation and norms of modesty. The United States has been labeled a sex negative culture, but we regard it as sex ambiguous. This approach considers both the negative and positive cultural dimensions of sexuality that give mixed and conflictive images to the individual. The changes in sexual norms and values as a result of the 1960s has certainly made some inroads into the negative estimation of sex in this society. Bumper stickers stating "make love not babies" signaled a changing of the guard in terms of sexual attitudes. This was enhanced, but not caused by, the widespread availability of contraceptives for women. However, the 1980s and early 1990s have seen a return of sexual conservatism given impetus by a decade of political conservatism and fear of AIDS. Amidst these changes ambivalence toward sex is rampant. The double standard for sex, while loosening up dramatically in the 1970s, has not changed much qualitatively from the 1950s. College sexual activity for women has in fact leveled off in the 1980s to about 50–60% for women (Leo 1984:77). This reflects

the "conditional double standard" " . . . males are allowed more freedom than females to engage in premarital sex, but females are
permitted to be sexually active as long as they are in affectionate
relationships (Reiss 1960 in Sprecher et al. 1991:87). The Clinton
administration with its more liberal approach will undoubtedly
impact the sexual climate in the U.S. through policy.

Everywhere one turns the media is displaying sex; in rock
videos, soap operas, weekly prime time television, popular maga-
zines, and of course in advertising. At the same time, there are
nationwide efforts to convince the young to remain celibate until
marriage—the "just say no to sex" approach of the 1980s era.
Monogamous sex is touted as the antidote to AIDS, yet rock videos portray images of young men with entourages of young women.
While the media portrays images of sexually provocative young
women in advertising, women are encouraged to repress their
sexuality. With the advent of the companionate marriage, this
same population is expected to unleash this repressed sexuality
with their spouse. Without belaboring the point further, ambivalence about sex may be generated by cultures that affirm and/or
sanction sex. Since few cultures are purely sex positive or negative, this ambivalence represents an assortment of rules around
the expression of sexuality throughout the lifecourse and in different contexts.

In the West, sexual dysfunctions ironically embody this ambivalence in two prevalent dysfunctions identified by the clinical
sector: sexual addictions and those persons experiencing low levels
of sexual arousal known as ISD or inhibited sexual desire (or
hypoactive sex desire). It has been estimated that as many as six
percent of Americans are sex addicts who use sex in the same way
the alcoholic uses alcohol. Such individuals may engage in sexual
behaviors that place them at risk for STDs. However, sociologists
Martin P. Levine and Richard Troiden (1988:347–363) warn that
the creation of sex addiction as a disease category represents the
convergence of two powerful forces: medicalization and politicalization. The erotically non-normative are labeled deviant and mentally ill and may then be treated so that they will conform to
dominant norms. Psychologists counter that sex addicts want and
need help with this serious psychological disorder.

According to some reports, inhibited sexual desire is also a
growing disorder in the client population of therapists from the
1980s through 1990s. This problem is clinically defined by disinterest in sexual gratification (Ficher and Eisenstein 1984:147). Some
therapists report that as high as 30–50% of their clients experience

this problem. Harold Lief has suggested that "20% of all adult Americans are afflicted with ISD's" (Leo 1984:83). Some sexologists attribute this to fallout from the sexual revolution of the 1960s. Caplan and Tripp regard sexual taboos as basic to sexual excitement, so that as the last sexual taboos were broken during the '60s and '70s so was the excitement associated with the forbidden (in Leo 1984:83). Irvine (1990) offers a socio-cultural analysis. Irvine regards ISD as "the medicalization of a simple non-pathological product of sexual boredom or indifference, due to the widespread problem of flagging sexual interest in marriage" (in Davis ND:13).

According to some studies, sexual functioning problems are widespread in the U.S. One study of one hundred happily married couples reported the following: 40% of the men experienced erectile or ejaculatory dysfunction, 63% of the women had arousal or orgasmic dysfunction, 50% of the men and 77% of the women complained of loss of sexual interest or an inability to relax (Goldberg 1985:147). According to Goldberg (1985), our cultural conditioning undoubtedly creates a situation in which "sex is at cross-purposes." The gendering process of sexual socialization creates mismatches between the sexes. Men are socialized to experience sex as conquest with the goal orientation of orgasm. This has been described as a recreational model of sex. Women are socialized to experience sex as intimacy and as a process of achieving closeness. This is termed a relational model of sex. Mismatches also occur in the gender roles. As the initiator, the male carries the entire responsibility for making sex fulfilling and, as a consequence, will experience guilt and resentment over this situation. In tandem, the female, in her role as reactor, comes to feel anger over being sexually controlled. Goldberg suggests other cultural oppositions occur as a result of different gender role expectations. These mismatches include cultural expectations of male freedom and female desire for monogamy, and in the role dichotomy of male as animal-female as madonna. In addition, the expression-repression mismatch is one in which males learn to regard sex as a "preoccupation" and females see "sex . . . [as] . . . something that may be enjoyed under special conditions of commitment and intimacy but, at the same time, something a women feels she can do without if the circumstances are not right" (Goldberg 1985:147).

While these trends may be exaggerated and not representative of the heterogeneity nor of ethnic diversity in sexual attitudes of the nineties, there is still some truth to these as cultural ideals for many middle class Caucasians. Much sexual dysfunction is due to the anxiety, guilt, and repression spawned by inequalities between the sexes and gender role disparities. Paradoxically, while the sexual revolution and the new age of sex research may have tempered some of these problems caused by gender role differences, a host of new fears associated with performance anxieties have arisen.

The anthropological perspective regards not only the cultural context in which sexual dysfunctions occur as worthy of interest, but the numerous ways in which the medical and mental health sectors define, label, categorize, diagnose, and treat sexual dysfunctions as well. For example, the terms we have for describing human sexuality such as "dysfunction" and "dyspareunia" focus on difficulties and pain. Medical terms incorporate a disease model and concentrate on the dysphoric aspects of sex rather than the euphoric. The clinical approach to diagnostic categories for sexual disorders has been critiqued as ethnocentric, classist, and sexist by an ample body of socio-cultural literature (Davis ND:1).

"Sexual dysfunctions can be defined as the various inabilities to experience sexual arousal to orgasm with a partner in a smooth, well-integrated manner" (Levine 1981:418 in Ficher and Eisenstein 1984:147). Problems may occur throughout the course of human sexual expression. Female dysfunctions among western women include the following: anorgasmia, dyspareunia, and vaginismus. Anorgasmia refers to lack of orgasm and sexual responsiveness, but the term "pre-orgasmic" is now supplanting this terminology. Primary orgasmic dysfunction is found in women who have never had an orgasm. Situational orgasmic dysfunction occurs in certain contexts but not others (Masters and Johnson 1970). Dyspareunia is defined as painful intercourse. This condition, as well as a low level of sexual desire and sex addiction, affect both males and females (Ficher and Eisenstein 1984). Vaginismus refers to the involuntary spasms of muscles around the vagina so that penetration is very painful. Kaplan and Owett (1993:3–24) have identified the "Female Androgen Deficiency Syndrome" substantiating the 1959 study of Waxenberg et al. that the loss of androgens in women can lead to loss of sexual desire, decrease in orgasms, and loss of orgasmic ability. This phenomenon was described among women who had undergone chemotherapy or oophorectomies (removal of the ovaries). Testosterone replacement therapy is the prescribed treatment. Although Sildenafil (Viagra), the very successful medication in the treatment of male impotence, has not yet been tested on women, it may prove beneficial in treatment of women's sexual dysfunction. ("In the Public Eye" 1998:35).

Males encounter difficulties with early ejaculation in which they cannot control and delay when they ejaculate. Delayed ejaculation refers to the opposite problem of taking too long to ejaculate. Impotence[2] is defined as the inability to obtain an erection. The concept of male potency does not have a counterpart in terminology for women. This is an obvious example of the dual and sexist symbolization of male sexuality (Richardson 1988:129). The term "preerectile" would be more appropriate for this condition which is " . . . usually the result of excessive fatigue, heavy drinking, and/or overanxiousness about sexual performance—all of which are characteristically associated with masculine activities" (Richardson 1988:129). Additional factors causing or contributing to impotence include: prescription medications, especially those for high blood pressure, diseases such as atherosclerosis and diabetes, damage to nerve fibers due to surgery, and low testosterone levels. According to the Harvard Health Letter ("Putting the Pill (for Men) in Perspective" 1998:4) " . . . about 85% of impotence cases are due to a physical, not a psychological cause."

Treatments for impotence have been revolutionized by Pfizer, the creators of Sildenafil (Viagra), whose clinical studies found Sidenafil produced erections in 70% of a test population of men with erectile problems (Ault 1988:1037; Handy 1998:54). Other treatments for erectile dysfunction include psychological therapy, hormone replacement therapy, vacuum devices, urethral supposities, penile injections, vascular surgery, and penile implants ("Putting the Pill (for Men) in Perspective" 1998:4). New oral medications for increasing blood flow to the penis are on the horizon.

While these problems are all too real for the individual enduring them, this should not blind us to recognizing that there are cultural biases in these "dysfunctions" (Davis ND:1–21). In the disorders associated with timing, there are cultural proclivities around time expectations that can vary with individuals by gender, ethnic group, and cross-culturally. For example, some heterosexual males may regard delayed ejaculation as more of a problem than their partners do. A women's anorgasmia may be related to her lack of information about sexual functioning, beliefs that sex is dirty or bad, or religious views that see sex as immoral. Or there is the situation where an individual's problem is caused by her/his partner's lack of knowledge or technique for arousal. However, very often as in the case of anorgasmia, it is the woman who is labeled with the problem, not her partner, reflecting gender bias in the clinical sector.

The causes of sexual dysfunction may be analyzed from the societal/cultural level to that of the individual. When discussing the psychosomatic aspects of sexual dysfunction that are associated with anxiety, the cultural perspective must be included so that the effects of cultural patterning may be considered. In addition, there are physiological factors that can affect sexual functioning including disease (e.g., cancer, bladder disorders, diabetes, structural damage) and response to drugs (e.g., diuretics, adrenal steroids) (Kaplan 1979). Davis (ND:1–35) has reviewed and critiqued *The Sexual Disorders in the Diagnostic and Statistical Manual of Mental Disorders IV* and has proposed that in order to maintain and understand the role of culture in our western diagnosis, we need to move beyond "categorizing and classifying sexual behavior to an emphasis on questioning and analyzing the constructions of the categories themselves, as culture bound" (Davis ND:11). This is particularly important in seeking to understand sexual dysfunction in other ethnic groups or cultures because the western clinical theories of sexual dysfunction ignore the role of culture and ethnicity.

Birth Control in the United States and Cross-Culturally

The world has faced dramatic changes in terms of overall population. It took the longest period, from *Homo Sapiens* to 1850, to reach the first billion people. In another 80 years the second billion was reached and then in another thirty years, the third billion. In only sixteen more years the world's population was four billion. The U.S. is growing at the rate of two million a year and it is estimated that by the year 2000, 70% of the predicted population of six billion will be residing in third world countries (Gordon and Snyder 1986:155). To assess population growth, both the birthrate and death rate must be considered. The world birthrate is about 27 per 1,000 people with the death rate at about 10 per 1,000, resulting in a growth rate of approximately 1.8%. Developing nations have a higher growth rate of 2.4% (vos Savant 1991:26).

Before venturing into our discussion of birth control in the U.S. and cross-culturally, it is important to define terminology related to the topic of contraception. **Contraception** refers to "any natural, barrier, hormonal, or surgical method used to prevent conception and pregnancy" (Francoeur 1991a:657). **Birth control** refers to any method whereby births are interfered with and includes methods for **controlling fertility** through birthspacing, late marriage, long post-partum sex taboos, herbs, abortions, and magical means, among other methods. **Population control** is an abstract concept that includes births, deaths, and migrations. It is used to discuss major demographic trends. "Fertility is a measure of the rate at which people are born. Mortality is a measure of the rate at which people die. Migration is the movement of people into or out of a geographical area" (Eshleman et al. 1988:591). Until approximately two hundred years ago, high death rates with high birthrates maintained a situation of worldwide population stability. In 1850, a major demographic transition occurred in which birthrates continued at high levels in conjunction with decreasing death rates that had begun in the 1800s. Death rates declined because of improved diet and advances in preventative medicine and treatment (Gordon 1978:513). This created a period of explosive population growth (Eshleman et al. 1988:595). However, gradually, family planning and birth control measures began to have an impact, and rapid growth has been reduced. It should be pointed out that while industrialized nations followed this pattern, the transition may not yet be completed in third world countries (Wells 1978:517–518).

If we assume fertility as the norm, the relative chances of conception in a year of unprotected intercourse are 85% to 90% (King et al. 1991:130). This is important information in the evaluation of the effectiveness of contraceptive methods over the course of a year in an unprotected couple. Two strategies for assessing contraceptive effectiveness are used. The **theoretical effectiveness** *rate* refers to the percentage of couples who would conceive using a particular technique correctly and systematically. The **actual effectiveness** *rate* is lower than the theoretical effectiveness rate since all couples do not use birth control methods properly or consistently. The effectiveness of a technique is reported in terms of the number of failures per 100 couples in a year of use called woman years, since women are the people who get pregnant. Over time the chances for pregnancy increase with any technique. For example, a technique with a 5% failure rate the first year will increase to between 23% and 40% over the course of ten years (King et al. 1991:130).

The problem of unwanted pregnancy is far more complex than just an issue of access to contraceptives and birth control. While the white middle and upper classes may be exposed to birth control methods through the availability of sex education information, this is not necessarily true for the lower socio-economic classes, various ethnic groups, and immigrant populations. Traditional sex education programs may not be culturally sensitive. For example, Taylor and Ward's (1991:129) study of ethnic issues in sex education found that the focus on individuality and autonomy in sexual decision making was culturally inappropriate for Vietnam. For Vietnamese, birth control is only used by married couples and this is a decision that is made with the input and approval of the couples' parents. For the Vietnamese then, contraception is a family concern, not an individual choice. There is class and cultural bias in terms of access to information, availability of birth control methods, and societal and economic support.

Other factors must also be considered in the examination of contraceptive techniques. For example, Hyde's (1985:264) study using a random sample of Denison University students revealed that 32% of the males and 21% of the females answered "don't know" or "nothing" in response to a question about what contraception was used during their most recent sexual encounter. Hyde reports that these findings concur with those at other universities. It is difficult to argue "the contraceptive ignorance theory" in light of the evidence from one study of women having abortions. Half had used prescription contraception and the majority had knowledge of birth control (Luker 1975 in Hyde 1985:264). Luker (1975) has suggested an alternative theory of "contraceptive risk taking." What happens, according to this theory, is that a woman engages in a cost and benefit analysis so that if the costs of contraception outweigh the perceived risk of pregnancy, then unprotected intercourse is likely to occur. The costs to women of using contraceptives are social in that their reputations are at risk in terms of self-concept and peer approval. The double standard, although modified, still prevails and affects the contraception decisions of females. A public "coming out" as sexually active may be expressed by purchasing over-the-counter contraceptives at a drug store or making an appointment to see a physician. This necessarily has emotional and psychological consequences for the individual and relations with peers and family.

Some forms of contraception reduce spontaneity. This may be perceived by women as a potential problem in their relationships with men. They may fear the possibility of being spurned by their partners should they ask them to use contraception (Hyde 1985:265). This also places them at a risk for HIV infection as well as pregnancy. Conversely, there are young women for whom pregnancy is regarded as a positive event, a quick ticket to adulthood and status for those who have not reckoned with the economic realities of childrearing as a single parent or as a young couple. It is important to consider the meaning of pregnancy, as well as its socio-cultural matrix including socio-economic factors, male-female relations, power inequities, and ethnic issues when evaluating contraceptive use.

A wide array of contraceptive options are available in the world today. These include: abstinence, natural methods, barrier contraceptives, spermicides, hormonal contraceptives, intrauterine devices, and surgical methods (Francoeur 1991a:249; King et al. 1991). The following table illustrates the prevalence of various kinds of contraceptive techniques found cross-culturally and in the West.

Table 11.1
Who Uses What Kind of Contraceptive?

Country/ Region	Tubal Ligation	Vasec- tomy	Pill	IUDs	Condoms	Other Methods
United States	23.2%	11.4%	30.0%	7.9%	12.9%	14.8%
China	37.5	12.9	4.8	41.1	2.0	1.6
India	40.0	40.0	2.9	8.6	5.7	2.9
Caribbean & Latin America	36.8	2.6	36.8	5.3	7.9	10.5
Middle East & Africa	14.3	0	57.1	14.3	7.1	7.1
All developed countries	13.0	7.4	26.9	11.1	24.1	17.6

(Francoeur 1991a:29)

In order to summarize the multiplicity of techniques that are available, we refer you to the table on pages 260–263 which describes first year failure rates of various methods of contraception as well as a review of how each of the techniques works to prevent conception. Information on medical problems associated with each method and noncontraceptive benefits are also included. Please note this table does not include abortion discussed later in this chapter.

We will briefly describe the most well-known methods of contraception available today in western developed nations. Abstinence is an effective choice for avoiding pregnancy altogether while prayer and wishing obviously are not. Two other ineffective methods are withdrawal, or coitus interruptus, where the male pulls out prior to ejaculation, and douching after intercourse.

Natural means include the rhythm methods consisting of three techniques: the calendrical method, the basal body temperature method, and the Billings method. The rhythm methods are based on identifying when the twenty-four hour fertility window occurs in a woman's cycle. The calendar method is based on a formula in which 18 is subtracted from the length of the woman's shortest cycle and 11 is subtracted from the length of the longest cycle recorded over at least an eight-month period. The time frame to avoid sex spans the sum from the shortest cycle to the sum of the longest.

Table 11.2
First Year Failure Rates of Birth Control Methods

First Year Failure Rates of Birth Control Methods*

Method	Lowest Observed Failure Rate in Users Who Use It Correctly and Consistently (%)	Failure Rate in Typical Users(%)	Primary Mechanism of Action	Major Health Complications	Noncontraceptive Benefits
Vasectomy	0.1	0.15	Removes part of male's vas deferens, thus blocking passage of sperm	Usual complications associated with surgery; form antibodies to sperm	—
Tubal Sterilization	0.2	0.4	Removes part of female's fallopian tubes, thus blocking passage of egg	Usual complications associated with surgery	—
Cervical cap (with spermicide)	6	18	Barrier	Possibly infection with prolonged use	Possibly offers some protection against gonorrhea and chlamydia
Diaphragm (with spermicide)	6	18	Barrier	Possible infection if left in for a prolonged period of time	Offers some protection against gonorrhea and chlamydia

Method					
Spermicides	3	21	Kill sperm	—	Offers some protection against gonorrhea, trichomoniasis, chlamydia, and herpes (and perhaps AIDS)
Sponge (with spermicide)	6 (women who have never given birth)	18	Barrier, kills sperm	—	Possibly offers protection against gonorrhea and chlamydia
	9 (women who have given birth)	28			
Rhythm			Abstain from sex during ovulation	Frustration due to long periods of abstinence	—
calendar	9	20			
Basal body temperature	2	20			
Billings	3	20			
Withdrawal	4	18	Abstain from ejaculating in vagina	—	—
Chance (unprotected intercourse)	85	85	—	—	—

(continued on next page)

Table 11.2 (continued)
First Year Failure Rates of Birth Control Methods

First Year Failure Rates of Birth Control Methods*

Method	Lowest Observed Failure Rate in Users Who Use It Correctly and Consistently (%)	Failure Rate in Typical Users(%)	Primary Mechanism of Action	Major Health Complications	Noncontraceptive Benefits
Combined birth control pill	0.1	3	Prevents ovulation	Cardiovascular problems (particularly in women who smoke, are over 35, and/or have other health problems such as diabetes or hypertension); nausea and/or breast tenderness; depression, eye problems; headaches; hypertension; weight gain	Reduces risk of uterine and ovarian cancer, ovarian cysts, rheumatoid arthritis, pelvic inflammatory disease. Alleviates premenstrual syndrome, menstrual bleeding, and menstrual pain
Progestin-only pill	0.5	3	Thickens cervical mucus, inhibits implantation, and inhibits ovulation	Irregular menstrual cycles	Relieves menstrual pain and bleeding in some users; probable protection against pelvic inflammatory disease

Norplant (hormone implants)					
Six capsules	0.04	0.04	Thickens cervical mucus and inhibits ovulation	Irregular menstrual cycles	May help to prevent anemia; probable protection against pelvic inflammatory disease
Two rods	0.03	0.03			
IUD					
Overall	3	3	Prevents implantation; copper and hormone coatings inhibit ovulation	Spotting, bleeding, cramps (except with Progestasert), IUD expulsion, infection if pregnancy should result; uterine perforation; pelvic inflammatory disease	Progestasert decreases menstrual blood loss and menstrual pains
Progestasert	2	3			
Copper T380A	0.8	3			
Condom (no spermicide)	2	12	Barrier	—	Highly effective in reducing the spread of sexually transmitted diseases, including AIDS

*Modified from Hatcher et al., 1990 and Trussell et al., 1990.

These sums represent the days in a woman's cycle reckoned from the beginning of her period that are unsafe (King et al. 1991:133). The basal metabolic method assesses when ovulation occurs by daily charting one's temperature upon rising. A woman's temperature rises by a few tenths of one degree Fahrenheit within 24 to 72 hours after ovulation. This approach suggests refraining from intercourse until after menstruation and until two to four days after the rise in temperature. The Billings method is based on the evaluation of vaginal mucous to determine when ovulation takes place (Francoeur 1991a:249). Mucous is usually cloudy. It becomes clear, slippery, and stretchy one to two days prior to ovulation. After ovulation the mucous returns to its cloudy pattern. The Billings method may be combined with the basal metabolic method to enhance the effectiveness of either technique.

Barrier methods include the condom, a thin cover for the penis made of latex or animal membrane, and the diaphragm, a thin latex barrier shaped like a dome that prevents sperm from entering the cervix. Spermicidal jelly or cream is used inside the diaphragm. The cervical cap is a much smaller barrier method for women that, like the diaphragm, fits over the cervix. Spermicides are chemical agents that kill sperm. They come in the form of suppositories, creams, and foam and are inserted prior to intercourse. These are most effective in combination with barrier methods rather than alone. The sponge is a relatively new addition to the barrier methods with a 90% effectiveness rating that combines a spermicide with a sponge for absorbing the ejaculate. The sponge has the advantage of being inserted up to 24 hours before penile-vaginal intercourse (Klugman 1993:70).

The most recent innovation in barrier methods is the female condom—Reality. It fits the female vulval area by covering the labia and the base of the penis as well. It is reckoned at a 95% effectiveness rating although it has not been proven to prevent sexually transmitted diseases (Klugman 1993:70; "HIV Prevention Strategy Highlights Women" 1994:26).

There are several options involving hormonal methods. The oral contraceptive pill, the OCP, traditionally combines estrogen and progesterone. This combination tricks the body into thinking it is pregnant and thereby prevents the release of the egg, or in the case of the mini-pill containing only progesterone, inhibits the development of the uterine lining. There are a variety of kinds of OCPs on the market. A new class of synthetic progesterones has been approved by the FDA (1993) and include two new kinds of progestins, Norgestimate and Desogestril. These function similarly to other OCPs, but reduce the side effects of weight gain, breast tenderness, acne, spotting, and nausea (Austin 1993:57). This class of hormones has been in use since 1981 (Klugman 1993:59). Currently there are conflicting reports as to whether OCPs increase the risk of breast or cervical cancer.

The morning after pill includes taking two doses of an oral contraceptive pill that contains ethinyl estradiol and dl-norgestrel (Percival-Smith and Abercrombie 1987). If the pill regimen is followed properly, there is little chance of conception (Klugman 1993:59). Ovral, an FDA-approved OCP, has been used as a morning after pill with a 90 percent effectiveness rating in preventing pregnancy. Two Ovral are taken within 72 hours of unprotected sex, followed by two more 12 hours later. Ovral works by preventing implantation of the fertilized egg (Evans 1993:52). Another hormonal method that has gained a great deal of press is RU 486 which is not yet available in the United States, although it is currently used in Europe. It was tested on 2,000 women in the U.S. in the fall of 1994 (American Health 1994:8). RU 486 works by inducing menstruation in women who suspect they may be pregnant by preventing implantation, and it may prompt a miscarriage in early stages of pregnancy (Dowie 1991:137–140; Klugman 1993:59; Hall 1989:44).

Norplant is one of the more modern hormonal methods of birth control. It is a surgically implanted and removed device, about three inches long, that is placed under the skin of the arm. It releases progestin and provides protection for up to five years and is 99% effective (Findlay 1991:126; Sivin et al. 1983; Klugman 1993:70).

Another hormonal method proposed for the U.S. is Depo-Provera, which was approved by the FDA as a contraceptive method in January 1992. Depo-Provera is an injectable synthetic progesterone that prevents the release of an ovum as well as interferes with the implantation of a released egg in the lining of the uterus. One injection prevents pregnancy for three months and is 99% effective. It has been used in ninety countries over the past twenty years (Welch 1992:48).

The intrauterine device, or IUD, flourished in the 1960s and 1970s. The IUD is a small metal or plastic device that is put in the uterus. However, medical follow-up revealed that some IUDs were associated with Pelvic Inflammatory Diseases and other health problems including infertility. Today, the IUD is primarily restricted to women outside the U.S. who have already had children.

The World Health Organization has studied testosterone enanthate (TE) injections as a form of contraception in 271 men in seven countries. The weekly injections had few side effects and a success rate of 99.2% which makes it more effective than OCPs (97%), IUDs (94%), and condoms (88%). After four months of weekly testosterone injections, a man's level of sperm was reduced enough to have a contraceptive effect. This was reversible after six and a half months ("Men's Shots Used as Contraceptive" 1990:B1; Prendergast 1990).

Another direction in contraceptive research is focused on the development of a vaccine without the side effects of hormonal methods. Such an approach includes the field of molecular biology and involves studies of the immune system. Tests on the contraceptive vaccine have been initiated on both men and women (Wheeler 1995:A9)

Surgical techniques include sterilization. In males, a vasectomy is performed in which the vas deferens are severed. In the tubal ligation, the woman's fallopian tubes are cut and tied, cauterized or banded. Both are permanent methods, although reversal is possible at great cost and without a high success rate. An injection vasectomy, used in China for 10 years, is under review in the U.S. This technique involves injecting a plastic polymer in the sperm ducts thereby sealing them. Chinese researchers are researching possible reversals in such a technique ("Injection Vasectomy" 1992:28).

Abortion is a highly controversial surgical technique for dealing with unwanted pregnancies in the United States. The arguments can be classified as residing in two camps. The right-to-life camp argues that the fetus is a human life and therefore has a right to life regardless of the societal and personal costs and risks to the individual. The other camp argues for *choice*, that a woman has the right to determine what to do with her body. Some research on women who have chosen abortions reveal that the experience is "not traumatic" and that "[m]ost women report feeling relieved and happy . . . [and] . . . [f]ewer than 10 percent . . . experience psychological problems afterward, and most of them had problems before the pregnancy and abortion" (Hyde 1985:266). In the 1992 *Gallup Poll* only 13% of the American public felt that abortion should be illegal under all circumstances, while 48% felt it should be legal under some circumstances and 34% felt it should be legal under all circumstances with 5% of no opinion (Gallup 1992).

Hyde (1985) points out that there are a number of methodological problems with the research on psychological adjustment and abortion. Post-abortion adjustment is related to how the woman reaches the decision to have an abortion to begin with. There appears to be a great deal of variation in how women decide to have an abortion. In fact:

> By studying abortion decisions and their consequences, we may come to see the whole process of making a decision to have an abortion (or to have the baby and give it up for adoption, or to keep the baby) as fostering psychological development and growth, at least if handled well. Women may emerge from the process being more mature and having better-developed moral sensitivities (Gilligan 1982 in Hyde 1985:268).

1973 marked an important year in terms of women's choices over unwanted pregnancies. In the now famous *Roe v. Wade* case, the Supreme Court in a 7-2 decision ruled that in the first trimester of pregnancy the decision to terminate was one to be made by a woman and her physician. This judgment also allowed for abortions during the second trimester of pregnancy (King et al. 1991:153). Ironically, the *Roe v. Wade* decision set a standard for future government interest in the rights of the fetus during the third trimester (Lacayo 1990:23). In July 1989, the Supreme Court ruled in favor of a " . . . Missouri law that further restricted public funds and facilities for abortions and that also required physicians to test for fetal viability (the potential for the fetus to survive) at 20 weeks" (King et al. 1991:153). This ruling, the Webster decision, paved the way for states to pass laws limiting and restricting abortions in a variety of ways (Carlson 1990:16). However, given the election of Clinton and a Democratic pro-choice platform, we have begun to see greater government support of the *Roe v. Wade* decision in recent years.

First trimester (the first twelve weeks) abortions use a method known as vacuum curettage (also called vacuum aspiration or suction). The majority (91%) of abortions take place during this period, with 50% prior to the eighth week, and 41% in the 8th–12th week period with .01% after the twenty-fourth week ("Who Gets Abortions" 1990:G1). This technique involves dilating the cervix with an instrument or laminaria rod, a dried seaweed that gently expands the cervix. A suction machine is attached to a tube and the tube is inserted into the uterus. The suction of the machine withdraws the fetal tissues from the uterus. The benefit of this method is that it takes only ten minutes and is done on an outpatient basis with a local anesthetic requiring only a few hours of recuperation in a clinic.

This method is considered safe up through the 20th week of the pregnancy (Hyde 1985:266). After the first 12 weeks, vacuum curettage is not enough. At this point, a technique known as dilation and evacuation or D and E, is the most widely used method of abortion for the 13th–16th week. Because suction alone is not enough to remove the unwanted cells, the uterine inner walls are scraped to ensure complete removal. After the 16th to 24th week, prostaglandin or saline induction is used. These techniques terminate fetus viability and induce contractions so that the woman gives birth to a non-living fetus.

In the United States, it is estimated that 3.4 million women become pregnant unintentionally. Of those, 47% have abortions (about one and half million), 40% give birth, and 13% miscarry. Under 25-year-old women account for 59% of the abortions ("Who Gets Abortions" 1990:G1). Of those having abortions, 81% were unmarried and 70% were performed on Caucasians (Henshaw 1987). If the abortion rate is broken down per 1,000 women, then African-Americans represent 53%, Hispanics 43%, and whites only 23% ("Who Gets Abortions" 1990:G1). Nine percent of the U.S. college women have had a minimum of one abortion (*1989 Gallup Survey*). In spite of the volume of abortions in the United States, health risks to the mother are very low. Deaths from pregnancy are 20 in 100,000, while deaths from medically monitored abortions are less than one in 100,000 and are safer than pregnancy or childbirth (Lebolt et al. 1982; Grimes and Cates 1980). Abortions do not increase a woman's future health risk for other pregnancies. There is no increased incidence of infertility or miscarriages ("More on Koop's Study on Abortion 1990).

Birth Control in Cultural Context

The following quote from Marvin Harris (1989:210) is a good way to begin our discussion of birth control in the cross-cultural context.

> Sex does not guarantee conception; conception does not lead relentlessly to birth; and birth does not compel the mother to nurse and protect the newborn. Cultures have evolved learned techniques and practices that can prevent each step in this process from occurring.

The cross-cultural context for birth control includes research on indigenous societies prior to western contact, societies in the process of culture change, as well as studies of third world and developing nations. It is vital that we stress the value of relativism in assessing various techniques of controlling birth. Some of these methods include abortion and infanticide and may coincide with preferences for one gender over the other. Amniocentesis and sonograms designed to detect anomalies in a fetus are used in India and Asian countries to select male children by aborting female fetuses ("Asia: Discarding Daughters" 1990:40). In South Korea, " . . . male births outnumber female births by 14%, in contrast to a worldwide average of 5%" (Burton 1990:36). Such an imbalance can have far-reaching consequences in terms of adult marriage patterns. China's policy of one child per couple, encouraged by severe financial sanctioning, has resulted in the continued practice of female infanticide and amniocentesis to identify and then abort female fetuses. In a society that historically favored males, females were considered an economic liability (Burton 1990:36).

In clear contrast is Ireland, the only European country with a complete ban against abortions (except where the mother's life is endangered). Abortions have never been allowed in Ireland; and, since 1983, this ban has been incorporated into the Constitution. Birth control is also illegal in Ireland. As a consequence, it is estimated that annually 4,000 Irish women have abortions in Britain (Johnson 1990: B9).

Cross-culturally about 40% of the world's population live in countries where abortion is legalized and available by request, while 23% of the world's people live in countries where it is illegal unless there is a medical/social reason. Some nations are more restrictive and do not allow abortions unless the mother's life is endangered by the pregnancy (African, Asian, and South American countries, Ireland among others). Abortion has been legalized since the 1930s in Eastern and Central Europe. Romania has the highest abortion rate of all European countries with a rate of 192 per 1,000 women aged 15–44 (as of 1990) believed to be related to low availability of contraceptives. The Netherlands has the lowest with a rate of 5.1 per 1,000 women aged 15–44 despite very liberal abortion laws, but contraceptive use is high (Terenzi 1992:54).

But lest we fall prey to creating an exotic "other," a review of abortion in our own history illustrates how our perspectives on abortion have changed over time. Until the nineteenth century, it was left to a woman to determine if and when to have an abortion. In other words, she was free to do with the fetus as she liked until she felt a "quickening" (fetal movements at about 4 months of pregnancy). This was related to a pre-nineteenth-century ideology that life began at quickening, not at conception (Crandon 1986:471). As a result, punitive action against women was negligible. "In 1800 there was not, so far as is known, a single statute in the U.S.

concerning abortion" ("Abortion in American History" 1990:8). Drugs for abortions were widely advertised and available. However, over the next one hundred years, abortion became illegal except as a lifesaving technique for the mother in every state. This may be partially attributed to the efforts of the newly formed American Medical Association's desire to legitimize and coopt this territory which had previously been under the control of women. According to the medical view, only physicians were sufficiently informed to know when abortions were necessary and justified (Crandon 1986:471). Until the 1960s, women were excluded from equal entry in medical schools, thereby ensuring that females had no voice in this issue until *Roe v. Wade* ("Abortion in American History" 1990:8).

In addition to abortion, industrializing nations may also include OCPs or other contraceptives in their birth control efforts. In the U.S., two-thirds of married women use some form of contraception. However, before we pat ourselves on the back, the U.S. Census Bureau has noted that contraception is higher among married women in Europe and in the following developing countries than in America: Taiwan 78%, China 77%, Mauritius 75%, Hong Kong 72%, South Korea 70%, Costa Rica 68%, Thailand 67% ("Birth-Control Popular Here" 1990:15).

We have focused on contraception and abortion in the West and have presented evidence from third world and developing nations that use reproductive technologies developed by western medicine. Bear in mind that societies worldwide and over long periods of time have been interfering with reproduction in a variety of ways prior to the advent of the condom, the pill, and other modern methods. A variety of methods are used among pre-industrial peoples to avoid impregnation and may include: masturbation, homosexuality, and coitus interruptus (Harris 1989:211). In East Bay, a Southwest Pacific people studied by William Davenport, abstinence from marital coitus was the preferred form of birth control used to facilitate birth spacing, while coitus interruptus was the technique used for extramarital relationships (Davenport 1976). According to Gregersen (1983:290–291), coitus interruptus is the predominant birth control method found cross-culturally. Other techniques include trying to expel the semen after intercourse as used by the Kaviondo and Zande who rely on body movement by standing and shaking (Gregersen 1983:290–291). This is not unlike the sexual folklore among some college students that standing up immediately after sex reduces the chance of conception.

The cross-cultural and historical record reveals numerous ethno-theories and practices including reports of herbs, medicines, and potions used as contraceptives and abortificants among the Greeks, Romans, Egyptians and others (Riddle et al. 1994:29–35; Kroeber 1953:248; Schneider 1968:385). Silphium was used by the Greeks from the seventh century B.C. to the fourth century A.D. when demand for it led to its extinction. Research by Riddle and his colleagues suggests that the ancients used a variety of herbs that inhibited conception and terminated pregnancy including Queen Anne's Lace, pennyroyal, willow, date palm, pomegranate, and acacia gum. The effectiveness of the chemical properties of these herbs in affecting fertility has been clinically demonstrated (1994:29–35). Plains Indians wore a contraceptive charm known as a "snake girdle" made of beaded leather over the navel (Gregersen 1983:291). In addition to these methods, effective interventions and approaches such as abortion and infanticide were also practiced.

Pre-industrial cultures have been using a number of techniques to engage in the control of births. As discussed earlier in this chapter, attitudes about abortion as well as the practices used, are related to economic, subsistence, political, social, historical, and religious factors. We shall present several examples to highlight the interaction of these variables in influencing methods for controlling birth. As we shall discuss, population and how people make a living are interrelated. With plant cultivation, horticulture but much more dramatically with agriculture, we see population growth occurring in an escalating system of sedentism, social organization complexity (ranking and hierarchy), centralization of authority, and increasing surplus. These features are related to issues of birth spacing, beliefs about family size, the value of children to the household economy, and gender preferences for children which, in turn, impact birth control beliefs and practices.

Foragers, or gatherers and hunters, are an excellent example of how culture shapes biology in relation to population. One widely accepted anthropological theory correlates sedentism, permanent residence due to plant cultivation, with population expansion. This line of reasoning is supported by data from the ethnographic spectrum and includes evidence from !Kung peoples of the Kalahari Desert. The settlement of some !Kung groups has provided an excellent opportunity to study the effects of sedentism on a previously foraging population and by implication the effect of sedentism on populations in terms of cultural evolution, adaptation, and demographics.

The !Kung, like tropical foragers in general, are known for small populations. A number of interacting factors are believed to contribute to this situation. Foragers have diets that are low in fat and high in protein and consequently, they have less body fat than do food cultivators. According to the critical fat hypothesis (Frisch 1978:22–30) this can impact fertility in several ways, most notably through adolescent sterility (introduced in chapter 10), and through the impact of extended lactation without supplemental food sources. In order to ovulate, women need a certain amount of body fat, minimally estimated at 15%. Surplus calories are associated with plant cultivation. Combined with reduced physical activity of agriculturalists, this can lead to a higher body fat ratio.

The critical level of fat storage necessary for menstruation and ovulation to occur is about 150,000 stored calories of energy, enough " . . . to permit a woman to lactate for 1 year or more without having to increase her prepregnancy caloric intake" (Lancaster 1985:18). A period of subfertility for several years after *menarche* exists among foragers but this declines in settled populations.

> Sedentism combined with high levels of caloric intake lead to early deposition of body fat in young girls and "fool" the body into early biological maturation long before cognitive and social maturity are reached (Lancaster 1985:18).

Foragers have other cultural practices that contribute to smaller populations. One method is birth spacing. As people who cover vast territories, gatherers cannot afford to nurse and carry more than one child who is not able to walk. !Kung, for example have solved this dilemma by spacing children about four years apart. Several mechanisms contribute to !Kung birth spacing. A long post-partum sex taboo that requires abstinence from coitus for a minimum of a year is a cultural method affecting the fertility of women. Foragers are well-known for long post-partum sex taboos for women as a vehicle for birth spacing and indirectly controlling births. There is a correlation between societies that have long post-partum sex taboos and those with low fertility rates (Nag 1962:79)

Foragers whose body to fat ratio is low to begin with can interrupt ovulation by prolonged nursing, which helps keep fat levels low. "It is now well established that the longer a mother nurses her baby without supplementary foods, the longer the mother is unlikely to start ovulating again" (Ember and Ember 1996:156). !Kung mothers nurse for two or three years without the additional sources of food available to settled peoples such as milk from domesticated animals or harvested grains. Lactating females, in populations where body fat is low, are less likely to become pregnant (Kottak 1991:203).

Two other techniques that are used by foragers to control births are infanticide and abortion. Hunters and gatherers practice abortion and are known for their knowledge of pharmacology in which animal and plant poisons are used to cause the fetus to miscarry (Gregersen 1983:291). Abortion is not limited to foragers and is practiced among a great many other peoples. Procedures include the use of herbal treatments and toxic substances from animals, hitting the stomach vigorously, manipulating it through massage or squeezing, and having someone jump on the abdomen (Riddle et al. 1994:29–35; Gregersen 1983:290–291; Devereux 1955; Sarvis and Rodman 1974). In addition, Devereux (1955) and Sarvis and Rodman (1974) report attempts to kill the fetus through strenuous activity or through the use of devices or substances that are inserted into the uterus. These techniques may all cause harm to the mother, are dangerous, and may not be particularly effective.

In regard to methods of abortion, David M. Schneider's (1968:383–406) report and analysis of abortion on Yap, a Caroline Island in Micronesia, is a good example of how beliefs, political economy, gender relations, and demographics intersect. Yap is an island that once supported 50,000 people. Yet, by 1945, the population had fallen dramatically to only 2,500. The culture is one that was geared to a much larger population; consequently, the population decline affected socio-political organization in that there were no longer enough people to fill positions and perform necessary political services. This resulted in a generalized Yapese concern and desire for more children. In spite of this, women continued to practice self-abortions. Self-induced abortions on Yap have, in fact, exacerbated the depopulation problem since these also tend to occur "during the maximum years of fecundity" (Schneider 1968:384). However, abortion may not be the sole cause of continued depopulation as gonorrhea or possibly some other diseases have undoubtedly contributed to the low fertility rate.

Given this situation, it is not surprising that self-induced abortions on Yap are negatively sanctioned and kept secret, especially from the men. Abortions are objected to on moral grounds. Yap women are under pressure lest they become known as aborters which could jeopardize their marriages or chances for marriage since men want and desire children. To understand the perpetuation of abortion in spite of depopulation, it is necessary to engage in a **holistic** analysis of Yapese culture; that is, one that considers all aspects of Yapese life.

There are three methods of self-induced abortions on Yap. We shall quote directly from Schneider in regard to these (1968:385):

> One [method] consists in a series of magical manipulations with little apparent efficacy. . . . The other two techniques are empirically

more effective. One of these is drinking boiled concentrated sea water. Women described the effect as a general feeling of illness accompanied by vomiting and severe cramps.

The other technique consists in introducing a thin rolled plug of hibiscus leaves (which expand when moist) into the mouth of the cervix and then injuring and scratching the mouth of the cervix with a bit of stick, stone, iron, fingernail, or other sharp object until blood is drawn. Women informants generally agreed that injuring the area about the mouth of the cervix was necessary in addition to inserting the plug; . . .

The latter method can lead to infection and associated reproductive problems, although medical reports suggest that the resulting infection is mild and not life threatening.

Historically, Yapese population decline may be traced back to a period when population had peaked. What is remarkable about this situation is that, despite the serious depopulation that followed, Yap culture in its traditional form has continued as an adaptation to a dense population; this is an example of culture lag. Abortion was a part of this earlier adaptation and represented an effective solution to an overpopulation problem at the time. The persistence of abortion in the face of decline in population, however, must be understood in the context of the totality of the existing culture, including gender relations and expectations.

Abortion is tied directly to differences in role expectations for women over the life course.

Women up to the age of thirty do not want children because they would no longer be free to fall in and out of love, to attract lovers, to have and break off affairs at will, to practice the elaborate games of love and sociability that appeal to young Yap men and women. They do not want to be tied to a child and to a husband when they are in the best position to gain and enjoy the rewards of being unattached. . . . On Yap the standard and available means of avoiding children is to induce abortion when pregnancy occurs (Schneider 1968:393).

After the age of 30 however, women's attitudes shift to desiring children. This coincides with the shift from youth to adult status.

In summary, abortion persists on Yap in spite of population decline because the period of youth is a time when women have access to the rewards and pleasures of love affairs in a system in which they will never achieve the prestige positions and rankings available to men.

Infanticide cross-culturally also illustrates the complexity and necessity for a holistic perspective in understanding birth control.

This should not be construed to mean that infanticide has not been practiced in the West as a method of regulating unwanted births. It has! Historically, Europeans favored indirect methods of infanticide such as overlaying (where a mother "accidently" suffocated her child by rolling over on it when in bed); wet nurses whose reputations for infant care literally guaranteed that the child would die; or foundling homes such as those in France between 1824–1833 where 336,297 children were abandoned. "Between 80% and 90% of the children in these institutions died during their first year of life" (Harris 1989:214).

Infanticide as a method to control birth is found in a variety of forms around the globe. It includes the indirect methods just discussed that are favored by the Europeans and others such as the Northeast Brazilians (Scheper-Hughes 1987:535–545), or may include conscious systematic neglect, starvation, and/or exposure as more direct approaches. In a cross-cultural study of pre-industrial societies, George Devereux found that 464 groups practiced abortion (1967:92–152). Cultures may openly condone or condemn such forms of birth control. Women are placed in conflicted situations in those societies that simultaneously prevent contraception and abortion. In such cases, surreptitious infanticide may be the only solution for the mother with an unwanted pregnancy (Harris 1989:211). Reports of direct infanticide in non-technologically complex societies suggest that between 53% to 76% of the societies allow for the practice of this method. In such situations, the cultural conception of being a person, a member of the family and the group is not given to newborns (Harris 1989:212, 214).

Marvin Harris's (1974, 1993) re-analysis of Napoleon Chagnon's study of the Yanomami of Venezuela and Brazil is a classic work that illustrates how infanticide relates to warfare and a male supremacy complex among tribal cultivators. The Yanomami represents a case study in a broader theory by Divale and Harris (1976) whose cross-cultural correlational study of the HRAF files proposes that warfare is the most common way of regulating population among tribal cultivators. This happens in an indirect way by leading to female infanticide, rather than directly by male deaths due to war. The thesis behind their argument is that in order for population size to be controlled, the limiting factor must be females and not males. The idea behind this is that one male can impregnate a number of females, therefore societies can afford to lose adult males without affecting their population.

The Yanomami are a tribal society of 15,000 living in 125 villages (Chagnon 1983). They were originally riverine Indians whose ancestors were pushed into a forest adaptation due to population pressure. These Yanomami became skilled hunters and engaged in shifting cultivation in the forest. About 400 years ago they began cultivating plantains and as a result of this semi-sedentary existence combined with additional calories supplied by the plantains, they experienced a rapid population spurt. This feature is a central part of the argument. Increased calories and sedentism offset natural mechanisms for birth spacing and low population that we have seen operating in gatherers and hunters such as the !Kung. Plant cultivation, which provides more carbohydrates and higher caloric intake than foraging, allows for earlier puberty, increased conception, and generally a longer childbearing period. Eighty-five percent of the Yanomami diet is from plantains and the bananas that they cultivate.

But shifting cultivation does not meet Yanomami needs for protein. As they became semi-settled cultivators, their high carbohydrate diet and increased population led them to put increasing pressure on the local game resources for protein. This, in turn, facilitated the development of a male supremacy complex in order to help create better hunters for the dwindling game. In addition, the growing populations who settled in villages came into conflict over protein as a scarce resource which contributed to and escalated warfare. This complex of warfare and hunting, therefore, placed a premium on males, rather than females. In the history of the world, it is rare for women to participate in warfare since this would make poor evolutionary sense. The premium on males ultimately led to female infanticide as a method for parents to select for sons over daughters. The Yanomami prefer that the firstborn be a son and may practice infanticide if a girl is the firstborn. Because Yanomami females do not fight, hunt, and bring home the protein, they are valued less than males. Female infanticide contributes to a population inequity in the ratio of male to females; there are 449 males to 391 females in seven villages studied by Chagnon (1983). This gender inequity perpetuates fighting and raiding to acquire women from other villages which in turn escalates the warfare even further. The Yanomami practice polygyny so that the best fighters and hunters may acquire several wives through the lure of protein and rank. This adds fuel to the fire by creating an even greater shortage of women. Twenty-five percent of the men are polygynous. In this situation warfare operates as a way to disperse populations and relocate them in the environment in order to temporarily relieve population pressure on resources. Among tribal cultivators, warfare and the infanticide it engenders serve as mechanisms to limit population (Kottak 1991; Harris 1974).

As we have illustrated in this chapter, both sexual practices as well as fertility control are intricately tied to the broader cultural system and articulate clearly with a number of cultural variables as well as ecological ones relating to demography, types of subsistance, and adaptations.

Summary

Chapter 11: Topics in Adult Sexuality: Human Sexual Response and Birth Control

1. Sex is defined as a bio-cultural phenomenon.

2. Mangaian and Tantric models of sexuality are contrasted with Western.

3. Western theories of sexuality are introduced in a critical and historical framework.

4. Prominent researchers of sexology are discussed including: Freud, Kinsey, Masters and Johnson, Kaplan, Singer and Singer, Whipple and Perry.

5. Problems in sexual response are addressed including sexual dysfunction and sex therapy. These are placed in a bio-cultural framework.

6. Definitions for concepts related to birth control are offered including: contraception, birth control, population control, theoretical effectiveness, and use effectiveness.

7. Currently available methods of birth control in the West are presented.

8. The abortion controversy is discussed.

9. Birth control is placed in cross-cultural context: nonwestern examples of developing nations and indigenous contraceptive and birth control practices are provided.

10. Pre-industrial examples of birth control include: coitus interruptus, lactation, abortion, and female infanticide.

OVERVIEW

Chapter 12

Topics in Adult Sexuality: Life Course Issues Related to Gender Identity, Gender Roles, and Aging

This chapter:

1. Examines gender role relative to concepts of psychological masculinity and femininity.

2. Incorporates an anthropological discussion of gender variance including transsexual (TS), transvestite (TV), and transgendered persons and the *Berdache* (two spirit).

3. Discusses concepts of androgyny, scripts, bonding and intimacy.

4. Applies concerns about intimacy to late twentieth century U.S. society.

5. Discusses the bio-cultural aspects of sexual aging cross-culturally.

Chapter 12

Topics in Adult Sexuality: Life Course Issues Related to Gender Identity, Gender Roles, and Aging

While concepts of **gender identity**—knowing that you are male, female, or **intersex** (Williams 1986), and **gender role**—adopting culturally defined male and female behavior, are probably widespread cross-culturally, their expression is highly culture specific. There is a large anthropological literature discussed in chapters 1 and 2 on the symbolism of maleness and femaleness and the diversity of gender role behavior among cultures; e.g., Margaret Mead's work is one body of classic, though recently challenged research on this topic (Mead 1928, 1949, 1935). We also know from the literature that culture change can impact on gender roles cross-culturally (Lurie 1973; Radin 1926; Sharp 1981). However, the extensive amount of energy vested in examining and trying to understand gender role behavior through the life cycle in the latter twentieth century is largely a function of middle class western and specifically U.S. culture.

In this chapter some of the current issues in male-female relationships, sexual aging, and gender roles in mainstream U.S. culture as well as subcultures within the larger society are addressed. As part of the "**sexual revolution**" of the 1960s, a tremendous amount of time, energy, and attention was directed toward defining and elaborating concepts of **gender role** and **gender identity**, masculinity and femininity.

The sexual revolution of the 1960s and 1970s was a sexual, social, economic, and political statement against the perceived negative aspects of traditional gender role behavior and expectations. In its most extreme form, anything defined as "traditional" was seen

281

as negative. There is some question as to whether there actually was a sexual revolution, which connotes radical or root change, qualitative widescale behavioral and attitudinal changes (Goldberg 1979, 1984; Farrell 1986; Elshtain 1989; Ehrenreich, Hess; and Jacobs 1986); or whether it was a reform movement, a rebellion in which behaviors became more overt and widespread, and attitudes remained relatively constant (cf. Hendrick and Hendrick 1987; McCabe 1987). Sexual behaviors have become more visible, more open and more discussed in this culture since the 1960s. There is more of certain kinds of behavior occurring. For example, we have the highest teenage pregnancy rate of any western, industrialized nation, including Japan (WHO 1988; Wattleton 1990). The sexual revolution introduced the concepts of liberated and **androgynous** gender role behaviors and expectations. Androgyny, to be discussed in more detail later, is behaving appropriately to the situation, as opposed to behaving according to a gender role stereotype. But whether there are qualitatively "new behaviors" and whether attitudes about sexual behavior have qualitatively changed is arguable. For example, a double standard of heterosexual nonmarital sexual behavior persists, even though more nonmarried people openly admit to being sexually active (Kelly 1988; Masters, Johnson, and Kolodny 1985).

To review, **gender identity** generally is the perception of oneself as a man or a woman. In a biomedical, western model it rests on phenotypic sex, i.e., the presence of male or female primary and secondary sex characteristics, as well as an early and deeply ingrained sense of "I am a boy or girl, man or woman." Gender identity is labeled at birth. **Gender role** is learned based on the gender identity. Gender role is the internalization of culturally recognized attitudes, behaviors, beliefs and values that compliment one's gender identity. It includes both verbal and nonverbal behavior and relates to concepts of masculinity and femininity.

Masculinity and **femininity** are patterned, learned, verbal and nonverbal signs, symbols and behaviors that reinforce socially defined concepts of maleness and femaleness. Culturally specific, they include speech and dress patterns, activities and affect, world views, and body language. The boundaries between innate and learned aspects of gender role/identity and masculinity and femininity are controversial and unresolved as the debates on various theories on left and right brain dominance illustrate.

Transsexualism and the Berdache

The cultural specificity of gender identity categories such as maleness and femaleness, masculinity and femininity are probably most dramatically represented by the **transsexual** (TS) identified person and the *Berdache* (two spirit). The transsexual identity and the Berdache as expressions of gender identity illustrate both the **culture-bound** (i.e., unique to a given culture) dimensions and cross-cultural continuities of defining one's gender within the context of society. Transsexualism, introduced in chapter 7 on "Fertility, Conception, and Embryology," is seen as a pathological state, something aberrant and "to be fixed" in this culture. While transsexualism is not labeled as such cross-culturally, various intergender and cross-gender behaviors and identities are recognized and accepted in some nonwestern societies and are granted legitimacy. These topics are expanded in this section.

It has been suggested by Benjamin (1966:17–19) that transvestism and transsexualism represent a continuum of cross-dressing behavior and identity. Based on research on over 200 clients, he suggested that **transsexualism** and **transvestism** represent variations on a theme of **gender identity** dissatisfaction. For the transsexual individual, however, the distinction lies in the fact that transsexuals are "normals" for whom nature has given the wrong bodies. They are not "deviant" men then, they are women with ersatz physiques. Their emic (insiders) view as a gender community reflects western culture's central tenants in the construction of gender (cf. Bolin 1988:13–14).

Bolin's fieldwork with transsexual people conducted from 1979–1981 illustrates how social constructions of identities are historically embedded and may change through time. Subsequent research in 1993 revealed changes in transsexual and transvestite social identities in response to several cultural features, including an increasing politicalization at the grass-roots level, and culture change in gender roles and conceptions of femininity (Bolin 1994:462–485). In *In Search of Eve: Transsexual Rites of Passage*, Bolin participated in the lives of sixteen preoperative transsexuals affiliated through a group she referred to as the Berdache Society (a fictive name). In her work, it became evident that transsexual gender identities weren't fixed, but transformed as they underwent changes in their roles and appearances. This view of identity is in the constructionist tradition in anthropology, which regards identities as socially created and brokered by gender cues, including cultural meanings associated with bodies and behaviors as masculine and feminine. Some examples of gender cues are body contours, degree of muscularity and of roundness or softness, body modifications, speech, including non-language vocalizations (paralanguage), language expression and communication patterns, and clothing styles (sartorial expression) among others. Bolin's research with transsexual people suggests that rather than being born with a "cross-sex identity," some individuals develop an identity as women through group interactions, social networks, and experience in the female role. The Berdache Society created its own *rites de passage*. Bolin found that the rites of passage created by the research population fostered the development of their identities as women. Part of the process whereby transsexuals learned to present themselves successfully as women in order to pass, embodies cultural meanings of physical, verbal, and nonverbal expressions. Learning to pass is crucial in the development of an identity as a woman in the male-to-female transsexual person. According to Devor (1989:140), "gender status is learned by displaying the culturally defined insignia of the gender category with which one identifies."

Transsexuals support the North American gender ideology that there are only two genders, that these are "natural," one is either male or female, not in-between and that these are determined by the genitalia (Garfinkel and Stoller 1967:116–85; Feinbloom 1976:150; Kessler and McKenna 1978:112–41). Male-to-female transsexuals are very aware of the "cultural reality" of gender. In the everyday process of living one's gender, gender is attributed on the basis of gender cues; or insignias that are visible, not hidden from view such as the genitalia. Transsexual gender theory incorporates two seeming contradictions. First, gender is a socially constructed phenomenon that transsexuals learn (Bolin maintains this is escalated through a rite of passage). Second, while gender is learned it is based on an ideology of the biology of genital differences. Therefore, subscribing to this logic, transsexuals must have the surgery because women are people with clitorises and vaginas just as men are people with penises and scrotums. While endorsing this view, transsexuals acknowledge that in the course of their physical feminization produced by female hormone therapy and learning the art of passing, they become, in fact, women with penises for a time (cf. Bolin 1987:50–52; Bolin 1988:73–105). Therefore, they know that the genitalia is not the ultimate determinant of gender. They learn to act as women; that is, they *learn* their new gender as women. In addition, they recognize that gender identity, the component transsexuals regard as the core criteria of gender, is independent of genitalia and gender roles, for their gender identities were not congruent with their physical selves. In other cultures the continuity between sex and gender role is far more flexible. Williams' discussion of Native American *Berdache* clearly illustrates this (1986).

Culture and Gender Variance

The anthropological literature offers another view on the subject of gender variance that also suggests the importance of the cultural overlay in understanding gender diversity. Taking a traditional anthropological perspective, the phenomenon of transsexualism is a culture-bound syndrome, one created and given definition by medico-psychiatric and mental health sectors and reflecting a broad western gender schema of sex dichotomization or polarization. A relativistic perspective challenges the "syndrome" of transsexualism by regarding it rather as rooted in the gender system whose rigidity may have potentiated it.

In contrast to traditional psychological views, an anthropological perspective on the subject of western transsexualism focuses on the cultural construction of gender and its learned parameters. For example, in Bolin's research, the central question of "how can males become females in a system where gender is regarded as an ascribed characteristic?" is part of understanding the western gender schema (or belief system about how males and females are distinguished). Following the theoretical model suggested by the ethnomethodological approach of Kessler and McKenna (1978), gender is considered a social construction that pretends to be biological. In this way culture impersonates nature and creates a set of beliefs about what is natural or biological about gender. For purposes of clarification it must be remembered that when referring to sex, we mean those biological attributes related to reproduction, while gender refers to those characteristics that are psychological and cultural: men and women, girls and boys, masculinity and femininity, femaleness and maleness. Central to the western gender schema is the notion that there are only two genders. This is regarded as inevitable, an "incorrigible proposition" about what is constant and unchangeable in reality (Kessler and McKenna 1978). Gender is regarded as not only dichotomous but also as oppositional in western societies. The schema posits only two options: male and female with nothing in-between or outside of this dichotomy. As we shall explore shortly, this is a rather rigid and polarized view of gender and one not shared by all peoples of the world.

Transsexuals interpret their own identity discomfort within this gender schema so that they regard themselves as people born into the wrong bodies. They paradoxically support the western bipolar gender ideology that women are people with vaginas. The male-to-female transsexual rationale is that since they are not men psychologically, they must therefore be women and pursue the sex change surgery. This provides a window on our gender schema so that we can see that social womanhood includes appropriate genitalia and secondary sexual characteristics, both of which can be acquired by surgical and hormonal reassignment. While none of us walk around with our handy-dandy gene scanners or technology to run a blood profile on those with whom we interact, we nevertheless attribute sex or assume certain biological concomitants, even though we do not have x-ray vision to see through clothing to make a genital assessment. Our gender model is different from those in other societies. For example, the kind of work one does may be a more important attribute of gender than are genitalia, once the initial assignment at birth is made on the basis of visual inspection. As we shall see, it is possible for a physiological female to become a social male by taking on the occupation of men. It is an interesting point, that since the gender revolution of the 1960s we in the West have begun to challenge the biological implications of gender roles, e.g., that women are more emotional and men more aggressive, but we have not gone so far as to challenge the system of gender as dichotomous and sex and gender as continuous (Kessler and McKenna 1978). The recent emergence of transgenderism (described earlier in this chapter and chapter 7) with nonoperative transsexualism as a social option along with the blending and blurring of boundaries between transvestism and transsexualism may in fact represent subversion of the dominant western gender ideology.

The fallout from this system is a group of individuals whose identity discomfort has led them to surgical resolution. The concept proposed by Ruth Benedict that cultures may select from a wide range of possibilities, is suggestive here (1959:52–53). While there are numerous individuals whose self-concept may place them in intermediary types of positions (as proposed by Benjamin's model discussed earlier), surgical resolution offers the only hope for a nonstigmatized life and a sense of "normalcy." Other options leave one vulnerable to stigma. In the group of male-to-female trans-sexuals Bolin researched with in the early eighties, covert pressure was applied to select an identity upon affiliation with the Berdache Society. Only two identities were available at that time: transsexual (TS) and male transvestite (TV). If one was TS, then pursuing surgery was regarded as inevitable. In the Berdache Society, the male transvestite identity was a heterosexual one. Transvestites were not defined by the group as having a gender identity conflict, but were regarded as people who cross-dressed for other reasons, including relief from male role strain or **fetishism**, a sexual compulsion toward an object, in this case arousal, through dressing in women's clothing. They were thought of as qualitatively different from transsexual people. However, given another culture and another time, such differences might be considered negligible and people we call transvestites and trans-sexuals might find themselves grouped together under a status designation such as that of the *Berdache* or the preferred "two spirit."

Recently, the community of gender variant persons, represented by diverse national, regional, and local organizations is involved in a process of revisioning gender variance. As mentioned in chapter 7, the term "transgenderism/ist" is a recent addition to western gender variance. Transgenderism is a community term indicating kinship among those whose identities are not satisfied by man or woman as these genders are currently conceived. An emerging state of collectivity has resulted in the recognition that gender variant identities exist on a continuum rather than in opposition. Transgenderism includes traditionally defined transvestism and transsexualism but it transcends those identities as well (Bolin 1993: 460–462).[1]

Holly Boswell defines transgenderism as a "middle ground" in her article, "The Transgender Alternative." She goes on to state that it is "a viable option between a cross-dresser and transsexual person, which also happens to have a firm foundation in the ancient tradition of androgyny" (Boswell 1991:29–31). This is supplemented by the following statements Dallas Denny has made about transgenderists when they are:

> described as persons who change gender roles, but do not plan to have reassignment surgery. They have alternatively been defined as persons who steer a middle course, living with the physical traits of both genders. Transgenderists may alter their anatomy with hormones or surgery, but they may purposefully retain many

of the characteristics of the gender to which they were originally assigned. Many lead part-time lives in both genders; most cultivate an androgynous appearance (Denny 1991:6).

Denny further states:

Even many of those who have chosen to alter their bodies with hormones and surgery, like myself, maintain a proud transgender identity rather than attempting to assimilate into the mainstream culture (personal communication).

Figures 12.1-9 illustrate the faces and voices of transgenderism. The term "two spirit" (formerly *Berdache*), is used commonly by anthropologists, but different cultures have different names such as the Omani *Xanith*, the Tahitian *Mahu* and the Chukchee *Softman*. The *Berdache* have been reported by anthropologists from as early as 1904 in the work of Bogoras among the Chuckchee, to as recently as Williams' historical and contemporary study of Native American Berdachism published in 1986 and Nanda's work among the Hijra's of India (1990).

We shall shift the discussion to a very brief review of the cross-cultural record of gender variance in order to further contrast anthropological interests with the clinical ones. The spectrum of gender variance cross-culturally is a wide one and includes a number of situations in which gendered behaviors (broadly conceived to include demeanor, dress, and activities) relegated to one gender are adopted by those of another in part or in whole. These include ritual or ceremonial occasions such as among the Iatmul of New Guinea in which males will mockingly dress as elderly matrons (Bateson 1958), or as among the Dani where women will don the garb of young warriors on ritual occasions.

Gender-crossing may also include the system of **woman-marriage**, a predominantly African institution, whereby two women marry one another. One becomes the "social male" in the relationship and the other woman is the "wife." The social male may then arrange for a progenitor as in the well-known Nuer case, and then become a social father. Evans-Pritchard (1951) has reported on this as a kin recruitment strategy for barren women. Whether woman-marriage is a cross-gender institution for the creation of social males, or whether it represents another alternative, an intermediate gender status or just an alternative role for women has not been resolved (Blackwood 1984). Whether sex between the wife and the social husband occurred is unclear as well (Blackwood 1984).

The Faces and Voices of Transgenderism

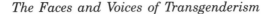

Figures 12.1 and 12.2 Dallas Denny

This is the same individual at 23 and 42 years of age. Differences in appearance are due to age, electrolysis, and the long-term use of estrogens (female hormones). Dallas Denny has had sex reassignment surgery, but this, of course, is not apparent from the photographs. Dallas, a leading authority and author on transgender and transsexual issues, is executive director of the American Educational Gender Information Service, P.O. Box 33724, Decatur, Georgia 30033-0724, telephone (404) 939-2128 and editor/publisher of *Chrysalis: The Journal of Transgressive Gender Identities.*

Biology is not destiny. The technology for changing my body was available, and I took advantage of it in the same way that other Americans use technology to improve their vision or repair physical defects.

I didn't ask to be transgendered. The feelings that my body and social role were not as they should be were always there, and they made it very difficult for me to function in society as a man. There was nothing about being male that was of value to me.

I made the decision to change my body only after careful thought. It was a moral choice based on my gender identity, and not my sexual orientation. I don't ask people to understand or to approve of my decision, nor do I apologize for who I am. I'm proud to be transgendered.

When I appeared to be a male, people who knew that I thought and felt as a woman weren't able to relate to me on any level as a female. Neither did they believe that I could have a viable future as a woman. Now people have difficulty bringing themselves to believe that I could ever have been a man. But only the outside has changed. I'm the same person I always was.

Figure 12.3 JoAnn

My name is JoAnn. I work for Caterpillar, Inc. as a Senior Systems Analyst. I am active with the American Business Women's Association and currently serve on the executive board of the local chapter as Treasurer. This photo was taken at the ABWA Spring Conference in June (1994) in Williamsburg, VA. My age at the time of the photo was 46. I am also active in my church, St. Patrick's Catholic Church. On the advice of some of the parish members, I am being urged to seek a position on the school board there later this spring. We'll see about that. For the past six years I've been heavily involved in fitness activities with aerobics and step. I am the music coordinator for one of the local fitness centers and have recently been asked to consider being an instructor, too. In conjunction with this, I have put together a workshop for 6th–10th grade girls as part of my State University's "Expanding Your Horizons" conference to help young girls become more aware of the career opportunities available to them. The title of our workshop is "Aerobics for Fun, Fitness, and Profit" and is one of about 70 in the conference.

Some vital statistics about myself—I'm 5'6" and weigh 120 pounds (and post-operative). I'm currently 50 years old (birthday in November) and single. I suppose someday I'd like to get married, but it isn't a high priority. I'm really quite happy now.

Figure 12.4 Gitte-Maria

I was born September 9, 1950 in Copenhagen. My parents baptized me Glenn. Already as a child at the age of 3 to 4 years, I found out that I was different from my big brother and the other boys. My feelings, my interests were different; but I was afraid to tell anybody about my feelings. As I grew up and became a teenager, my feelings to be a girl grew stronger (if that was possible, because they were very, very strong all the time). Of course, I dressed when I had the possibility. I lived the life of a closet-TV, that was no way acceptable. I wanted to be and to live the full life of a girl/woman, I just had this terrible fear that everybody would consider me insane if I told them. Therefore I kept my secret for myself and tried to live a life of a "normal" boy/man. I got married in 1973, had two beautiful daughters (one just made me a grandmother). In 1979, I got divorced.

In 1990 I couldn't go on living this double life, a life in disguise. I wanted to fulfill my dreams, make them come true. I had now been thinking over the consequences for over ten years and I found that if I should get something out of the rest of my life, I had to change now and have a Sex Reassignment Surgery, no matter what it might cost me psychologically, socially, my relationship to family and friends and any other costs. At that time I had succeeded to hide my life for my children, they knew nothing about my secret life though they knew that I was different just not in what way and now I had to tell them. They are really wonderful daughters. When I told them and showed them some pictures they just said: We want to see you for real. I think they expected to see a Frankenstein so when they saw me they were very surprised and accepted me at once. Of course we had long talks, conversations about what would happen to me in my life but they were very understanding and supporting. Since that day they have never seen me as a man, nobody has. My daughters still call me dad when we are at home. I will of course always be their father but when we are walking the streets they call me Gitte. I am now waiting to have my surgery.

Figure 12.5 Gwendolyn

I am a male-to-female pre-operative transsexual, living full-time as a woman. Just a brief biography: I am 43 years old; the oldest of nine children, I have five sisters and three brothers; enlisted in the Marines (served two months) and the Air Force (more than 12 years as a journalist); retired from the Army (eight years as a military police sergeant); served in both Vietnam and Saudi Arabia and Kuwait (during the Gulf War); have an Associate of Arts degree; attended Rhode Island College majoring in secondary education, history, and psychology, but did not complete degree requirements; am employed by the Navy Exchange Service and have been married twice.

Figures 12.6 and 12.7 Wendi

Wendi is pictured in her naval uniform and in casual dress. She explains:

I am an out post-operative male-to-female transsexual, currently active in the political arena, as I am one of the four major players in our continuing education and awareness protests at the Michigan Women's Music Festival. Also, I am the sponsor of the New Woman's Conference, a retreat for post-operative male-to-female transsexuals that I hold in September each year.

Figure 12.8 Merissa Sherrill Lynn

Most people would refer to me as a male-to-female post-operative transsexual, or a "new woman." I don't buy into any such categories. Instead, I think of myself as a person who found it necessary to make some biological alterations in order to live, and enable the rest of the world to relate to me properly. My history is simple. In 7/7/42, I was born a Downeast Yankee, and a male with a female identity. I spent the first thirteen years expecting God to fix it. After realizing God was not going to fix it, I spent twenty years looking for an acceptable way to run away and die. I've spent the remainder of my life fixing it myself, and empowering others to do the same.

Although I am religiously dead-set against organized religion, I am an intensely spiritual person, and a philosopher by education and nature. Who and what I am is a God-given gift, and I have an obligation to use that gift in a good way. I believe God (for lack of a better word) made me this way to bring balance to people, the natural world, and to spirit. Everything I have done, and everything I will do, is based on that purpose, to make this a better world in which to live, not just for myself, but for all people.

Everything I founded, the Tiffany Club, the TV/TS Tapestry Journal, the Coming Together Convention, and the International Foundation for Gender Education, was created to fulfill that purpose. The objective was to build a happier, safer, more balanced world in which to live. The way to do that was to help people learn respect for each other, and to make the compromises necessary to enable us to live together. In order to teach people, we needed an army of educators and leaders. We needed a community of people willing to work together for that common purpose. Since our primary issues related to gender identity and gender expression, transvestism and transsexualism was our focus and our foundation. The Tiffany Club, the Tapestry, the Convention, and IFGE were all created to build that community of educators and leaders, and provide that community with the tools that would enable them to build a better world.

The Tiffany Club was one of the original and most successful social/support organizations, and served as the model for dozens of other groups. The Tapestry (now innocuously called Transgender) remains the most widely read and influential transgender publication. The Convention was the only event specifically designed for transgender leadership development, and to address the issues to the transgender community. IFGE (The International Foundation for Gender Education, P.O. Box 367, Wayland, MA 01778, telephone: (617) 899-2212/894-8340) remains the largest and most effective organization addressing transgender issues.

In early 1995, I retired from a leadership role within these organizations and services, to pursue other projects. My intention now is to write the books, tell the tales, and do everything I can to ensure the objective and message doesn't get lost in the cacophony of effort.

Figure 12.9 Michelle

I am a 40-year-old post-operative transsexual woman.

The cross-cultural record reveals the need for a reappraisal of terminology. For example, it is all but impossible from some of the early ethnographic reports to determine the distinction between an alternative gender status or position and a role variant for a gender status. In western terms this might be like trying to distinguish between a female-to-male transsexual and a tomboy. The alternative gender status is referred to by some researchers as a gender transformed status, one in which a new gender status may be created by crossing over or blending statuses. This has also been termed by Martin and Voorhies (1975) as a supernumerary gender, or an additional gender.

In this regard the North Piegan "manly hearts" provide an interesting example. The **"manly hearts"** is an institution of "macho women" characterized by "aggression, independence, ambition, boldness and sexuality," traits associated with the male role in a culture with high gender role distinctions. Manly hearts are typically married and are wealthy. Although they may "act like men" and are considered a blended status by some anthropologists (Martin and Voorhies 1975), it is proposed here that the manly heart is more likely an alternative role for women, in something similar to, but necessarily different from, the western tomboy (Bolin 1996).

The gender-transformed status has been referred to by anthropologists as the *Berdache*. The *Berdache* (two spirit) is usually described in the literature as a genetic male who dresses partially or completely as a female, adopts the female role to various degrees and may or may not engage in heterosexual, homosexual, or bisexual behaviors. The female version of the *Berdache* has been referred to by some as a cross-gender status since the term *Berdache* is derived from the Arabic *bardaj*; a boy slave with sexual duties. Walter Williams has referred to it as the "Amazon" (1986), although this has been contested by some anthropologists. The anthropological literature on this phenomenon for women reflects gender bias in reporting so that accounts of male *Berdachism* far exceed those of females with the exception of Evelyn Blackwood's work on female gender crossing and sexuality among others.

Reports of the male *Berdache* (two spirit) institution reveal how widespread it is. Westermarck (1956) offered the first ethnological overview of the *Berdache* in 1906. Steward recorded its presence among Native Americans for Kroeber's *Culture Element Distributions* (1943, 1973), Devereux (1937) among the Mohave, Hoebel (1940) and Lowie (1935) reported it for Plains Indians; Hill noted it among the Navajo (1935) and Pima (1938), Bogoras for Chuckchee (1904–1909), Evans-Pritchard for the Azande (1970), among others.

Anthropologists are interested in the cultural context for this sort of gender variation and offer a variety of explanations derived from the ethnographic context and through the cross-cultural correlational method in which data from numerous societies may be quantified and compared.

Hoebel (1949) proposed that the *Berdache* (two spirit) represented male protest in societies in which the extant warfare complex demanded excessive male aggression, but Goldberg's test of this hypothesis found no correlation between warfare and *Berdachism* Munroe, Whiting and Hally (1969). Downie and Hally (1961), and Munroe, Whiting, and Hally (1969), correlated Berdachism with societies that had low sex role disparity (differences) rather than high as previously suggested in Hoebel's research. They hypothesized that societies with few differences in sex roles would facilitate the gender reversal of the predisposed individual. Such societies would be more tolerant of those that did opt for *Berdache* status. Levy (1973) in his work with the Tahitian *Mahu* (*Berdache*) makes the same argument but interprets it differently. According to Levy, it is because Tahitian culture has few gender role differences, that the *Mahu* is vital for making differences apparent, lest blurring of genders occur. Wikan's study of Omani *xanith* regards this Middle Eastern version of the *Berdache* as a third gender option that may be chosen intermittently as an avenue for economic opportunity (via homosexual prostitution) and whose secondary function is to preserve female honor. Thayer's (1980) analysis of the *Berdache* among Northern Plains Indians proposed it is an alternative and intermediary status between the secular and the sacred (in Bolin 1987).

Williams' *The Spirit and the Flesh: Sexual Diversity in American Indian Culture* (1986), is noteworthy as the first contemporary study of Native American *Berdachism* in which participant-observation was employed as the field method. Williams argues against views of the *Berdache* as a gender-crossing role (Callender and Kochems 1983), and asserts that the *Berdache* is an alternative gender. Williams notes that this reflects the American Indian gender ideology wherein one's spirit is the salient characteristic of gender rather than the polarity suggested by the western genital ideology of gender. Williams cites the enhanced status of the *Berdache* among Native American groups whose gender relations include a high valuation of women. He suggests that "[b]ecause women are valued, androgyny is allowed, but if women are devalued, are regarded as polluting, female characteristics in males will be denied" (Williams 1986:268–269).

This overview of the *Berdache* has been presented to clarify the distinctions between the anthropological and the clinical approaches to what may be loosely termed "transsexualism." Both anthropologists and clinicians are concerned with etiology but have two very different perspectives. While the clinical is distinguished by its search for biological and psychological factors, the socio-cultural perspective is interested in the cultural construction of gender and how gender variance is embedded in broader systems of gender ideologies and relations within the cultural context. The clinical literature typically focuses on the individual or the family as primary in the formation of the transsexual identity in contrast to the anthropological research which looks to the cultural meanings and gender role expressions in the totality of a gender system. "While the clinical perspective focuses on transsexualism as a syndrome to be explicated and treated", the socio-cultural perspective regards the transsexualism social identity " . . . as a category of people who are stigmatized in our culture . . . " (Bolin 1987:59). Finally, anthropologists question the view of the transsexual identity as a syndrome and rather prefer to analyze the system of relations that spawns gender variance within our own culture and cross-culturally.

Gender Roles Cross-Culturally: A Synopsis

Transsexualism and the *Berdache* are extreme examples of the cultural variety and response to gender identity. Our fascination with gender identity extends to various beliefs concerning gender role, changing gender role behaviors, and how to manage male-female relations in the late twentieth century. Specifically these beliefs will be examined relative to concepts of androgyny and intimacy, as primarily culture-bound phenomenon.

Traditional gender roles refer to those preferred behaviors and expectations that are clearly gender specific and are associated with male and female middle class behavior from the late 1800s to the mid-1960s. These include affective and behavioral characteristics based on gender which tend to emphasize female passive-aggressiveness, overt displays of emotion other than anger; nurturance, intuition, gentleness, and softness. Male behavior is sharply contrasted: aggression, emotional constriction other than expression of anger, rational-logical interpretations, decisiveness, hardness. In fact, in a now-classic study, Broverman et al., 1970, clearly relates culturally defined gender role appropriate behavior with definitions of adulthood and mental health in this society. Basically a mentally healthy adult is a phenotypic male who expresses appropriate gender role behavior. Culturally, appropriate female gender role behavior and expectations generally are considered less healthy. It is a clear statement that what was (is) valued in this culture is male behavior and female behavior is viewed ambivalently at best. For example, motherhood is both exalted and seen as innate, natural, not something which is learned, defined and expressed as the role is experienced through the child's life.

Androgyny is a Greek word combining the words for male *andros* and female *gyny*. It refers to a state of being in which an individual behaves, thinks and emotes in response to the situation, regardless of gender-defined characteristics. As a situation-specific response, it synthesizes culturally defined categories of male and female behavior. For example, fear is a common human reaction to a threatening or dangerous situation. Traditional U.S. male behavior would show no fear; U.S. females would express fear. An androgynous individual would express fear as the appropriate response to the situation regardless of gender. Androgyny extends from behavorial-affective situations to include modes of dress, speech, and demeanor in popular U.S. culture; e.g., gender neutral clothes or colors. Culturally, androgynous behavior and affect are still **androcentric** or male oriented and easier to achieve for females than for males. For example, while both men and women now wear earrings, this was a difficult and not totally resolved barrier for men to cross. Men's earrings tend to be less ornate, smaller and simpler than women's, and worn in one ear only. In contrast, women can freely shop in men's clothing departments and be considered stylish. There is little fear or chance of reprisal or discrimination toward women who do this.

The concept of androgyny may be very culture specific, even within western cultures. Some cross-cultural examples may help to illustrate this. Within other societies which have roots in western traditions such as Latin America, there is a very clear recognition and acceptance of men's culture and women's culture.

Among *Latinos*, men and women know who they are; they accept, sometimes with anger and resentment that men and women are different. They behave very differently in each other's presence, particularly in public, than they do when they are with their own gender. These changes can be subtle—a shift in facial expression or body posture, or more overt—such as changes in the volume, inflection or tone of voice, vocabulary used, or the topics considered appropriate for discussion in a mixed group. Through affinal ties,

one of your authors, Whelehan, belongs to an extended Peruvian
kinship system which encompasses three continents and islands in
the Caribbean. These households are middle class and female centered. However, the presence of a male, particularly an older one,
shifts attention to his needs and away from the females. Females
create their own space in several ways when men appear. Women
can attend to the man's desire for food and drink in order to "take
care of him and get him out of the way." Women can move to
another area of the house or engage in a "female activity" such as
cooking, or going shopping, which takes them out of the house
(Whelehan ND).

An even clearer example is found in Muslim societies in the
Mideast. *Purdah* is the veiling and seclusion of women. It is perceived by many westerners as an extreme situation of female sub-
mission to and oppression by men. However, in *Guests of the Sheik*
(Fernea 1965), Fernea discusses the gradual changes in her perception of *purdah*, and her eventual acceptance of it for herself and
the women with whom she lived. Veiling, which is done in public
and in the presence of nonkin-based men, provides women with a
sense of privacy, space, and protection. Women are not veiled in
their own space at home, which is considered one of the most
important sectors of Muslim society. There they make decisions
that are integral to daily life and the welfare of the family. Veiling
separates them from a male world which they do not necessarily
want to join nor which they necessarily define as superior to their
own. They know that their work and effort maintain the kin group,
a key survival unit in society. To them, androgyny is a very strange
concept in societies of clearly known men's and women's cultures.

Intimacy

The concepts of psychological masculinity and femininity have
been widely discussed in the U.S. since the 1970s by writers such
as Hite (1976, 1980, 1981, 1987), Cassell (1984), Farrell (1974,
1986), Zilbergeld (1978, 1992), and Goldberg (1976, 1980, 1984). In
essence, this research states that men and women have very similar needs and wants in relationships. However, their means of expressing and getting their needs met are different, and are not
necessarily well communicated to or well understood by the other
gender. For example, both men and women state they want emotional bonding and depth in relationships and that trust, being
able to be oneself, and honesty are important. They also state that

they value these needs over genital sexuality, per se. However, according to the work by the researchers cited above, culturally defined ways of meeting these needs and perceptions of the other gender either impede or enhance need fulfillment. In general, men and women attitudinally and culturally still fulfill their **scripts**; those socially defined roles of masculinity and femininity (Gagnon 1973). People's deep sense of who and what they are, regardless of their overt behavior, rests on fairly well ingrained pre-1960's ideas of masculinity and femininity. This is expressed in women when they value relationships, communication, and being physically attractive over sexuality per se and when they assume primary responsibility for the relationship (Hite 1976, 1981, 1987; Cassell 1984). Men express this by using sex as an example of emotional caring, by defining the quality of a relationship sexually rather than through verbal or affective means, or choosing partners primarily on physical characteristics (Farrell 1974, 1986; Zilbergeld 1978, 1992; Goldberg; 1976, 1980; Hite 1981).

The result is that while needs for closeness, trust, and bonding exist, the means people use to express and meet these needs may not achieve this purpose. In addition, pre-1960's rules about male-female interaction have changed. New ones have not been culturally recognized and accepted socially on a wide scale. While some individuals have achieved relationship satisfaction, there is much confusion, anxiety, and miscommunication on a cultural level about male-female relationships. There is also a wider variety of relationships more openly and visibly present in this society today (Blumstein and Schwartz 1983; McWhirter and Mattison 1984).

One aspect of couple relationships that receives much attention is **intimacy** or social and emotional bonding. Development of intimacy or bonding is a hominid and primate characteristic and rests on biobehavorial interaction (Perper 1985). Intimacy develops in stages that include verbal and nonverbal cueing, kinesics, and interaction (Perper 1985). Perper suggests that the most elementary steps toward intimacy may rest in forms of cueing that are universal. Intimacy draws on our social evolution as primates in relation to our need for continuous social interaction and recognition from members of our own kind, species, and group. Paths to intimacy are culture specific and defined. For example, in many nonwestern societies, adult social intimacy is found with members of one's own gender through men's and women's groups, initiation ceremonies, voluntary associations, or extended kin group participation (Murphy and Murphy 1974; Turnbull 1961, 1972; Frayser 1985; Gregersen 1983).

In U.S. culture, intimacy is an elusive goal. There are a variety of books, talk shows, and self-help groups to help us achieve intimacy (e.g., Hite 1987; Goldberg 1976, 1980, 1984). Both men and women in this culture express a strong desire for intimacy in their social relationships and interactions with each other, but seem to have a difficult time achieving it (McGill 1987; Hite 1987; Farrell 1986). In part, this is due to the changes in socio-sexual rules and the lack of new, culture-wide rules to guide male-female interaction. Clear, well-defined roles for male and female behavior and affect no longer exist in this society. At the same time **culture lag** exists. Culture lag occurs when behavior changes faster than the belief systems that support it. There is a lag or gap between how people behave and the consonant belief system that underlies the behavior. So, people behave one way and may hold beliefs or values that do not fit comfortably with the behavior. This may be exemplified by the number of people in counseling for sexual and relationship problems, and the discrepancy between people's sexual behavior and the comfort level and attitudes that accompany one's behavior (Kaplan 1974; Allgeier and Allgeier 1991; Whelehan and Moynihan 1984). Intimacy necessitates an acceptance of interdependency. U.S. values on independence and individuality carried to its extremes in the 1970s and 1980s exemplified in the phenomenon of Yuppies, work against intimacy. At this point, we are ambivalent about issues of intimacy and independence; masculinity/femininity as defined before the "sexual revolution" of the 1960s; androgyny, commitment, and autonomy. Concerns with AIDS and other STDs that impact on fertility and the quality of life intensify this ambivalence and confusion.

Biocultural Aspect of Aging Sexuality in Cross-Cultural Perspective

Dealing with gender identity, gender role concerns, and sexual behavior carry across the life cycle. As we age (which is a **"normal"** physiological process for all animal species!), our physical and sociopsychological expressions of sexuality change as well.

Through the aging process certain physiological changes occur that will affect how people express and experience their sexuality. But we must bear in mind that the aging process of sexuality is as much shaped by cultural norms, values, and expectations as it is by the physiology of the aging process. In addition, class, status, ethnicity, and gender are all intervening variables which may intersect with illness factors which in turn may influence the way sexuality changes through the life course. Therefore cultural features intervene in the biological process of aging.

Unfortunately, there are a number of methodological issues in studying the aging process, particularly as it applies to the elderly. One problem is in defining what we mean by elderly. In the United States elderly is defined as 65 or older and mandated by government issues of subsidy and retirement. But studies of the elderly include a broader scope and may define elderly as beginning at sixty years old (Kinsey et al. 1948). Ours is a temporal definition inflecting U.S. concern with time (Clark and Anderson 1975:335). Cross-culturally, it may be difficult to ascertain the age of a consultant (informant) who may not know their chronological age; and old may be defined in ways other than chronological. The cross-cultural record reveals that old is defined in a variety of ways. On the Irish island of Inis Beag, Messenger notes, "a man is a 'boy' or 'lad' until forty, an adult until sixty, middle-aged until eighty, and aged after that..." (1971:33). According to Messenger, for Mangaians old age is not experienced as a separate phase but a continuation (1971:145). Even our approach to life stages as childhood, adolescence, adulthood, and maturity is a culture specific one (Brandes 1987).

Issues of the definition of sexuality in the aging process are problematic as well. How one defines what is sexual activity may differ from young through middle age and old age, with coitus often a central defining feature of sexuality. This leads to a model of sexuality that eliminates the great variety of nonintercourse behaviors in what Riportella-Muller (1989:213) refers to as an all or none paradigm. In addition, the biological baseline for changes in sexual functioning are based on data from American society which may reflect cultural expectations about the elderly rather than physiological capability.

Although the aging process for women's sexual and reproductive life is clearly marked by **menopause**, the male **climacteric** is not so clearly identified. Sperm production gradually declines through the male life course so that by age 75 a male may be producing only 10% of the sperm he produced before age thirty (Kelly 1990:58). While females lose significant amounts of estrogen and progesterone as they age, male hormone levels of testosterone do decline but how much and with what impact this has on his virility and sexuality is debatable (Angier 1992; Marino 1993).

While there may be no documented major changes in hormone levels, this is not to say there is no male **climacteric**. A male climacteric or mid-life crisis is a culturally acknowledged period in the western male's life cycle where he may experience a variety of symptoms including anxiety and depression (Moss 1978; Henker 1977 in Kelly 1990:58; Masters, Johnson, and Kolodny 1982:170). The "male menopause" as it is erroneously referred to in U.S. culture may in fact be a response to the cultural conception of aging in America. This is a period where the male, whose patriarchal culture has encouraged, advantaged, and celebrated him, confronts a mitigating cultural feature: the youth orientation of North American society. The aging male must face the inevitability of aging and the profound changes in status and prestige that may accompany it. The "male menopause" may well be a response to this male dilemma, for ours is not a society that venerates the elderly. We expect this to vary cross-culturally depending on the cultural meanings embedded in the particular society's ideology of aging.

Kinsey et al. (1948, 1953) found lower levels of sexual activity after sixty, but later studies by Masters and Johnson (1985) and Greeley (1992 in Cross 1993) found much higher levels in the U.S. This may reflect changes in societal attitudes toward sex in general and sex for the elderly. However, there are some real problems in sampling bias in sex research with the elderly. Some researchers have relied on reports from senior centers and others only from married partners (Riportella-Muller 1989:214). There is also a bias in the literature on Caucasians. Masters, Johnson, and Kolodny (1982:170) report the following changes in men over age fifty-five:

> (1) it usually takes a longer time and more direct stimulation for the penis to become erect; (2) the erection is less firm; (3) the amount of semen is reduced, and the intensity of ejaculation is lessened; (4) there is usually less physical need to ejaculate; and (5) the refractory period—the time interval after ejaculation when the male is unable to ejaculate again—becomes longer.

It is very important for individuals to have some awareness of these physiological changes in the aging process. This process and sexual expression are tied into the general features of the physiology of aging in which the body experiences an overall slowing down. This slowing down of sexuality should not be looked upon as an end to sex or as a dysfunction, but a change in one's sexual expression and behavior. It is our societal attitudes about "dirty old men" that may do much harm to the aging man. Western ideal cultural standards set a very narrow age range for socially approved periods of sexual behavior. Sexuality is denied to the young as well as the old in the United States in the ideal cultural norm. This reflects the traditional value on sex for procreation. If our model of sexuality was recreational, elderly sexuality would probably be championed. The elderly man as well as woman who wishes to continue to experience his or her sexuality may be the brunt of jokes and other sanctions among western Anglos. It is as if sex is something that should be outgrown among the elderly. This damage may be first expressed through the male climacteric in anticipation of this cultural expectation.

Between the ages of 45 and 55, women undergo the climacteric. This refers to the time period surrounding **menopause**. It includes the period of **perimenopause**, the five years prior and subsequent to the last menstrual cycle, covering a period of about ten years (The Association of Reproductive Health Professionals 1993:3). Menopause is the cessation of menstruation that occurs as a result of decreasing hormone production of estrogen and progesterone by the ovaries. Over a period of time, a woman's menstrual cycle becomes increasingly irregular, although there are some women whose menstrual cycle just stops. However, because the climacteric is generally a period of irregularity, using birth control for a year after the cessation of menstruation is advised (North American Menopause Society 1993:18,21–23).

"The change of life" in western Anglo populations is said to be associated with a variety of psychological changes in addition to the hot flashes and other physiological changes discussed below. These include reports of radical mood swings and disorders, depression, and greater emotionality (Vliet 1993:14–16). However the cross-cultural record, while sparse, indicates that women's experience of menopause is largely determined by the cultural perception of it and is linked to other factors such as women's status, the meaning of aging, sex, and reproduction. The Inis Beageans, an Irish folk community noted for their repressive sexual attitudes, believe that one of the consequences of menopause is mental illness, so that some of the physiological symptoms reported among western women of hot flashes and mood swings are regarded as signs of insanity. According to Messenger who studied this community . . . some women have "retired from life in their mid-forties and, in a few cases, have confined themselves to bed until death, years later" (1971:15). In contrast, Marshall's study of Mangaia reveals that women experience no shifting of moods and are "good tempered" through the course of menopause (1971:145).

Physiological changes in female sexuality in the latter half of the life course are related to the hormonal changes that occur with menopause. Breast and uterus size is decreased (Kelly 1990:55).

The thinning of the vaginal wall may occur as well due to the reduction of vaginal lubrication as a result of the decrease in estrogen. Both these factors may result in painful intercourse. However, this is not inevitable and treatments are available. External lubricants or application of an estrogen cream may be very helpful in such situations where additional lubrication is needed (Riportella-Muller 1989:216; Mayo Foundation for Medical Education and Research 1993). According to the Boston Women's Health Collective, the only universal physiological changes in menopause are loss of fertility, vaginal thinness and drying, and vasomotor changes or "hot flashes"(1992; also Greer 1992). In 1964 Dr. Robert Wilson began a treatment regimen to offset some of the effects of menopause which includes a loss of the youthful appearance of the skin as well as other reported effects. This regimen is known as ERT (estrogen replacement therapy) or HRT (hormonal replacement therapy) and is today considered highly controversial because of the possible link of ERT with breast and uterine cancer (Utian et al. 1993:11; Boston Women's Health Collective 1992). Other regimens are currently being tested and used including the addition of progesterone or testosterone to ERT. Progestin may be used alone to reduce hot flashes. Presently in the U.S., less than 15% of post-menopausal women are using HRT/ERT (Utian et al. 1993:16).

According to some research on western women, some decline in sexual interest is indicated. However, Butler et al. (1988) note that less than 20% of the post-menopausal women in their study experienced any significant drop in sexual interest, while Rienzo (1985) and Marron (1982) found that women's sexuality was enhanced by not having to be concerned about pregnancy at a time when the "nest" has been emptied (Riportella-Muller 1989:216). Again the problems in sampling methodology as well as ethnicity must be considered before conclusions can be reached. Post-menopausal age and for U.S. women, the presence of a sexually attentive partner, are also factors (Brooks 1993:27–28). In a study of Greek and Mayan peasant women, Beyene (1989 in Browner 1990:569–570) found none of the symptoms associated with menopause in the West. Both Greek and Mayan women looked forward to it as an end to fertility and "reported more interest in sex and improved sexual relations with their husbands" (Browner 1990:570).

Women, like men in western society, experience a dramatic drop in status associated with aging. While the U.S. focus on youth affects both aging women and men, unlike men who may be compensated for the effect of aging on their appearance by an increase in power, women as they age will become less attractive since beauty is equated with youthfulness. Currently, there is no model of elderly or mature beauty. This is related to gender differences in social roles for women and men. It is well documented that in the West gender roles break down into a gender biased dichotomy of action orientation for men and expressiveness and relationship orientation for women. While gender roles are changing, the cultural messages supporting this dichotomy still prevail in a variety of arenas. We can summarize this dichotomy by saying men "do" while women "display." Women have come to be primarily defined in terms of beauty and allure, highly visible adornments and demonstrations of men's success (Bolin 1992:79-99).

This loss of status for women as they age can weigh dramatically upon women's self-concepts as being attractive and sexual. In addition, the same age and sex norms are applied to women that are applied to men as they mature toward the elderly age category. While there is no "dirty old woman" concept in the United States, there is the notion of the "old bag" and the prim and proper matron. Both are regarded as unattractive and sexually unappealing archetypes of the aging women. Like men, women are also stigmatized for their interest in mature sexuality. Homosexual men may also face a similar stigma due to the youth orientation of the bar culture, although the degree to which this is true for those outside this scene and lesbians remains to be determined (Solnick and Corby 1984 in Riportella-Muller 1989:222).

Elderly women may also buy into cultural notions that sex is something they should give up. Kaas (1978) has identified this as the "geriatric sexuality breakdown syndrome" which is a self-fulfilling prophecy for many elderly who are responding to societal norms against sex. The aging individual may come to feel guilty or even that they are deviant for their continued interest in sexuality (Riportella-Muller 1989:219; Brooks 1993:27, 31, 34). However, the medical evidence contradicts this norm. In fact the reverse is true according to Cross (1993:177) who quips one must "use it or lose it". Sex is good for the sexual organs which will shrink in size if not utilized. In fact, women who continued to experience their sexuality after sixty had less trouble with self-lubrication of the vagina (Riportella-Muller 1989:216).

Because women live about eight years longer than men, this can have ramifications upon the sex lives of heterosexual women who are married to men of the same age or older. There is an "imbalance in the sex ratio for those over 65; there are more women than men, and more women left without partners" (Riportella-Muller 1989:212). This problem may be compounded by institutionalization, where sex may be discouraged among unmarried residents by formal administrative policies and by informal practices of employees in such institutions. It should be remembered that only about 5% of the elderly are in homes for the aged at any given time (Riportella-Muller 1989:212), although this should not dissuade us from concern over the effects of institutionalization on the health of the elderly. We will turn now to the cross-cultural record to examine some of the evidence around the issue of aging and sexuality from the middle years through later years of the life cycle.

Cross-culturally the period of the middle years is defined temporarily as the period in which one is not yet old and defined functionally as a period when one's children have reached adulthood (Brown and Kerns 1985). Although cross-cultural research is limited on this period, Judith K. Brown and Virginia Kerns' edited volume *In Her Prime: A New View of Middle-Aged Women* offers an excellent overview of this subject. Brown and Kerns' work contains articles focusing primarily on women. There are unfortunately no "systematic cross-cultural studies of men in their middle years" (Oswalt 1986:161). However, Stanley Brandes' (1987), *Forty: The Age and the Symbol*, offers an important cultural analysis of the meaning of forty in western society. According to Brandes (1987:85), one of the weaknesses in the adult life span literature is its focus on universals to the detriment of class, cross-cultural (author's addition), and ethnic differences. The perception of when one is aging and at what point transitions and stages are demarcated is largely a cultural construct overlying a biological continuum of changes. Thus, Glascock and Feinman (1981 in Oswalt 1986:165), note that maturity and aging are not clearly identified by physiological changes, but rather by other changes in one's life. These could include changes in occupation and work effort, status of children, passing on of inheritance, etc.

Pre-industrial societies do not show the expected variation in the middle years experience of women. In fact "[t]he changes in a woman's life brought about by the onset of middle age appear to be somewhat positive in non-industrialized societies" (Brown and Kerns 1985:2). Three changes accompany transition into the middle years: restrictions may be lifted, authority over certain younger relatives may be expected, and women may become eligible for special non-domestic statuses (Brown and Kerns 1985:2–3).

When women undergo menopause in societies in which menstruation is regarded as polluting or taboo, the post-menopausal woman may gain a great deal more freedom of movement and flexibility in interaction. For example, they may be free to talk with non-kin males and act in indecorous ways in groups in which propriety in young women is demanded (Brown and Kerns 1985:3).

Richard Lee (1985:23–35) reports that !Kung women between the ages of 20–40 years old are required to project a non-sexual image of "shy sweetness." After age 40, !Kung women are given much more sexual freedom. An older woman may have an affair with a young man that may be common knowledge or she may engage in open sexual joking with men (if over about 50 years old). Women's status among the !Kung, which is high to begin with, becomes increasingly higher as they age so they have greater influence in arranging marriages, participating in gift exchanges, and acting in the role of kinship expert.

Cross-culturally older women are given the opportunity to be more influential and exert more authority. This may include access to the labor of children and their spouses as well as a more managerial role in food getting and distribution activities. Control over the distribution of food is one way that informal power of older women is expressed. Finally, aging may provide women access to extra domestic positions such as that of shaman, holy or sacred roles, ceremonial planner, and midwife, among others (Brown and Kerns 1985:4–5).

The cross-cultural spectrum is broad concerning the issue of sexuality among older women and men. For example, Vatuk (1985:147–148) notes in her research in Western Uttar Paradesh and Delhi that men and women are expected to give up sexual relations upon the marriage of the son. In contrast, for the !Kung, a healthy sexuality is accorded even more leeway for the aging woman. Lee notes an interesting marriage pattern of older !Kung women and younger men, a pattern sanctioned negatively in North American society. Approximately about 20% of all marriages at /Xai/xai waterhole are between older women and young men. Following divorce or widowhood it is not uncommon for an older woman to take a younger man as a husband (Lee 1985:30). It would be interesting to explore the beliefs about older women as sexual partners in these kinds of relationships.

Information on sexuality among the elderly cross-culturally is not extensive and subject to the same methodological dilemmas of sex research in general. For example, while human sexuality textbooks may include a cross-cultural discussion of sexuality in childhood and adolescence, like the subject of "middle age," there is little available on the elderly. This is true for the anthropological literature as well, although there is a growing body of research on this subject.

In this regard Davenport (1976:115–163) cites some intriguing evidence from the peasants of Abkhasia who live in the Caucasus region. These people are known for their longevity and continued sexual functioning "long after 70, and even after 100" (1976:118). This group illustrates the relationship between biology and culture in the aging process, reproduction, and in sexual expression. The Abkhasians represent an enclaved genetic population, so there are obviously genetic factors involved in their longevity. Notably, 13% of the women continue menstruating after age 55. According to Davenport (1976:118), "[o]ldsters continue to work, enjoy their food and have heterosexual relations in diminishing amounts well beyond ages at which Western Europeans and North Americans consider such activities to be almost impossible." However, the cultural factors are very important in understanding the Abkhasian sexual vigor at advanced ages. These peasants have no concept of retirement and change at old age. People continue through life doing everything they have always done, including having sex, but to a lesser extent. There are no specific sanctions around sex among the elderly in contrast to the West. The variety of perspectives on sexuality among older people continues to illustrate the richness and diversity of human sexuality throughout the life cycle and the importance of understanding the effect of culture upon behavior and biology.

Summary

Chapter 12: Topics in Adult Sexuality: Life-Course Issues Related to Gender Identity, Gender Roles, and Aging

1. Gender role is an expression of gender identity.

2. Gender role is culturally defined and expressed.

3. The transexual identity and the *Berdache* are explained bioculturally.

4. The concepts of androgyny and scripts were developed in the 1970s to explain gender role expression in U.S. culture. The concept of and desire for androgyny may be culture specific.

5. Intimacy, bonding, and male-female sexual relationships are of great interest and concern to many researchers in late twentieth century U.S. culture.

6. The bio-cultural aspects of aging sexuality are presented cross-culturally. As with other aspects of sexuality, the cultural responses to an older person's sexuality vary widely and differ from those in the U.S.

OVERVIEW

Chapter 13

Sexual Orientations, Behaviors, and Life-Styles

This chapter:

1. Defines and describes the currently known types of sexual orientations.

2. Distinguishes between sexual orientation and sexual behavior.

3. Defines heterosexism and homophobia.

4. Attributes normalcy to the known sexual behaviors as a continuation of what is found in the mammalian and primate world.

5. Does not assume that a heterosexual orientation needs no explanation.

6. Presents various theories which attempt to explain nonheterosexual orientation.

7. Presents a range of partnering relationship forms in the U.S. and cross-culturally.

8. Discusses forms of pre- and post-sexual revolution relationships.

9. Discusses gender role behavior: including liberated, androgynous, and traditional gender roles.

Chapter 13

Sexual Orientations, Behaviors, and Life-Styles

Orientations

It is almost a given in this culture that research and writing on sexuality assumes **heterosexuality** as the norm and starts from there. For those of us who are heterosexual, this seems "normal and natural." For those of us who are identified as **gay, bisexual**, or **lesbian**, however, this appears as a bias and **heterosexist ethnocentrism**. Assuming that as a species we are sexual beings, this chapter discusses a variety of sexual orientations, their possible expressions, and relationship forms.

Some basic definitions are needed. **Sexual orientation** refers to one's attraction to sexual and romantic love partners. Currently, this orientation is perceived in U.S. culture as being **homosexual, bisexual**, or **heterosexual**. A homosexual orientation denotes sexual and romantic attraction toward individuals of one's own gender. A bisexual orientation denotes sexual and romantic attraction toward both one's own and the other gender; this attraction may also be referred to as **ambisexual**. A heterosexual orientation is sexual and romantic attraction toward individuals of the other gender. Given our culture's insistence that people identify and behave heterosexually, it may take self-identified homosexuals and bisexuals time to accept and become comfortable with their orientation. Comments relative to this situation include: "I always knew I was different." "Something wasn't 'right or normal,' but for a long time I didn't know what it was." "I always had girlfriends, but I was always interested in and attracted to boys" (Author's files).

Sexual orientation is not synonymous with **sexual behavior** and these are discrete entities. As with orientation, sexual behavior may be homosexual—sexual involvement with same gender individuals; bisexual—sexual involvement with individuals of one's own and the other gender; and heterosexual—sexual involvement with individuals of the other gender. Relatively nonjudgmental terms used to describe each of these orientations include "**straight**" for male and female heterosexuals; "**gay**" for male homosexuals, particularly for those who are open or "**out**" about their orientation; "**lesbian**" for female homosexuals, and "**bi**" for male and female bisexuals. These terms will be used in this chapter.

A person's sexual orientation and behavior may or may not be consonant. For example, this can occur when one's sexual partner of choice is not available, as in gender-segregated institutionalized populations such as prisons or all boys' or all girls' schools, or where one's gender choice is culturally proscribed. This latter situation frequently occurs in U.S. culture which is overtly homophobic. **Homophobia** is the fear, prejudice, and negative acting-out behavior toward people who self-identify or are believed to have a homosexual orientation. Researchers such as Boswell (1980) and Greenberg (1988) believe that the presence of homophobia in current western cultures is a continuation of practices and beliefs instituted by the Catholic church and found in Christianity since the twelfth century A.D. In part, these homophobic positions are a reaction against the nonreproductive aspects of same gender sexual relations and, the accompanying sex-for-pleasure aspects of them. The repercussions of this prejudice have been felt politically, economically, socially, and religiously to the present. In a homophobic society such as ours, for example, homosexuals' behavior may be homosexual and hidden—they "pass." Their behavior may be heterosexual and they pass; or a homosexual orientation may be expressed openly in communities where they can find support and relative degrees of acceptance and safety—"gay communities" (Kirk and Madsen 1989; Kelly 1988; Blumenfeld and Raymond 1989).

The cause of anyone's sexual orientation is unknown. During the past hundred years in western culture, volumes have been written in the professional and lay literature to explain the roots of one's sexual orientation, particularly if it is homosexual or bi (*Time* 1993). Heterosexuality is assumed to be "normal" and therefore needs no causal explanation (Allgeier and Allgeier 1991). If we examine sexual behavior and orientation from cross-cultural, evolutionary, and interspecies perspectives, we find a wide variety of sexual expression (Vance in *SOLGAN* 1992). Ford and Beach (1951) and others document a range of sexual expression in the mammal and nonhuman primate world that includes hetero, bi, and homosexual behavior (Weinrich 1987). Ford and Beach (1951), Herdt (1981, 1983, 1984 a and b), Williams (1986), Gregersen (1983, 1992), Marshall and Suggs (1971), Frayser (1985), and other anthropologists document the widespread nature of homosexual, bisexual, and heterosexual orientations and behavior cross-culturally and through time (Vance in SOLGAN 1992).

Ford and Beach have documented at least 76 out of 141 societies where homosexuality is acknowledged and receives varying degrees of acceptance for those who identify or behave as such (1951). In addition, Williams' research clearly shows that **third-sex**, **intersex**, or other forms of sexual identities are well integrated into some cultures worldwide. These identities do not translate well into a western, U.S. world view because of our homophobia, heterosexism and cognitive rigidity in the formation of sexual identity boundaries. These identities, such as the *Berdache*, discussed in chapter 12, have a meaningful, respected role in their own cultures (1986).

A variety of Melanesian groups provide a wealth of data that contradicts our western notions of sexual identity, orientation, and behavior. In Herdt's (1984) edited volume on *Ritualized Homosexuality in Melanesia,* ritualized homosexual behavior is examined from spiritual, social, male identity, and gender relations perspectives. Our ideas about sexual orientation and behavior may be culture-bound when compared with sexual behavior that is perceived to be related more to concepts of spirituality, generativity (perpetuating oneself and the group) adult male-female systems of balance and order in the world, and the cycle of life and death. The Sambia from New Guinea will once again be used as an example of an alternative view on sexuality and male-female relations.

Data on the Sambia in New Guinea reveal a radically different approach to homosexuality than in our culture. The horticultural Sambia live in an impacted habitat of perceived limited resources of which ejaculate is also seen as a scarce, precious commodity. As with other "spermatic economies" (Barker-Benfield 1975), semen (more accurately ejaculate), a vital life fluid, is believed to exist in finite quantities. Since semen is seen as life enhancing and a source of male strength, preadolescent and adolescent boys engage in fellatio with other older males to nourish and build their strength and vitality. Male-to-male fellatio is seen as essential for healthy male psychosexual and physical development and preparation for heterosexual marriage and procreation (Herdt 1984, 1993). Women, however, for a number of reasons, are seen as a potential drain to this vital life essence. Therefore, male-female sexual contact, particularly P-V intercourse is carefully controlled and channelled to protect the male from "losing" his energy and to ensure the healthy development of the fetus (Herdt 1981,1993).

As with research done on sexual orientation in the U.S., little cross-cultural research addresses bisexuality and lesbianism (SOLGAN 1992:9–10). This may be due to several reasons. First, male researchers rarely have direct access to women and their daily, intimate lives cross-culturally. The potential for genital sexual behavior that could disrupt indigenous patrilineal and bilineal descent systems makes this access highly taboo. Secondly, there is more interest in male homosexuality than either female homosexuality or bisexuality in western cultures. Thirdly, gender role boundaries are more rigid for males than for females in western culture. For example, the same behavior engaged in by two men as by two women in this culture is more likely to be interpreted as "homosexual" for the men but not for the women. Two men walking arm in arm solicit different labels and responses in this culture than two women walking arm in arm. A similar kind of tunnel vision may be operative in examining homosexuality cross-culturally.

Minimally, the nature of our sexual orientation and behavior is a complex interaction of a number of factors—socio-cultural, psychological, and biological. Our sexual behavior is probably one of the more plastic or malleable behaviors we engage in as a species. Numerous theories in this culture propose biological, genetic, in utero, psychoanalytic, and socio/environmental arguments to explain sexual orientation and behavior that is *not* heterosexual. We have a difficult time accepting that we are sexual beings, and that a variety of sexual orientations and behaviors are part of the human sexual dimension. The only difference in other and same gender sexual behavior is one of reproductive success: other gender sexual behavior may lead to production of viable offspring; same gender behavior does not. However, as discussed in chapter 3, with the loss of estrus not all other gender sexual behavior leads to reproductive success. Witness: kissing, mutual masturbation, oral genital contact, anal penetration, and effective birth control with penile-vaginal intercourse do not result in conception. At various times in this and some other cultures, all these behaviors have been seen as "unnatural," "abnormal," or as a "sin" in certain religious contexts (Bullough 1976).

Biological theories to explain sexual orientation examine hormone levels pre- and post-natally, differences in the size of brain structures as well as the sexual differentiation process in utero (e.g., Kelly 1988; Money 1988a; Green 1987; Gladue et al. 1984; LeVay 1991, SOLGAN 1992). Most recently, a genetic basis for *male* homosexuality has been proposed (*Time* 1993). All of these theories attempt to explain the cause of homosexual orientation, particularly for men. These theories speculate on delays or differences in release of androgens, LH, relative levels of estrogen and testosterone pre- and post-natally, genetic predispositions carried on the X chromosome, or the size of the hypothalamus as predisposers of homosexual orientation. These theories gain some level of support from the self-descriptions of gays who believed from an early age they were more comfortable with and more attracted to members of their own gender. They are biased in that they only look at male homosexual orientation, not orientation as a concept, other orientations, or female behavior (see also Vance's comments in *SOLGAN* 1992).

The psychoanalytic theories often are misinterpretations of Freud's view that we are essentially bisexual in nature, only to have that nature suppressed and channeled heterosexually by society (Freud 1920 and 1959; Kelly 1988). These theories frequently posit a bi- or homosexual orientation as deviant. It is depicted as "arrested psychosexual development" or viewed as a function of a poor parent-child relationship. The American Psychiatric Association (APA), after intensive lobbying efforts by members of the gay communities, removed homosexuality from the list of personality-sexual disorders in 1973. While a label can be removed or included in the DSM-III by the APA, this step alone does not immediately change deeply held beliefs about the behavior labeled. Homosexual orientation and behavior still are not seen as "normal" by much of the medical and lay community (APA 1973; Bjorklund and Bjorklund 1988).

Socio-psychological, environmental, and learned behavior theories on sexual orientation also focus on homosexuality, thereby assuming heterosexuality as the norm. They primarily discuss gay men, not lesbians or bisexuals. This kind of bias implies a negative difference if one is anything except straight in behavior and orientation. These culture-specific theories relate sexual orientation and behavior to childhood and adolescent sexual experiences, role models, learned pleasurable and unpleasant sexual activities (Kelly 1988).

While acknowledging and recognizing these causal explanations for our sexuality, this chapter posits that human sexuality is innate, that it is an interaction of biological and learned experiences whose boundaries are currently unknown to us. Our orientation is part of our socio-psychological and biological make-up. Increasingly, we will probably learn that orientation itself has an innate component (Money 1988a; Small 1993). Each culture channels sexual behavior into defined appropriate outlets. From here, we will examine straight, gay, bi, and lesbian life-styles as functions of their cultural-milieus[1] and as dimensions of the richness and diversity of the human sexual repertoire. Like other behaviors, sexuality is plastic and adaptive given demands of the immediate context.

Bisexuality

There is relatively little research on bisexuality compared to what exists for gays and straights (Klein 1978 and Paul 1984). A bisexual orientation is a romantic and sexual attraction toward people of your own and the other gender. Both men and women can self-identify as bisexual. Behaviorally, they can express either or both aspects of their orientation simultaneously, serially, or only express one dimension of their orientation. The following statements serve as examples. A self-identified male bisexual states: "I am a self-identified bisexual who generally acts out on my attraction to men. However, every once in a while there is a woman to whom I'm strongly emotionally and sexually attracted and I decide to pursue that interest"; or "I believe everyone is pansexual (open to a variety of sexual experiences). I am attracted to both men and women and don't care about society's labels." (female); or "It's natural. It's an extension and expression of the love I feel for some of the people in my life, both men and women." (female) (Whelehan's notes).

Of all the sexual orientations discussed, bisexuality appears to receive the least acceptance, even though relative to behavior its incidence rate has been estimated to be as high as 45% (Klein 1978). On Kinsey's scale which is a measure of behavior not orientation, bi's are a 2-5, clustering in the 3-4 range (Kinsey et al. 1948, 1953; see scale this chapter, figure 13.1). Bi's tend to be very closeted-opening up primarily to other bi's since both the straight and homosexual communities have difficulty understanding and accepting them. Bi's are pressured to choose either a straight, gay, or lesbian behavior and identity, possibly so that the non-bi's are more comfortable. Frequently, gays and lesbians perceive bi's as closeted homosexuals who "pass." The homosexual community politically may perceive bi's as similar to "oreos" in the black community (Bolin 1974). Straights can see bi's as "playing a game," "going through a stage," as being indecisive, or as being homosexual. The fact that some closeted gays and lesbians do engage in heterosexual contact as part of being closeted reinforces their misunderstanding of bisexuals. In addition, since this culture only accepts a heterosexual orientation and behavior as "normal," people can be confused trying to sort out their feelings and behavior. They may experiment sexually with members of their own and the other gender as part of psychosexual growth (Klein 1978). One of the more unfortunate aspects of bisexuality which is also shared with homosexuality is the hiding and passing some bi's engage in because of nonacceptance by the larger society. As Paul posits, bisexuality is part of the range of human sexual behavior (1984). It is probably the most inclusive of the orientations and one that can directly contribute to reproductive success.

Homosexualities

A homosexual orientation is the romantic and sexual attraction to members of one's own gender, frequently called gay when the attraction is openly acknowledged between men and lesbian when it occurs between women. Kinsey et al. estimate that at least 10% of the U.S. population is exclusively homosexual (1948). While Kinsey et al.'s figure may be under-reported since their scale does not clearly distinguish between orientation and behavior, research from other sources including cross-cultural material would support at least that percentage (Walters 1986; Bell and Weinberg 1978; Ford and Beach 1951; Kirk and Madsen 1989). Same gender sexual experience or behavior is higher, probably 50% for boys by the time they are 18 years old and 33% for girls (Elias and Gebhard 1969). A more recent survey from the Kinsey Institute also indicates the incidence to be closer to 50% for boys and girls collectively (*Wellness Letter* 1991:4). Research released in 1993 stating that homosexuals comprise only 2% of the population is questionable since orientation and behavior were confused and criteria by one of the researchers included that his subjects were out to family, friends, and the researcher (*Time* 1993:39).

As part of culture change since the 1960s relative to relationships, homosexuals have become more open and more desirous of formal social recognition (Kirk and Madsen 1989; Bell and Weinberg 1978). These include both gay and lesbian relationships. Gays and lesbians want to live and work openly with their partners, be part of extended kin groups as couples, and receive comparable economic and social recognition and acceptance as do straight couples. As non-accepted people in our culture, gays and lesbians have also established flexible, adaptable gender roles and division of labor patterns relative to economics, household management, and social behavior. Straight couples struggling to redefine gender role expectations and behaviors may be able to use the flexibility of stable gay and lesbian relationships as models (Blumstein and Schwartz 1983; Kirk and Madsen 1989; Bryant and Demian 1990). As a recent survey indicates, a large number of homosexual couples live in stable, satisfying long-term relationships. They have worked out economic, social, and sexual issues. Similar to many heterosexual couples, homosexual couples can have the same types of problems with communication (Bryant and Demian 1990).

Since the Stonewall riots in June 1969 (Kelly 1988; Kirk and Madsen 1989), homosexuals as a group in this culture have become more organized and vocal in demanding equal legal, economic, and social treatment in their personal and professional lives. This involves an end to discrimination and harassment relative to service in the military which currently involves a heated dispute, housing, employment, medical care and in their relationships (Lambda Newsletter 1990). It encompasses respect for them as individuals and the range of roles they fulfill in society, including their roles as parents, friends, and family members.

Lesbianism

While there is a growing awareness and increasing body of literature in this culture on lesbianism (cf. Blumenfeld and Raymond 1989), and cross-cultural documentation of lesbian relationships (Blackwood 1986), lesbian relationships and life-styles, like bisexual relationships, are less well known and discussed than those for gays (SOLGAN 1992 cites). There may be several reasons for this bias. Lesbian relationships tend to be less formalized and ritualized. Females in general have greater flexibility to form female-female bonds and be demonstrative than do males. Thus, lesbians may "pass" intentionally or unintentionally more readily than gays or bi males. Thirdly, female sexuality is not as accepted as male sexuality. It is not seen as having the same kind of force, power, visibility and possible threat as is male sexuality (Kelly 1988; Blumenfeld and Raymond 1989; Blackwood 1986).

Cross-culturally and in the U.S., lesbian relationships manifest a great deal of flexibility and tend to emphasize the interpersonal dimensions of the interactions (Weil 1990; Blumenfeld and Raymond 1989; Blumstein and Schwartz 1983; Bryant and Demian 1990; Herdt 1984b). As with gays, lesbians are parents with well-adjusted children, friends, colleagues, and neighbors. They show, overall, no greater psychological problems than do straight women. In fact, there is research that homosexuals who accept and are comfortable with each other have high levels of self-esteem and may have more stable relationships than average straights (Blumenfeld and Raymond 1989; Blumstein and Schwartz 1983; Kirk and Madsen 1989; Weil 1990; Bryant and Demian 1990; Herdt 1984b). Lesbians and gays may have had to learn to be psychologically strong to confront in a healthy way the discrimination they experience from the larger U.S. culture.

Homosexual Youth

Compared to adolescent Sambian males, homosexual adolescents in the U.S. live in a non-accepting, potentially hostile and rejecting environment if they self-identify or are perceived to be homosexual. Homosexual adolescents in this culture have issues to resolve as well. Not only do they share common adolescent concerns about appearance, peer acceptability, sexual and drug decision making, communicating with adult authority figures, but they also need to recognize and accept their sexual orientation. In a society which places as much emphasis on conformity and normalcy as ours does, bringing a same gender date to a public event, not concealing pronouns when referring to boyfriends and girlfriends, and not wanting to appear "queer" or "faggy" can create considerable stress for a gay or lesbian adolescent. Our homosexual youth have a relatively high suicide attempt rate (Whitaker 1990). Parental and other adult support, validation of the worth of the person, location of support, and homosexual adolescent groups[2] can help gay and lesbian youth accept and flourish as who they are. One benefit of gay liberation is that homosexual youth are beginning to have their own support groups and organizations. These are "safe" places to be themselves and interact (Sullivan 1998). Given this support, homosexual youth do accept themselves and can develop into well-adjusted, functioning adults (Boxer and Cohler 1989).

Heterosexuality

The degree of heterosexual bias in our culture is illustrated by the number of heterosexual inventories that exist and the assumption that people here are heterosexual in both orientation and behavior. We have trouble recognizing different life-styles in our own and other cultures. As Vance has stated, our concept of homosexuality is "only found in modern, Western societies" (SOLGAN

1992:9). For example, in Williams' book on nonheterosexual orientation and behaviors cross-culturally, it is difficult to find vocabulary to label and describe the *Berdache* among Plains Indians, the
nadle among the Navaho, and other forms of nonheterosexual identity and behavior that is common, accepted, and valued in some
other cultures (1986).

In post-World War II, the ideal adult sexual standard was that
of a heterosexual, monogamously married couple with a minimum
of two children—a boy and girl, in that order of gender preference.
The couple owned their own home where the woman worked full
time without monetary compensation and the man worked outside
the home with paid full time employment (Frayser 1985; Blumstein
and Schwartz 1983; Kinsey et al. 1948, 1953). This ideal permeated
until the "sexual revolution" of the mid-1960s (discussed in chapter
12) when these values and behaviors labeled as "traditional" were
behaviorally challenged and questioned. During the sexual revolution many people's behaviors but not necessarily their attitudes
changed. The changes in behavior but not attitudes led some researchers to question whether there was a sexual revolution (Weil
1990; Kelly 1988).

Behavioral changes include more open nonmarital sexuality,
cohabitation, open marriage (O'Neill and O'Neill 1972), higher divorce rates as well as the continuation of the traditional marriage
(Blumstein and Schwartz 1983). As the twentieth century ends,
there are numerous relationship patterns:

1. while over 90% of the adult population marries at least
   once, the traditional marriage comprises about 13% of
   relationships;

2. over 10% of the people cohabitate;

3. about 50% of marriages end in divorce with about 75%
   of these people remarrying;

4. over 25% of the households are single-parented, generally female-headed;

5. serial monogamy is the dominant relationship pattern
   (U.S. Bureau of Labor 1985; U.S. Census 1980, 1990;
   Blumstein and Schwartz 1983; Weil 1990).

There was a slight drop in the divorce rate in the late 1980s,
attributable to the fear of AIDS and other sexually transmitted
diseases; the economic benefits of staying married contrasted to
the economic hardship of separation, divorce, single parenting,

and child support; and the realization that being an older single adult can lead to social-sexual isolation (Weil 1990). Ironically, this drop in divorce rates is not attributable presently to couple's love for each other or their desire to be together as a socio-psychological unit.

While heterosexual marriage continues as a statistical norm, more people, greater than 10%, choose to be single. People are marrying slightly older—in their mid-twenties for both men and women; and there is a bimodal distribution relative to the onset of pregnancy—one in the teens and one in the thirties (U.S. Census 1980, 1990). These variations can be attributed to greater educational, career, and economic opportunity and flexibility for both men and women; behavioral changes in gender role expression and expectations, greater materialism; and a generation of children of divorce who are now adults and who may be postponing marital commitment based on their experiences as children in custodial situations.

Interestingly, when men and women, regardless of their sexual orientation in this culture are asked how sexuality fits into their life and relationships, there tends to be intragender consistency and intergender diversity through time. Women, both lesbian and straight, tend to see sexuality as part of and an expression of the relationship. Men, both gay and straight, tend to see sexuality as a physical pleasure, and a release of sexual tension (Critchlow-Leigh 1989; Blumstein and Schwartz 1983; Shilts 1987; Kinsey et al. 1948 and 1953; Hite 1976, 1981, 1987). Both men and women are orgasmic, enjoy sexual release, and enjoy sex in the context of a love relationship regardless of orientation (Hite 1976, 1981, 1987; Blumstein and Schwartz 1983, Critchlow-Leigh 1989; Goldberg 1984; Farrell 1986; Blumenfeld and Raymond 1989; Bryant and Demian 1990).

Communes

Other alternative relationship styles during the sexual revolution (mid-1960s through late-1970s) included attempts at group marriage and communal living (Crooks and Baur 1990; Weil 1990). Group marriage, more than two adults who agreed to economic, sexual, social, and emotional sharing, did not receive larger culture support and did not last very long. While many of the economic, political, social, and spiritual group-living or communal arrangements that emerged during the 1960s did not endure, some of the older, more established ones continue presently (e.g., in Virginia). These arrangements are intergenerational, include marital and cohabiting couples, and have various rules about sexual exclusiveness. Probably some of the more successful communes in this culture are those formed by various subcultures in this society: Mormons, Amish, Mennonites, and Hutterites. The Mormons no longer legally practice polygyny, more than one wife. All these groups, however, value **endogamy** or in-group marriage, and practice it fairly consistently. They live in extended kinship situations of shared social and economic arrangements, and have worked out a balance with the larger society in order to maintain their identity. Ethnicity is maintained through: common language, shared world view, endogamy, and a relatively homogeneous gene pool reinforced by endogamy (Barth 1969). These communes are highly successful in that they have adapted and survived for over 100 years within a larger society which has experienced rapid, intense, socioeconomic, and sexual change.

Two of these groups, the Hutterites and Amish, are flourishing. The Hutterites have been studied extensively by Hostetler and Huntington (1980) and Olsen (1976). Hutterite communities, farming settlements generally located on the Great Plains of the U.S., maintain a great deal of cohesiveness by carefully selecting and incorporating aspects from the larger society which help them to adapt. These include some farming equipment and minimal levels of electricity but exclude extensive social contact or use of recreational material culture such as radios, movies, or televisions. Family size is large, 8–10 children. Core values of monogamy, extended kin support and obligations, shunning to control and avoid deviant behavior, and minimal socio-sexual involvement with the larger culture are consistently and strongly reinforced. Values placed on hard work, honesty, and respect are shared with the larger society and grant them tolerance (grudgingly at times) from their non-Hutterite neighbors.

An Amish community is located near one of your authors. In an ecologically harsh environment of short growing seasons and six-month winters, this group survives through cottage industries such as quilting and cheese production. Women are deeply involved in both of these activities, as well as taking care of hearth and home. As with the Hutterites, monogamy and large families are encouraged, with minimal social contact with the "outside" (Whelehan observations).

Parenting Styles

Regardless of group structure and sexual orientation, the socialization of children is an ongoing concern in all societies. Parenting within mainstream U.S. culture is undergoing rapid behavioral and attitudinal changes. The average marriage in this culture lasts seven years, comparable to medieval western European marriages. However, our marriages usually end through separation or divorce in contrast to the death of a spouse as occurred in medieval marriages (Stone 1977). At the same time, our longevity continues to increase, with an average life expectancy of 72 years for a middle-class white male and about 76–78 years for a middle-class white female (Metropolitan Life Insurance Tables 1987). In addition, nuclear family size is decreasing largely due to economic and subsistence reasons: it's expensive to raise children to adulthood in a highly technical, post-industrial society. The demography of parenting is also shifting. A bimodal age for onset of parenting exists. At one end are young (under 15 years old) female teenaged parents who encounter socio-economic problems since teenagers are not recognized as legal, social, or economic adults. At the other end are "older" adult parents, women who are in their mid-30s with their first pregnancy. The overall average age of first pregnancy in this culture is about 24 years.

The ideal post-World War II nuclear family is realized in only 13% of our households. Over 25% of households are single parented. While over one million men are full-time single parents, the majority of single parents are women, either by circumstance—divorce, separation, or desertion—or by choice (U.S. Bureau of Labor 1985; Hochschild 1989; U.S. Census Bureau 1980, 1990).

Currently, in two-parent households, the mode is that both parents work outside the home full time. This creates a need for child care—largely in the form of paid, non-kin based arrangements such as day-care centers based on class not ethnicity. In-home division of labor and time spent parenting is highly variable in these situations. Frequently, the presence of a child restructures the adult relationship along more "traditional" lines, where the female assumes primary responsibility for housework and child care responsibilities in addition to full time paid involvement. While fathers are emotionally and socially bonded to their children, their in-home responsibilities usually are not equally shared with their female partners (Hochschild 1989). Hochschild's research, *The Second Shift*, indicates that women in two-income families work an extra month a year with their additional house-hold responsibilities (1989).

Parenting and sex role options are becoming more flexible as indicated by the greater availability of artificial insemination—donor (AI-D), more educational and career opportunities for women, a rise in divorce rates and greater numbers of single heads-of-households (AAUW 1989:5; Department of Labor 1985). With increased tolerance and visibility of alternative life-styles, more homosexual men and women are more openly involved in parenting. Simultaneously it is extremely difficult for gays or lesbians to adopt children, obtain custody or reasonable visiting rights of children from their marriages (Whelehan 1987; Boston Women's Health Collective, 1976, 1984, 1992; Lambda Legal Newsletter 1990; Douglas 1990). However, a few legal decisions have allowed gay and lesbian parents to maintain contact with their minor children. In Virginia recently, a lesbian mother was awarded custody (*Lambda Legal Newsletter* 1990; *New York Times* 1990; *SOLGAN* 1992; NPR Morning Edition June 1994). Lesbians also have the legal option of AI-D, which is available to them through clinics in urban areas or more informally through arrangements made with male friends. Lesbians stand a greater chance of parenting their biological children than do gays. For gays to do so, they must either "pass" in a heterosexual marriage, hope to receive visitation rights if they divorce, or find a woman willing to bear their child and let them raise the child. This latter option, known as surrogate mothering, is a difficult situation for gays or straights, married, or single men. It is probably not a viable option for most gay, bi, or straight men in this culture currently.

From research conducted in the 1970s (e.g., Green 1978, 1979a, b), children of homosexual parents are as likely to be homosexual as are children of heterosexual parents. More current research, which surely needs more follow-up, suggests possibly a familial tendency toward male homosexuality (*Time* 1993). Lesbians do not create man-hating daughters and sexually-confused sons anymore than gays create misogynist sons and sexually-confused daughters. One gay says of his 14-year-old daughter and 10-year-old son, "they're so normal," a statement many of us who are parents would like to make about our children. His children are proud and accepting of who their father is. Their conflicts, shared closeness, and confidences are what many parents in U.S. society hope to achieve with their children (Whelehan's notes).

Single-parent households, which are primarily headed by women, frequently function under severe economic constrictions unless the woman is a middle class career person who chose single parenting (Hochschild 1989; Crooks and Baur 1987). In female-headed single-parent households, there are a variety of parent-child interactions and male role models available to the children in the form of extended kin relations such as mother's brothers or mother's father, non-sexual male friends, and sexual partners (Dugan 1988). This is a well-established, adaptive pattern in many lower income African-American and *Latino* households (Dugan 1988; Stack 1974).

In single parent, male-headed households, the economic situation is usually stabler, more secure, and more comfortable (AAUW 1989:5). Men earn approximately 28–34% more than women in comparable positions, and after divorce experience about a 43% increase in disposable, or available income, while women often experience a drop in disposable income of almost 33% (Faludi 1991:22).[3] It is also more socially acceptable for a male as a single person in this society to seek a female relationship than for a female to seek a relationship with a man. As with women, men rely on paid child care, extended kin, and non-sexual friends of both genders for support and female role models for their children. Use of extended kin and non-kin as a means of socioeconomic support and socialization of children is a continuation of our hominid behavior of **alloparenting**, where nonhuman primate "sisters" and "aunts" care for the young as well as the biological parents.

Since a significant proportion (25%) of families in this culture are single-parent households, this involves qualitative shifts in parenting and child care similar to those mentioned for dual income families. In nonwestern societies, child care can be managed within extended kin and clan relationships. For middle class U.S. society at least, extended kin relationships have been attenuated since the beginning of the century. Increasingly, middle class parents, both single and dual, turn to paid, non-kin based child care, professional child care "experts"—counselors, educators, pediatricians, how-to-advisors and literature for information, support and help in raising children. Blue collar and lower class kin groups tend to tie into social service agencies and extended kin for help in child raising (Rubin 1976).

The present generation of children is also the first generation as a large group to be raised in either single parent or **blended** families, i.e., stepparent households on a wide scale basis. They grew up as children in kin groups who were living through and with the sexual revolution, changed gender roles, sexual behaviors, and divorce. It will be interesting to observe what decisions these children make as adults about finances and careers, marriage, childbearing and rearing, and relationships.

Teenaged Parents

A parenting situation defined in this culture as highly problematic is teenaged parenting, particularly for those teens under 15. We have the highest rate of teenaged pregnancy of any western society (World Health Organization 1988; Wattleton 1990). Over one million teenagers get pregnant each year in this culture. Yet, teenagers in the U.S. are not culturally recognized adults physically, socio-psychologically, economically, legally, or politically. Most of these teenagers have their children and many decide to keep and raise them. It is a cultural situation of children having children. These teenagers have a high pregnancy complication rate of premature births, low birth weight babies, or spontaneous abortions. Often, the fathers of the babies, many of whom are men in their twenties, not teens, cannot and do not economically and socially support the nuclear families they have created (Males 1992:525).

The younger teenaged parents often become dependent on the social service system, which varies greatly by state relative to how much economic, social, and emotional resources are available. These teens can have a difficult time gaining the culturally appropriate skills such as job-training, high school, or college diplomas to help themselves economically or socially. Survival mechanisms include support from extended kin networks particularly in economically poorer, White, Black and Hispanic (*Latino*) situations, and reliance on aid for families with dependent children (AFDC). Larger societal institutions perceive teenaged pregnancy and parenting as an unresolved crisis situation (Wattleton 1990).

Gender Roles

Parenting and relationship forms are based on cultural patterns of gender role behaviors and expectations. Gender role behaviors and expectations are learned, patterned, and symbolic. They are given cultural validity through both verbal and nonverbal forms of expressions. They relate to the culturally appropriate affect, behavior and perceptions ascribed to males and females, comprising in large part what we define as masculine or feminine. Cultural patterning of gender roles is ethnographic, i.e., culture specific. Each culture defines what is acceptable male and female behavior; thus there is much variety as well as continuity cross-culturally. For example, while women are the primary child caretakers from infancy through early childhood, illustrating interculture continuity, the forms of dress, mannerism, and speech patterns expected of males and females can vary greatly from one culture to another, and within cultures from one point in their history to another. Margaret Mead's work in this area, as discussed in chapter 1, is particularly explicative of the variety of gender role behaviors and expressions of masculinity and femininity (e.g., 1935, 1949). (See also chapter 12).

Over the past thirty years in this culture, much research has addressed gender role behaviors and expectations. This research has attempted to isolate causes of gender roles such as the persistent spurious nature-nurture controversy, discuss the range of behaviors, reasons for variations intra and interculturally, and explain continuity, change, and deviations from expected behavior. The research is done by men and women, is interdisciplinary, and is composed of male-female teams from the social and biological sciences. There is a large men's and women's literature available in this western society on this topic (e.g., Blumstein and Schwartz 1983; Bem 1974; Sherfey 1972; Symons 1979). Conceptually, gender roles are often categorized as **traditional, liberated,** or **androgynous.** These terms are value laden and take on political connotations.

The scale of traditional, liberated, and androgynous gender roles can roughly be conceptualized as a fluid continuum with traditional and liberated as more opposed than symmetrical to each other, and androgynous as comprising a wide, gray midsection as explained in chapter 12. It may be analogous to Kinsey's scale of sexual behavior relative to labeling it as homosexual, heterosexual, or bisexual (Kinsey et al. 1948). The following charts compare gender role categories on a continuum to Kinsey's sexual (behavior) orientation continuum.

Gender Roles			Sexual Behavior		
Traditional	Androgynous	Liberated	Heterosexual	Bisexual or Ambisexual	Homosexual
1 Exclusively	2-5	6 Exclusively	1	2-5 Exclusively	6 Exclusively

Figure 13.1 Gender role categories.

Perception is probably as important in interpreting this continuum as is the individual's behavior. Again, behavior (what you do) does not equal orientation (how you feel) relative to Kinsey's scale.

As with traditional gender roles, liberated and androgynous refer to both gender role behaviors and attitudes. The men's and women's literature gives full breadth to what constitutes a "liberated" individual (ref: Hite 1976, 1981, 1987; Goldberg 1978, 1984; Farrell 1975, 1986; Zilbergeld 1978; Greer 1971). It is a highly relative term, most consistently correlated with the concept of choice, a highly westernized, middle class phenomenon. Liberated is a rebellious term. It refers to a state of being culture-free of negatively perceived gender role expectations; that is, traditional roles, regardless of one's gender. It is individually oriented and defined, another phenomenon characteristic of U.S. culture. Since we are all creatures of our culture, liberated is probably more realized in theory than in actuality. It is a state of mind which may be reflected behaviorally and attitudinally by choosing gender role behaviors that are positive for oneself and one's relationships. Liberated is difficult to measure objectively; however, various scales and measures exist to assess androgyny (Bem 1974; chapter 12).

Regardless of the form of gender role behavior, there are several consistent variables. These exist as preferred states, and there may be considerable variance behaviorally and attitudinally from the preferred state. Few individuals consistently fit neatly into one ideal state: traditional, liberated, or androgynous. **Culture lag**, a situation where behavior changes more rapidly than the supporting belief system, can be found with each of these gender role patterns. A person may act one way yet feel very differently about the appropriateness of their actions; for example, a traditional male may be uncomfortable with the degree of emotional restrictiveness that is culturally expected of him. Not any one ideal state is inherently positive or negative. Depending on the situation, the actualization of any of the preferred states may have positive or negative consequences for the individual and group.

These categories of traditional, liberated, and androgynous overlap. There is a wide middle ground in our behavior and affect that incorporates all three states. As theoretical constructs, these may be useful analytic tools; behaviorally, people live the reality of their everyday lives, not theory. Lastly, these categories are culturebound, the context for their expression varies. While all cultures have concepts of maleness and femaleness, their actual definitions and expectations of male and female behaviors vary widely intra and interculturally (see chapter 12).

This chapter emphasizes the diversity in human sexual orientations, behaviors, and life-styles. It does so within the context of the range of known sexual behaviors among mammals, primates, and humans. It seeks an understanding, appreciation, and acceptance of the forms of sexual identities and their expression as inherent in human sexuality. Sexual orientation and behavior, in its most positive sense, enhances who we are and allows for talents in other areas of our lives to be developed. Finally, the chapter looks at relationships as an attempt to meet a larger primate need for bonding and connection, regardless of one's gender choice of partner.

Summary

Chapter 13: Sexual Orientations, Behaviors, and Life-Styles

1. There are several forms of sexual orientation—homosexual (gay or lesbian), bisexual, and heterosexual.

2. Sexual orientation is not synonymous with sexual behavior.

3. We do not know what causes anyone's sexual orientation. However, heterosexuality is assumed to be "normal" in the U.S., and thus is unexplained. In contrast, homosexuality and bisexuality often are seen as stigmatized or variant orientations and have been explained by a number of theories.

4. There is a wide continuum of sexual behaviors in the animal and human worlds.

5. Heterosexism and homophobia are widespread in U.S. culture but are not universal.

6. There exists a wide range of partnering relationships found within U.S. society and cross-culturally.

7. Female and single-parent households are increasing in U.S. culture.

8. Teenaged pregnancy is a common and serious problem in U.S. culture.

9. Blended families or stepparent families are increasing in U.S. culture.

10. Reproductive technology is changing the forms of parenting options available.

11. The sexual revolution of the 1960s in U.S. society was a time of testing traditional sexual boundaries and rules.

12. Culture lag exists in U.S. sexual behavior and values, creating confusion for individuals and groups and giving rise to a variety of gender role behaviors labeled liberated, androgynous, and traditional.

OVERVIEW

Chapter 14

HIV Infection and AIDS

This chapter:

1. Defines HIV infection, AIDS, and opportunistic infections.

2. Describes the characteristics of AIDS as a disease and illness.

3. Discusses incidence and prevalence of HIV infection and AIDS worldwide.

4. Discusses recent developments in treatment of HIV infection and subsequent restructuring of our attitudes toward the disease.

5. Discusses means of transmission of HIV.

6. Discusses education, prevention, and risk reduction of contracting or spreading HIV.

7. Presents a cross-cultural perspective on HIV infection using case studies from three specific populations: U.S. college students, Latin America, and subsaharan Africa.

8. Discusses issues related to HIV testing.

9. Puts HIV infection and AIDS in a human perspective.

Chapter 14

HIV Infection and AIDS

Introduction

Please note that the material in this chapter is current through June 1998. While some information about **HIV infection** and **AIDS** such as modes of transmission, risk reduction strategies, and safer sex has remained relatively constant over the past several years, other aspects such as treatment modalities, incidence rates, and prevalence patterns are changing rapidly. Please contact your local Department of Public Health AIDS task force, campus AIDS coordinator, or CDC (Centers for Disease Control and Prevention) reports for the most updated information.[1]

Easily within the past 1,000 years of human history, there have been several diseases such as the bubonic plague during the Middle Ages, syphillis, and leprosy that have engendered strong physical, social, economic, political, and psychological responses in those individuals and groups where they appear. In the late twentieth century, as a species we are again confronted with a disease which has extensive global and culture-specific ramifications for groups and individuals. This disease, first identified in U.S. culture in 1981 as **GRID**, or **gay related immuno-deficiency disease**, a misnomer, is now identified as **AIDS, Acquired Immuno Deficiency Syndrome**.[2] Currently believed to be caused by the retrovirus **HIV-I** in this country, and **HIV-2** in parts of Africa, **human immunodeficiency virus**, **HIV**, has spread to all continents except Antarctica through at least three different demographic patterns. It is one of our most serious health threats.

HIV infection and AIDS have numerous dimensions and are distinct from one another as will be discussed. They are similar to and different from other acute, chronic, and terminal illnesses as

335

defined in the following section. Depending on whom is infected and the route of transmission, HIV infection may be a sexually transmitted disease (STD), a disease of needle users and sharers, and a disease which can be passed in utero from mother to fetus, or post-natally through breastmilk by an infected woman to a nursing infant (Kennedy, Fortney and Sokal 1989; CDC AIDS Monthly Surveillance Report 1990; CDC AIDS Statistical Surveillance 1994; Stine 1993; IXth International AIDS Conference 1993; Altman 1998). It is important to stress that HIV infection and AIDS is a disease of behavior, not groups, although the prevalence of this disease is seen in some populations more so than in others.

At this point, we believe that the HIV virus is transmitted in five ways: through infected semen, vaginal fluids, blood—including menstrual blood and blood products, perinatally from mother to fetus, and through breastmilk. As a blood-borne, in contrast to airborne viruses such as those causing colds or measles, this virus must pass through skin and mucous membrane barriers to reach the blood. Once in the blood stream, the virus slowly acts to destroy the individual's immune system by eliminating the T-cells and infecting the macrophages which fight infection and maintain resistance to pathogens (Stine 1993). This is HIV infection. Over a period of time, the body is no longer able to resist infection from a variety of pathogens and the person becomes susceptible to **opportunistic infections** or **OI's**. OI's are pathogens such as **Kaposi's sarcoma** (KS), a form of cancer rarely found in young adults, and now occurring less often as the "face of AIDS," i.e., who has the disease, is changing; **Pneumocytis carinii pneumonia** (PCP), which generally does not affect a healthy immune system, and other diseases such as **tuberculosis** (TB), candidiasis or thrush, a yeast infection. These OI's are common in someone with AIDS.

An anthropological perspective—relativistic, cross-cultural, and holistic—is valuable for looking at HIV infection and AIDS as a human, species-wide phenomenon. HIV infection and AIDS is a disease as well as an illness (Fabrega 1974). A disease is the physical manifestation of infection by a pathogen—fever, limb impairment, sores—whatever the clinical signs and symptoms express. Illness is the social response to the disease which can involve psychoemotional, political, economic, behavioral, and symbolic reactions by both the group and individual involved. **People with AIDS**, referred to as **PWAs**, can become physically ill with fevers, chills, or an OI. This is the disease aspect of their situation. The disease is progressively debilitating, which means that eventually the person needs social and medical support from others, health care providers, kin groups, and friends to help the person function. In addition, the modes of transmitting the virus evoke their own social reaction. These factors—the effects of the progression of the disease and the transmission of the virus—are the illness characteristics of AIDS. Since the disease is pan-human, it is not only a cross-cultural, species-wide phenomenon, but a culture-specific one as well, as each society draws on its own means of responding to this disease and illness.

While the transmission of the HIV virus is behavioral, three demographic patterns of infectivity are identified (Mann et al. 1988). Pattern I is found largely in Europe, North America, parts of Latin and South America, and the Caribbean. Many people infected with the virus in this pattern are men who have sex with men and men who have sex with both men and women,[3] needle sharers, hemophiliacs, and the sex partners of these individuals. Pattern II is largely found in Africa, particularly subsaharan coastal areas and a belt across Central Africa—along trade routes and major rural-urban migration routes. It is also becoming more common in parts of Latin America, particularly Brazil (Brooke 1993). In this pattern, many of the people infected are behaviorally heterosexual men and women, their sex partners and the young offspring of infected women (Mann et al. 1988; WHO 1989; IXth International Conference on AIDS, Plenary 1993). Pattern III found in Asia, the subcontinent, Micronesia, and Melanesia generally involves people who have been infected by those people from countries in Patterns I and II. However, given the mutability of the virus—there are at least five strains of the virus currently in the U.S. (ATFCNY 1993), the difficulty in creating the necesary behavioral and attitudinal change to prevent the disease, the stigma of AIDS, and the long incubation period from infection to appearance of symptoms, currently estimated at nearly 11 years (Stockholm Conference 1988; ACHA Conference April 1989; IXth International AIDS Conference, Plenary 1993), these demographic patterns are shifting and are likely to change in the future. As better diagnosis and reporting become available, these patterns may also change or new ones appear.

Incidence and Prevalence

Incidence refers to the number of new cases of a disease or behavior in a population within a given period of time. **Prevalence** refers to how dispersed or widespread a disease or behavior is within a population. As of 1998, HIV infection and AIDS are global concerns. They are found worldwide in over 162 countries as the disease continues to spread (ACHA/NAFSA 1989; Parker, Herdt & Carbello 1991; IXth International AID Conference 1993; Mann and Tarantola 1998: 82–83). As of July 1998, the World Health Organization (WHO) estimates that there are more than **40 million** people worldwide who are infected with HIV and more than 12 million people worldwide who either have "full-blown" AIDS or who have died from it (CDC AIDS Statistical Surveillance 1994; NBC Nightly News 1994; Mann and Tarantola 1998: 82–83). As we stated, epidemiologists have noticed three patterns of prevalence. Pattern I, found largely in North America, parts of Latin America and Europe presently involves an eight (8) to one (1) male-to-female ratio of people with AIDS. Pattern II has a 1:1 male-to-female ratio and currently is largely found among men and women who have sex with each other. Pattern III is replicating Patterns I and II. It is believed to occur in these areas as a result of unsafe sexual and needle use contact with people from areas I and II[4] (Mann et al. 1988; WHO 1990; IXth International AIDS Conference, Plenary 1993; Mann and Tarantola 1998: 82–83).

In the U.S. the ratio of men to women with AIDS has slowly changed from 10:1 to 6:1. Currently, the incidence rate in the U.S. is starting to shift from middle class white gays to needle users and sharers and their sex partners, women, and adolescents. As a group, women in the U.S. are showing one of the three fastest increases in the incidence of HIV infection (Rosser 1993; Nevid 1993; CDC Reports 1993, 1994, 1998). It is estimated that by the year 2000, 60% of the people in the U.S. with AIDS will be women (CDC Reports 1993, 1998). Consequently, their concerns will be addressed throughout this chapter. From 1988 to 1990 the number of adolescents diagnosed with AIDS increased from 200 to 2,000 (MMWR 1988, 1990). While there can be a number of reasons for these shifts which include better diagnosis and reporting, as well as changing definitions of what constitutes AIDS, the reality is that HIV infection and AIDS are not isolated geographically, ethnically, or by any other category.

Several variables make tracking the incidence and prevalence of the virus and AIDS difficult. One, the virus mutates rapidly within and between populations and even within the infected individual. As stated, there are five known strains of the virus in the U.S., and a new strain of HIV-1, group O, has been discovered in France (ATFCNY 1993; Medical Alert 1994:8). The virus can "hide" in the lymph nodes for indeterminate periods of time before it can be detected in the blood, making earlier diagnosis of infection more difficult (Stine 1993:62). The illnesses resulting in an AIDS diagnosis which may develop from these strains include **AIDS dementia**

complex, wasting syndrome, KS (Kaposi's sarcoma), or PCP (Pneumocytis carinii pneumonia). Secondly, while AIDS is a reportable disease, HIV infection is not mandatorily reportable in all 50 states in this country. Not all cases of HIV infection and AIDS are known or accurately diagnosed. Symptoms of HIV infection and AIDS can mimic numerous other diseases such as a bad case of the flu, mononucleosis, or Epstein-Barr Virus Syndrome. Misdiagnoses are easy to make particularly for health care workers not well experienced in the signs and symptoms of the disease. Until January 1993, gynecologic disorders were not included in an AIDS diagnosis. A number of women with HIV infection and AIDS who *only* developed gynecological problems *never,* until recently, were diagnosed with AIDS and thus were unable to receive the treatment and supportive services that can accompany an AIDS diagnosis (CDC 1993; FOCUS 1993). They also did not appear in AIDS statistics. Fourth, the stigma associated with the disease and the very real economic, social, and medical discrimination that people along the HIV+ continuum experience forces many who are infected underground. Issues relating to confidentiality and anonymity discussed later under "Testing," the "right" of partners, family members, or colleagues/ coworkers to know vis-à-vis the individual's right to privacy involve legal, ethical, and personal dilemmas for the infected individual.

People delay seeking diagnosis and treatment for fear of reprisal as well as the fear of knowing they may be infected; denial is a major characteristic of the disease. For example, the Castro, the gay district in San Francisco, has been targeted by insurance companies as a "high-risk" district. People living there have a difficult time getting or keeping health and life insurance (Shilts 1987). A dramatic example of discrimination occurred in December 1987 in Florida. Three hemophiliac sons in a family are HIV+. The family's house was burned, the children were not allowed to attend school; and the family had to relocate (*New York Times* December 1987 in Stine 1993). Similar instances of HIV infection discrimination occur across the country (Hevesi 1990; OpEd 1990; Stine 1993). Fifth, people do not die of AIDS per se. They die of OI's or heart or kidney failure due to the virus' destruction of the immune system. Unless individuals and their families want to disclose an AIDS diagnosis on the death certificate, AIDS goes unreported. Again, the stigma associated with this disease promotes hiding its presence. Sixth, outside the U.S., particularly in Third World countries, diagnosis, treatment, and reporting are even more difficult due to limited resources and isolation, as well as the stigma associated with the disease (IXth International AIDS Conference 1993; Mann and Tarantola 1998: 82–83). Statistics from these areas may be even more skewed and under-represented than in the U.S. (Patton 1992). For example, by May 1, 1989, there were 1,892 cases of reported AIDS in Zambia. This is believed to be massive underreporting (WHO 1989). Relative to infection with HIV, as of July 1998, it is believed that there are 6,000,000 people infected with HIV in South and Southeast Asia; 3–5,000 in India; over 1.3 million infected in Latin America, and over 860,000 infected in North America (IXth International AIDS Conference, Plenary 1993; Maticka-Tyndale et al. 1994:206; Mann and Tarantula 1998: 82–83).

In the U.S. as of 1990, the most rapid increase in the incidence rate is among children, adolescents, women, and needle sharers. AIDS is the major cause of death for women 24–29 years old in New York City (CDC 1990; Altman 1989; Klass 1989; Rathus, Nevid, and Fichner-Rathus 1993).[4] It is the fifth major cause of death for women between 25–44 years old and the leading cause of death for men 25–44 years old in the U.S. (ATFCNY 1993; Kelly 1994). It is estimated that 1 in 500 college students in the U.S. is HIV+, a statistic which has remained constant since 1990 (Keeling 1990; ACHA conference April 1989). As of June 1998, 641,086 cases of AIDS have been diagnosed in the U.S. (CDC HIV/AIDS Statistical Surveillance Reports 1998a). People of color in the U.S. are disproportionately infected with the virus. Over 80% of the women with AIDS in this country are Black or Latina (Crawford et al. 1989; CDC Hotline 1993; Rathus, Nevid, and Fichner-Rathus 1993). Though Latinos comprise 8% of the U.S. population, they comprise 15% of the reported AIDS cases to the CDC in 1988 (Singer et al. 1990; CDC 1989). These incidence rates continue to increase as of July 1998 (CDC HIV/AIDS Statistical Surveillance Reports 1998a).

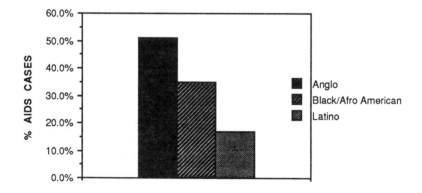

Figure 14.1   Ethnic distribution of AIDS in the U.S. as of 1990.

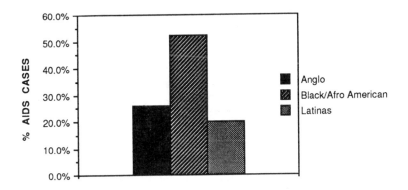

Figure 14.2   Women and AIDS: Distribution by ethnicity in the U.S.

Blacks and Latinos represent about 12–15% and 8% of the total population, respectively. Since Blacks and Latinos have differential access to health care in this society, these figures probably reflect underreporting.

Contrary to predictions made by some researchers, it does not appear that there will be an explosion of HIV infection and AIDS in the U.S. among self-identified heterosexuals (Masters, Johnson, and Kolodny 1987). Rather, the incidence within this population shows a slow but steady increase of 1–2% a year (Padian 1987; Padian et al. 1987; *Newsweek* 1990). This is insidious. By the time the rates are "statistically significant," the prevalence will be devastating. The disease could wreak the same havoc as it has among self-identified gays and bisexual men in the U.S., and among heterosexuals in Africa if people wait to change their behavior until the incidence and prevalence rates are statistically "meaningful." In addition, these statistics are misleading since, as with any stigmatized behavior and disease, there is under-reporting due to people not seeking help or diagnosis and due to misdiagnosis.

ASIDE: It is interesting (and upsetting) that while most First World, western presenters at the IXth International AIDS Conference recognized and accepted that increasing numbers of men and women worldwide, who have unprotected sex with each other are at risk for HIV infection, that the prevention emphases continue to be directed toward women who have sex with men and men who have sex with men. Only one paper at the conference specifically addressed strategies to encourage men who have sex with women to use condoms. There seems to be a shift in perception and consequently a placement of responsibility on women alone as opposed to *both* men and women who are sexually active to consistently and responsibly practice risk reduction.

We do not have accurate data on the number of women who have sex with women who have HIV infection or AIDS. As of July 1998, the CDC does not have an "exposure category" (i.e., means of transmission) that includes women who have sex with women (CDC Statistical Surveillance Report 1994, 1998a). The heterosexism is acute in this situation, particularly since there is a potentially dangerous assumption and belief held by some people that self-identified lesbians are "God's chosen" or "safe" from HIV infection. It is neither your identity nor group membership that makes you susceptible to HIV; it is your behavior. Lesbians who have receptive penetrative sex with men for whatever reason or who share vaginal fluids, menstrual blood, or other blood through sex toys, S/M activities, or other risky behavior with each other may become infected. A 1993 study of lesbian and bisexual women living in the Bay Area by the San Francisco Department of Health revealed high risk sexual and needle using behavior by these women with other women and men. It was not possible to tell whether their HIV infection rate was due to sexual or needle using behavior as these women engaged in multiple risky behaviors (SFDH 1993).

Research appearing between 1988 through mid-1989 reveals other trends in the disease. We have noted the difference in pediatric and adult AIDS. By gender, it appears that women have a more rapid progression of the disease than men; they die sooner after diagnosis than do men. Whether this is due to delays in seeking health care, hormonal effects of estrogen, a more compromised immune system prior to infection, or a combination of these variables is unknown (e.g., WAN Newsletter 1990; Klass 1989; Altman 1989).

Age is another variable. People over 40 with AIDS appear to die sooner than people in their twenties. People of color tend not to live as long as Caucasians (Singer et al. 1990). People diagnosed with AIDS in their teens appear to live the longest. Again, reasons for this are unknown (Williams 1989:1, 8)). It is known that chances for developing AIDS increase the longer one is HIV+. The average time from seroconversion to HIV+ status, where antibodies to the virus in your body are sufficient to be detected, to development of symptoms currently is about 11 years. In the late 1980s through the mid-1990s, the average life span from AIDS diagnosis until death was about four years for white middle class gay men in the U.S. With the development of the "cocktail," i.e., combination drug therapies, in the mid-1990s, there is hope that HIV infection/AIDS will become a manageable, though chronic and currently incurable illness. Depending on one's access to these new drugs, tolerance for their side effects, and one's overall state of psycho-physical health, the longevity rate currently is highly variable. Since 1996, the death rate from AIDS in the U.S. has declined 44%. The decline is attributed to the effects of these drugs (Bartlett and Moore 1998: 84–85).

As a balance, risk reduction strategies can prevent infection, thus decreasing the seroconversion rate and thereby reducing the incidence rate. In a study of gays in San Francisco who practiced safer sex consistently, the incidence rate has decreased steadily over the past several years. In this group, there was a zero seroconversion rate from HIV- to HIV+ status in 1987 and 1988 (ACHA Conference April 1989; Stall, Coates, and Hoff 1988; Winkelstein et al. 1987). Follow-up to this research presented at the IXth International AIDS Conference and current research indicate that consistent use of risk reduction leads to less infection and engaging in sexual risk taking can result in an increase in the infection rate (1993 JAMA HIV/AIDS Information Center 1998). Since behavior change is our only prevention at this point, it is important to recognize that adoption of safe and safer sex practices is effective. It is important that these practices be internalized and acted on consistently to prevent the further spread of infection. A widespread preventative vaccine probably will not be available until well into the next century (Baltimore and Heilman 1998:103). A cure appears to be even more futuristic (Matthews and Bolognesi 1988; IXth International AIDS Conference 1993). HIV infection can be prevented until that time through conscious sexual attitude and behavior change.

Transmission

The means of transmission of the virus stigmatizes the disease and those infected with HIV. As stated, there are five routes of transmission: through contact with infected semen and vaginal fluids, infected blood and blood products, infected breastmilk, and through perinatal transmission from infected mother to fetus (Shilts 1987; Mann et al. 1988; Heyward and Curran 1988; Kloser 1989; Kennedy, Fortney, and Sokal 1989).[5] The virus is transmitted through specific kinds of sexual behavior, sharing needles of any kind, and from mother to fetus/child perinatally or post-natally through breastmilk. The sexual behaviors which can transmit the virus include unprotected penetrative sex: penile-anal, penile-vaginal, oral-genital (both cunnilingus and fellatio), fisting or the insertion of (ungloved) fingers or hands into the vagina or rectum; sharing of sex toys such as dildoes and vibrators or any other forms of sexual behavior in which semen, blood, including menstrual blood, and vaginal fluids may cross tissue membranes and enter the blood stream. The **insertee** is more vulnerable to infection in any penetrative sex behavior than the **inserter** for reasons to be discussed under safe/safer sex. This means that in penetrative sexual encounters between men and women, the woman is more vulnerable to infection, particularly in unprotected receptive P/A and P/V intercourse, though men are also at risk. Because sexual behavior in general is culturally regulated with a variety of rules about communication, honesty, disclosure, openness, and acceptable practices, keeping oneself sexually healthy may be problematic. In U.S. culture, for example, there are many visual and verbal cues about sexuality, a lot of penetrative sexual behavior maritally and nonmaritally across the life span, but not a lot of openness about discussing one's sexual needs, desires, and limits, or communication about pregnancy, STDs, and HIV protection. People generally are not socialized in this culture to feel comfortable and knowledgeable about their sexual and reproductive selves and how to talk about this aspect of their lives. For example, we are inundated with ads and programs in the media which sell both sex and various products. Simultaneously, there continues to be opposition to sex and AIDS education in public schools. The incidence of unplanned pregnancies, particularly among younger (under 15) teenagers, is endemic. The incidence of incest and child sexual abuse in this culture is at least 10% for children under 18. There are a number of people in sexual and relationship counseling. There are available a plethora of self-help books on sexuality, relationships, and being attractive. These issues form the basis of many of our television talk shows (Atwood 1990; Planned Parenthood 1988–89; DHEW 1983; CDC 1990).

Compared to other societies across space and through time, we are rather restrictive in the kinds of sexual relationships we accept (Bullough 1976; Williams 1986; Gregersen 1983). In "post-sexual revolution" USA, we continue to be most accepting and comfortable with adult marital, heterosexual, monogamous bonding and those relationships which simulate this model (Blumstein and Schwartz 1983). We become progressively less comfortable with open digressions from this model: child and adolescent sexuality, lesbian, gay and bisexuality, the never married, and sex workers or prostitutes (Kelly 1988; Blumstein and Schwartz 1983; Williams 1986; Alexander and Delacoste 1987; Pheterson, ed. 1989). When initially categorized, AIDS was called GRID—Gay Related Immuno Deficiency Disease, a label that not only stigmatized the disease and those with it, but associated the disease with groups, not behavior. This stigma and association of the virus with groups, not behavior persists, an association which makes prevention and behavior change difficult. In reality, the sexual behaviors which put people at risk are engaged in by individual human beings, regardless of their sexual orientation. Some populations reflect a higher prevalence of HIV infection than do others.

Blood sharing, through contaminated needles or blood products, is another major means of transmitting HIV. Initially, injection drug users (IDUS) were labeled as the high-risk needle using group. While IDU definitely can transmit the virus through "dirty" or contaminated needles ("works"), this is as misleading a category as labeling AIDS a gay disease was with the acronym GRID. Needle sharing of any kind that carries the virus through blood either in the needle or the needle/syringe can spread the virus. This includes IDU such as occurs with heroin, speed, or crack; ear and body piercing with shared dirty (i.e., contaminated) needles; acupuncture with dirty needles; shooting steroids which is found in about 6% of high school males, as well as in 11–18% of male college athletes (ACHA 1989; McCammon, Knox, and Schacht 1993:248); going for "blood" brothers or sisters where friends make a small cut in their fingers and press the wounds together as a form of bonding; tattooing, or other activities where the blood of one person can be transferred through needles to another.

> **ASIDE:** One of your authors knows of a practice on a college campus where some fraternity members carve their fraternity letters into their arms in a shared ritual.

In this vein, blood donating is absolutely safe. A clean needle is used each time a person gives blood. Receiving blood through transfusions or blood products such as plasma or platelets as in situations of hemophilia, has been made much safer since 1985, when tests were developed to detect antibodies to the virus in the blood. As of April 1989, receiving blood or blood products is relatively risk free. About 300-400 pints of contaminated blood exist in a given year (Nevid 1993). As of July 1998, it was estimated that the risk of receiving HIV infected blood from one transfusion is about 1 in 450,000–600,000 units (CDC 1998b).

The last major means of known HIV transmission is from infected mother to fetus in utero, during birth, or through infected breastmilk to a nursing infant (Kloser 1989; Kennedy, Fortney, and

Sokal 1989; Altman 1998). Maternal-fetal transmission may occur before or during the birth process, regardless of whether the birth is vaginal or caesaerean (Vermund et al. 1991; Heyward and Curran 1988; IXth International AIDS Conference 1993). Neonatal AIDS is much different than older pediatric AIDS or adult AIDS. Anywhere from 67–85% of the children born HIV+ seroconvert to HIV- status by age two. Somewhere between 16–33% of babies born to HIV+ mothers are actually HIV+. A child is born with the mother's antibodies, an important characteristic because these antibodies can help an infant successfully ward off pathogens in its new environment, the external world. Thus, an HIV+ infant born to an HIV+ mother may be carrying her antibodies or may be actually infected. Monthly or bimonthly blood tests to assess the baby's serostatus will eventually determine whether the child is actually infected or merely carrying the mother's antibodies. An infected child will no longer retain the mother's antibodies between 18–24 months and will test HIV+. However, devastating this news is for the families of the 16–33% children who are HIV+, this incidence rate is lower than previously believed (*New York Times* March/April 1993). This incidence rate has definite implications for HIV+ women who are considering becoming pregnant, who are pregnant and deciding whether to risk giving birth to an HIV+ baby and for the counselors/health care personnel who are involved in her decision making and pregnancy. Of those babies who are HIV+, most die by the time they are nine, a significant increase in longevity from 1987 when most HIV+ infants and children died by age two. Breastmilk transmission of HIV occurs. Whether or not an HIV infected woman should breastfeed an infant is a complicated decision. Currently, breastfeeding is enmeshed in international controversy about ethics, economics, politics, cultural definitions of womanhood, and the availability of potable water (Altman 1998) (ref. Kennedy, Fortney, and Sokal 1989). Perinatal and pediatric AIDS is one of the more puzzling and less well understood aspects of this disease. Pediatric AIDS occurs in people less than 13 years old (CDC Hotline 1994). It is also an area of great interest and concern to researchers and health care workers since the incidence rate of neonatal and pediatric AIDS is rising relatively rapidly. For example, in the greater New York City area, it is estimated that an HIV infected baby is born about every 11 minutes in the larger city hospitals (Altman 1989; CDC Monthly Surveillance Report 1990). As of March 31, 1993, there were 1,192 cases of pediatric AIDS in New York State and 5,432 in the U.S. as of December 31, 1993. The following sections discuss each of these modes of transmission in detail.

Prevention and Risk Reduction

HIV infection and AIDS are hard to get. HIV infection, as a blood-borne disease, is not spread by casual contact such as being in the same room with an infected person, hugging or touching the person, or sharing a meal with an infected person. Since the virus is transmitted only through blood including menstrual blood and blood products, semen, vaginal fluids and breastmilk, if you do not exchange these fluids with another person, you will not spread or contract the virus. This information has remained constant over several years. Exchanging these fluids is largely a matter of choice and conscious decision making. Since a cure or vaccine is not expected until the twenty-first century (Stall et al. 1990; Baltimore and Heilman 1998), prevention through education and accompanying behavior and attitude change are our only means of not spreading the disease. HIV infection is totally preventable.

Prevention involves decision-making, assessment of individual areas of risk taking, communication and negotiation skills, and relatively high levels of self-esteem, or belief in your own worth and integrity. Prevention is all we have to halt this disease; prevention can be life enhancing and enjoyable. Prevention works to slow the incidence of HIV infection or number of new cases, and prevalence or how common the disease is in a population. Studies in San Francisco, California, an epicenter of the disease, on self-identified gays who engage in safer sex behaviors, show a dramatic decrease in the incidence of HIV infection (Stall et al. 1990; Stall, Coates, and Hoff 1988; Winkelstein et al. 1987).

Safer Sex and STDs

AIDS is a multifaceted disease. Since one route of transmission is sexual, AIDS is classified as an STD. The following chart on pages 348 to 352 chart compares AIDS with other STDs (taken from Crooks & Baur, chart on STDs).

Safer Sex

Sensuality is experiencing the world through one's senses: sight, sound, touch, smell, taste. Sensuous behaviors can enhance one's sense of sexuality, increase one's body awareness and intensify sexual and non-sexual pleasure. Sensuality is sexually safe. One cannot give or receive sexually transmitted diseases, HIV infection, impregnate, or be impregnated through sensuality. **Safe sex** involves all those behaviors that keep one disease- and pregnancy-free. These

Table 14.3
Common Sexually Transmitted Diseases (STDS):
Mode of Transmission, Symptoms, and Treatment

STD	Transmissions	Symptoms	Treatment
Bacterial vaginosis	The most common causative agent, the *Gardnerella vaginalis* bacterium, is transmitted primarily by coitus.	In women, a fish or musty smelling, thin discharge, like flour paste in consistency and usually gray. Most men are asymptomatic.	Metronidazole (Flagyl), ampicillin, or amoxicillin
Candidiasis (yeast infection)	The *Candida albicans* fungus may accerlerate growth when the chemical balance of the vagina is disturbed; it may also be transmitted through sexual interaction.	White, "cheesy" discharge; irritation of vaginal and vulvar tissue.	Vaginal suppositories or cream, such as clotrimazole, nystatin, and miconazole
Trichomoniasis	The protozoan parasite *Trichomonas vaginalis* is passed through genital sexual contact or less frequently by towels, toilet seats, or bathtubs used by an infected person.	White or yellow vaginal discharge with an unpleasant odor; vulva is sore and irritated.	Metronidazole (Flagyl), effective for both sexes

| Chlamydial infection | The *Chlamydia trachomatis* bacterium is transmitted primarily through sexual contact. It may also be spread by fingers from body site to another. | In men, chlamydial infection of the urethra may cause a discharge and burning during urination. *Chlamydia*-caused epididymitis may produce a sense of heaviness in the affected testicle(s), inflammation of the scrotal skin, and painful swelling at the bottom of the testicle. In women, PID caused by *Chlamydia* may include disrupted menstrual periods, abdominal pain, elevated temperature, nausea, vomiting, and headache. | Tetracycline, doxycycline, erythromycin, or trimethoprim-sulfamethoxazole |
| Gonorrhea ("clap") | The *Neisseria gonorrhoeae* bacterium ("gonococcus") is spread through genital, oral-genital, or genital-anal contact. | Most common symptoms in men are a cloudy discharge from the penis and burning sensations during urination. If disease is untreated, complications may include inflammation of scrotal skin and swelling at base of the testicle. In women, some green or yellowish discharge is produced but commonly remains undetected. At a later stage, PID may develop. | Tetracycline or doxycycline is usually effective |

(continued on next page)

Table 14.3 (continued)
Common Sexually Transmitted Diseases (STDS):
Mode of Transmission, Symptoms, and Treatment

STD	Transmissions	Symptoms	Treatment
Nongonococcal urethritis (NGU)	Primary causes are believed to be bacteria *Chlamydia trachomatis* and *Ureaplasma urealyticum*, most commonly transmitted through coitus. Some NGU may result from allergic reactions or from *Trichomonas* infection.	Inflammation of the urethral tube. A man has a discharge from the penis during urination. A woman may have a mild discharge of pus from the vagina but often shows no symptoms.	Tetracycline, doxycycline, or erythromycin
Syphilis	The *Treponema pallidum* bacterium ("spirochete") is transmitted from open lesions during genital, oral-genital, or genital-anal contact.	*Primary stage:* A painless chancre appears at the site where the spirochetes entered the body. *Secondary stage:* The chancre disappears and a generalized skin rash develops. *Latent stage:* There may be no observable symptoms. *Tertiary stage:* Heart failure, blindness, mental disturbance, and many other symptoms may occur. Death may result.	Benzathine penicillin, tetracycline, or erythromycin

Pubic lice ("crabs")	*Phthirus pubis*, the pubic louse, is spread easily through body contact or through shared clothing or bedding.	Persistent itching. Lice are visible and may often be located in pubic hair or other body hair.	Preparations such as A-200 pyrinate or Kwell (gamma benzene hexachloride)
Herpes	The genital herpes virus (HSV-2) appears to be transmitted primarily by vaginal, oral-genital, or anal sexual intercourse. The oral herpes virus (HSV-1) is transmitted primarily by kissing	Small, red, painful bumps (papules) appear in the region of the genitals (genital herpes) or mouth (oral herpes). The papules become painful blisters that eventually rupture to form wet, open sores.	No known cure; a variety of treatment may reduce symptoms; oral acyclovir (Zovirax) promotes healing and suppresses recurrent outbreaks.
Viral hepatitis	The hepatitis B virus may be transmitted by blood, semen, vaginal secretions, and saliva. Manual, oral, or penile stimulation of the anus are strongly associated with the spread of this virus. Hepatitis A seems to be primarily spread via the fecal-oral route. Oral-anal sexual contact is a common mode for sexual transmission of hepatitis A.	Vary from nonexistent to mild, flulike symptoms to an incapacitating illness characterized by high fever, vomiting, and severe abdominal pain.	No specific therapy; treatment generally consists of bed rest and adequate fluid intake.

(continued on next page)

Table 14.3 (continued)
Common Sexually Transmitted Diseases (STDS):
Mode of Transmission, Symptoms, and Treatment

STD	Transmissions	Symptoms	Treatment
Genital warts (venereal warts)	The virus is spread primarily through genital, anal, or oral-genital interaction.	Warts are hard and yellow-gray on dry skin areas; soft pinkish red, and cauliflower on moist areas.	Topical agents like podophyllin: cauterization; freezing, surgical removal; or vaporization by carbon dioxide laser
Acquired immuno-deficiency syndrome (AIDS)	Blood and semen are the major vehicles for transmitting the AIDs virus, which attacks the immune system. It appears to be passed primarily through sexual contact or needle sharing among IV drug abusers.	Vary with the type of cancer or opportunistic infections that inflict an AIDS virus victim. Common symptoms include fevers, night sweats, weight loss, loss of appetite, fatigue, swollen lymph nodes, diarrhea and/or bloody stools, atypical bruising or bleeding, skin rashes, headache, chronic cough, a whitish coating on the tongue or throat.	At present, therapy focuses on specific treatments of opportunistic infections and tumors. Some antiviral drugs, such as AZT, slow progression of AIDS and extend patients lives.

behaviors include activities such as abstinence or lack of genital contact with a partner, an entirely acceptable option; audio, visual, or verbal fantasies experienced by oneself or with a partner; involvement with erotic materials; masturbation; massages or bathing with oils and perfumes alone or with a partner; hugging, holding and non-penetrative kissing; manual, visual, and oral exploration of external body surfaces other than the genitals and anal region; sensuous feeding, and sleeping together without genital contact. The range of sensuous activities which do not exchange blood, semen, and vaginal fluids are limited only by one's imagination.

A guaranteed monogamous sexual relationship with partners known to be uninfected will also prevent STD and HIV infection. Note well: Sexual partners **KNOWN** to be HIV- (see Testing) can safely engage in any sexual behavior without risk of HIV infection. This behavior, however, would *NOT* guarantee being risk-free of other STDs or pregnancy. A lighthearted definition of a monogamous relationship is that:

> You have sex only with another and another has sex only with you. You have sex with no other other than your other and your other has sex with no one other than his/her other, i.e., you. You have had no sex with any other before your other and will have no sex with any other after your other. Your other has had no sex with any other before you and will have no sex with any other after you (ACHA Conference; Keeling 1989).

Monogamous relationships can be a very fulfilling, rewarding way of committing to, communicating with and loving another person. In the human and nonhuman primate world, they are also relatively rare. Gibbons, a great ape, are the only hominoid which pair bonds for life. In modern human behavior, monogamy is preferred by about twenty percent of human societies. In monogamous societies, the behavior may be preferred, but is not necessarily practiced (Frayser 1985; HRAF 1964; Atwater 1982; Kinsey et al. 1948, 1953; Hunt 1975). As stated in the introductory chapters, sex can occur both within and outside a marriage. The reality of sexual behavior in late twentieth century U.S. society is **serial monogamy** which may be a hominid characteristic (Fisher 1989; 1992). Serial monogamy is a series of sexual relationships, each one of which is monogamous for its duration. Relative to HIV infection, serial monogamy means that you and your partner are having sex with each other and everyone else you and your partner have ever been sexual with. As commonly stated, "It's a crowded bed."

Once one leaves the realm of safe sex, one enters the world of **safer sex** and risk-taking behavior. No one chooses to become HIV infected. People do, however, choose risks. Safer sex exists along a continuum from low risk to high risk behaviors. The safer sex continuum is defined relative to the risk of exchanging infected semen, blood, including menstrual blood, and vaginal fluids. Negotiating safe and safer sex behavior requires honesty, trust, communication between partners, and the ability to clearly state and act on your needs and limits. Knowing your own risk taking behavior is important in keeping yourself healthy and safe. In consensual sexual relationships, your behavior is voluntary and conscious. Practicing safer sex behaviors can enhance your own body awareness, strengthen your self-esteem, increase your pleasure, and reduce the chances of contracting an STD, HIV infection, or pregnancy.

Any behavior which exchanges semen, blood, or vaginal fluids increases one's risk of HIV infection. This includes any unprotected penetrative sexual activity whether it is anal, vaginal, or oral (particularly if this activity is receptive); and sharing sex toys such as dildoes or vibrators.

> ASIDE: "Any form of unprotected sex is risky, including oral sex . . . " "The risk of getting HIV from oral sex is not as high as from anal or vaginal sex, but there is a risk" (Author's bold) (CDC Health Resources and Services Administration 1993:6).

"Outercourse," interfemoral intercourse or what, in your author's generation was known as "dry docking" or "dry humping," is a relatively safer sensuous-sexual alternative to P-V or P-A intercourse. The penis of one partner is rubbed between the thighs of the other partner. Use of lubricants (saliva, lotions) increases the sensations, as does having the bottom partner cross his/her legs at the ankle, creating a tighter feeling. However, if either partner has sores or lesions on the upper inner thigh (for example, Herpes simplex 2 lesions can appear here), a latex barrier needs to be placed over that partner's thighs as protection.

Although undocumented, wet, deep, or French kissing where the virus could possibly pass through cuts in the mouth caused by flossing, toothbrushing, sores, or infection might transmit the virus. To date, there has been no documented cases of HIV through "deep kissing," and it is considered low risk.

Table 14.4
Activities that Cover the Safer Sex Continuum

Low Risk	Riskier	Riskiest
deep, wet or French kissing	multiple partners using latex unlubricated condoms, vaginal dams, or protective barriers for P/A or P/V intercourse	receptive penetrative anal, vaginal, or oral sex without properly using a lubricated, spermicidal condom
mutual manual stimulation of genitals without gloves		
outercourse	external water sports, i.e., playing with urine if intact skin cannot be guaranteed	receptive unprotected oral sex on menustrating women
penetrative oral-genital sex using unlubricated condoms[6]		sharing sex toys
	receptive oral sex	fisting, i.e., inserting a fist or a hand into vagina or rectum
		internal water sports, i.e., playing with urine inside a body cavity
		multiple partners without properly using condoms, vaginal dams, or other effective barriers
		improperly used condoms or vaginal dams

These are examples of behaviors that span the safer sex continuum from least risky to riskiest behaviors.

It is important to remember several things when making sexual decisions in the age of HIV infection and AIDS:

1. The **only** guarantee of not becoming infected through sexual behavior is to either be **abstinent** or to be involved in an **assured monogamous** relationship with a partner known to be **uninfected** as defined and described previously. Unfortunately, it is very difficult to guarantee monogamy and non-infection in your partner.

2. Once you decide to engage in genital sexual activity with one or more partners, you move from safe behaviors to risky behaviors.

3. If you decide to have genital sexual activity with a partner, it is important to practice consistently low risk behaviors that will reduce, but not eliminate the chances of infection.

4. Generally, the receiver/insertee in any penetrative sexual encounter—oral, vaginal, or anal—is at greater risk for HIV infection than the inserter, regardless of gender. This is because it is the receiver's mucosa/mucous membranes which are at greater risk for tearing or abrading than the inserter's. In male-female penetrative sex, for example, the female is at greater risk in P-A and P-V intercourse, or fellatio than the male. It is also believed that in P-V intercourse, the longer HIV infected semen remains in the vagina, the more HIV concentrated the semen becomes (Mariposa Newsletter 1992).

5. Consistent practice of low risk behaviors entails the correct use of condoms, spermicides and vaginal dams to decrease the chance of infection. Latex condoms are preferable to skin condoms.[6] HIV can pass through the pores in skin condoms. As far as we know, the pores in the latex condoms are too small for the virus to pass through (Consumer Reports 1989; Mariposa Newsletter 1992).

Properly used condoms plus spermicides provide greater protection than either spermicides or condoms alone (IASHS 1987). Condoms alone do not constitute safer sex. Properly using as many forms of risk reduction as possible in each penetrative sexual encounter increases the safety of your sexual behaviors and may lessen your chance of infection. As a reference, the September 1992 issue of *Mariposa Foundation Newsletter* rates condom effectiveness and aesthetics on a number of dimensions. The CDC, updating this report, believes most latex condoms to be equally effective *if* used correctly and consistently (CDC Annual Report 1993; JAMA 1998).

Vaginal dams are thin, rectangular shaped, colored, and flavored squares of latex used in cunnilingus ("going down on"), oral-anal ("rimming"), and oral-perineal activities. They are placed over the vulva, perianal, and perineal area for any oral contact with the female's genitals or male and female perianum. Figure 14.5 illustrates a safer sex brochure, the correct use of condoms, and vaginal dams.

Adopting safer sex practices as part of your regular sexual behavior is important in keeping yourself and your partner healthy and free of HIV infection, as well as other STDs. It requires knowledge and belief in one's needs and limits, respect and love of oneself, ability to stay within one's comfort level, and ability to resist pressure to engage in uncomfortable and risky behaviors. It requires good communication skills with a partner and mutual respect. It requires consistent use of safer sex practices—to incorporate them into your own sexuality so that they become as regular a part of your health and routine as are your other health practices such as brushing your teeth, or showering, for example.

It may be helpful to familiarize yourself and become comfortable with the proper use of condoms and vaginal dams before you engage in sexual contact with a partner. That way you have a level of confidence that may ease the negotiation of safer sex behaviors with your partner. In practicing safer sex, advantages other than infection reduction occur. They include a deeper awareness of your own and your partner's sexuality, an expansion of sensuality, improved communication, and possibly an increase in self-esteem. Practicing safe and safer sex are the only preventive measures we have at this time to prevent the sexual transmission of the HIV.

ASIDE: Much denial exists about one's risk for HIV infection. This denial is found among college students as well as in other segments of the U.S. population. College students are at definite risk for HIV infection. The situation of Alison Gertz, a 23-year-old, white, non-needle using, straight, upper middle class college student is a clear example of young adult risk taking. Alison Gertz knew her partner for two years. On the pill, she had one unprotected sexual encounter with him. He was HIV positive and has since died from AIDS. She was diagnosed with AIDS after having endured a bout with Pneumocistis carinii pneumonia (PCP), a common OI associated with AIDS. She and her family decided to devote the rest of her life to educating high school and college students about the risk of HIV infection. Alison Gertz died of complications from AIDS on August 8, 1992. (20/20 News report ABC 5/19/89; *New York Times* Obit Col. August 17, 1992).

Needle Sharing / Blood Contact

Blood is another medium through which HIV infection is spread. As discussed, blood can be shared sexually as with unprotected penetrative vaginal sex with a menstruating female or through sharing sex toys or through unprotected penetrative anal sex. A

SAFER SEX...

It's fun, it's easy, and it may save your life.

Don't Die of Embarrassment

Don't be too embarrassed to talk about condoms. Condoms can help prevent AIDS. Insist on the use of a condom if you have sex with a person whose health and drug history cannot be guaranteed to be HIV negative.

The materials in this kit are provided to help you maintain your health. The pamphlet is graphic and explicit in explaining ways of reducing your risk of contracting or transmitting the HIV virus. These materials are not meant to shock anyone. The goal is to encourage responsible behavior and decision-making relative to your sexuality and to keep you and your partner(s) safe and infection free.

AIDS HOTLINE
1-800-541-AIDS
Toll-free and confidential

For information, referral and support call your regional hotline and ask for the HIV counselor. This is a confidential service.

REGIONAL AIDS HOTLINES

Rochester area	716-423-8081
Syracuse area	315-475-2430
Buffalo area	716-847-4520
Nassau County	516-385-2437
Suffolk County	800-462-6786
New Rochelle	914-632-4133
	Ext. 360
Northeastern NY	800-962-5065
Bronx	212-447-8200
Brooklyn	718-230-4519
Queens	718-262-9100
Harlem	212-369-6378

Condoms, water-based lubricants, spermicides, latex gloves, and fingercots can be purchased in drug stores and some grocery stores, near the band-aid section.

Condoms can be obtained free from Student Health Services on campuses.

SAFER SEX

Safer sex involves reducing your risk of contracting the HIV virus as well as enhancing your sexuality. Once one decides to have genital sexual contact with a partner, one moves from safe sex to the safer-to-risky continuum. Only you can decide what level of risk and trust are acceptable to your health. If a person lies to you about their HIV status, your life is at risk. Remember, **PEOPLE LIE TO GET LAID**, unfortunately. You are solely responsible for your decisions.

If you decide to become sexually active with a partner in which semen, blood or vaginal fluids could be exchanged, your behavior enters the realm of risk-taking. Safer sex is for anyone—male or female—who is sexually active with males and/or females. Safer sex practices to reduce the risk of infection are encouraged. It is recommended to use as many layers of protection as possible: spermicidal lubricated condoms for penetrative anal and vaginal intercourse; unlubricated, nonspermicidal condoms for fellatio (oral sex on a male) and vaginal dams for cunnilingus (oral sex on a female) or any oral-perineal or perianal activity.

Cuts, scratches, scrapes on fingers, hands, lips—any broken skin surface may be a potential entry for infection. Therefore, finger cots and gloves are recommended in this situation. A glove can be turned into a vaginal/perineal dam by leaving the thumb intact for a finger cot and cutting open the rest of the glove. An unused unlubricated condom can also be used as a similar barrier by cutting it lengthwise.

The following is a list of those behaviors that are considered high-risk and absolutely unsafe. Most of the behaviors on the list you will probably recognize, but there are some that may be unfamiliar to you. By listing these, the editors of this pamphlet are neither condoning nor condemning them. But to be safe, you must have as much information as possible. It is important to remember that you must choose behaviors that are right for you, and you need not participate in any activity that you are uncomfortable with. Consider your own values carefully so you know how you feel before you find yourself in a situation that requires a decision about sex.

ABSTINENCE IS A PERFECTLY ACCEPTABLE, HEALTHY CHOICE TO MAKE.

Figure 14.5 Safer sex kit brochure.

PROTECTING YOURSELF

The following information from:
1989 San Francisco AIDS Foundation, San Francisco;
Our Bodies, Ourselves; Boston Women's Health Collective.

CONDOM USAGE

Befriend your condom. You might want to buy some cheap ones and play with them. Get used to how they look, feel, smell and taste and how to open the package before you want to use them. They are awkward at first, but so is riding a bike, learning how to use tampons, shaving — anything that is new and different.

How to put the condom on

Do retract (pull back) the foreskin if you are uncircumcised (uncut) before putting on the condom.

Do remove rolled condom from the package.

Do roll condom down penis as soon as it is hard, before you start to make love (foreplay).

Do leave 1/4-1/2 inch extra space at the tip of the condom to catch the ejaculate if the condom has no nipple.

Don't unroll condom before using it. Instead carefully roll the condom down the erect penis toward the base.

Don't wait to put the condom on until you are ready to enter your partner—it may be too late. Drops of semen—precum—may drip from the uncovered penis before ejaculation, and may infect or impregnate your partner.

Don't twist or bite or prick condom with a pin—this will damage it and allow fluid to leak out, possibly infecting your partner.

After ejaculating, hold the condom at the base of the penis and withdraw immediately. Unroll the condom from the penis and discard it.

A NEW CONDOM MUST BE USED EACH TIME ANAL OR GENITAL CONTACT OCCURS. A USED CONDOM OFFERS NO PROTECTION FROM AIDS OR OTHER SEXUALLY TRANSMITTED DISEASES.

Courtesy of AIDS Project Los Angeles

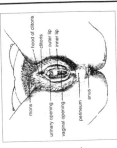

VAGINAL DAM USAGE

Vaginal dams are difficult to find here. They come in a variety of sizes, thicknesses, colors, tastes, and smells. Generally, they take more accommodation than do condoms. Vaginal dams are needed when performing oral sex on a woman, oral-perineal stimulation, and any kind of oral-anal contact.

Vaginal dams are placed over the genitals, perineum, or anal area before and during any oral contact. Stimulation is through the vaginal dam. In place of a vaginal dam, clear plastic wrap (nonmicrowaveable), an unlubricated condom or cut latex glove can be used.

A NEW VAGINAL DAM MUST BE USED EACH TIME ANAL OR GENITAL CONTACT OCCURS.

HIGH RISK, UNSAFE SEX PRACTICES

- Multiple genital partners without using condoms or vaginal dams
- A single genital sex partner without using condoms or vaginal dams unless monogamy and HIV-status can be guaranteed now and in the future
- Oral genital (male or female) contact, anal penetration, vaginal penetration, without condoms and vaginal dams; or sharing sex toys
- Rimming (oral-anal contact) without a vaginal dam
- Fisting (finger-hand vaginal-anal penetration)
- Oral-perineal contact without a vaginal dam

(the perineum is the soft skin between the genitals and anus)
- S/M (sado-masochism), B/D (bondage-discipline), tying up, spanking, M/S (master-slave) behavior that breaks the surface of the skin or deeply bruises it
- Sharing sex toys such as vibrators or dildoes with a partner
- Water sports (playing with urine) on broken skin and, possibly, internal water sports

Engaging in any or all of the behaviors in this group even once jeopardizes your own and your partner's health.

CO-FACTORS: NON-SEXUAL BEHAVIORS WHICH MAY INCREASE YOUR RISK

Using alcohol and other drugs may impair your judgment or lower your inhibitions and may cause you to behave unsafely.

SAFER SEX GUIDELINES

Always use lubricated, spermicidal condoms for anal and vaginal sex.

Don't get semen or vaginal fluids in your mouth. Use a condom or vaginal dam for any oral sex. Unlubricated condoms may be used, if preferred, for oral sex.

Don't have mouth to rectum contact without using a latex barrier.

Figure 14.5 (continued)

common non-sexual means of blood transmission of the virus is through needle sharing. To reiterate, donating blood is a safe activity. A clean needle is used for each donor and the needle is destroyed immediately after use. There is NO risk of contracting or spreading HIV infection by donating blood. There are, however, needle use behaviors that can be risky for HIV transmission.

Contaminated blood can be found in either the needle or syringe in situations of injection drug use (e.g., shooting speed, heroin, or cocaine); injecting steroids among athletes; acupuncture and tattooing; ear and body piercing; "going for" blood brothers or blood sisters; and any other situations where needles are used and can be shared. If a needle containing infected blood is shared, the virus can be transmitted. Means of prevention include not using IV drugs such as heroin, speed, or injectable cocaine, or getting off them and into treatment if you are a current user; not sharing needles with anyone for any reason; and properly cleaning the needles and syringes before use. The latter is the least desirable alternative and recommended only for people still committed to needle use and sharing.

Since 1985, screening the blood supply for the presence of HIV has improved dramatically. People receiving transfusions and using blood products such as hemophiliacs are at much less risk for infection at this time. While autologous donation where you donate your own blood is still recommended for elective (i.e., non-emergency) surgery, the current risk for infection through transfused blood is now considered statistically insignificant (ACHA 1989; CDC 1998b).

In the U.S., there is definite concern about the risk of contracting HIV infection in dental, medical, and surgical healthcare situations. The concern exists in two directions: the risk posed to healthcare workers who may be exposed to a patient's/client's HIV infected blood during a procedure (e.g., needle stick injuries or from blood splatters during surgery) as well as the risk posed to patients/clients from an infected provider. To date, the risk is very small (but real) to either the patient or provider, but the fear level on both sides is high. About 52 healthcare workers have become infected from procedures involving contact with HIV infected blood (Stine 1993 and 1998; Nevid 1993). Current recommendations to healthcare workers exposed to HIV are to begin a prophylactic course of treatment with several anti-HIV drugs within 72 hours of exposure to reduce the chances of infection (MMWR 1998; Buchbinder 1998). There remains a greater risk of HIV transmission from infected client/patient to caretaker than from infected caretaker to patient in medical settings (IXth International AIDS Conference 1993).

Five to six patients of Dr. Acer, a dentist who died of AIDS, became HIV infected while under his care. How these people became infected is still unknown; but the situation was dramatically and tragically represented by one of his patients, Ms. Kimberly Bergalis, who died of AIDS in 1992 after pleas to the government to pass legislation regarding HIV status disclosure among health care workers (Nevid 1993).[7] Without diminishing either the reality or agony of these situations, it is important to remember that the risks are small on either side of the patient-provider position and that adoption of universal health care precautions (e.g., gloving when performing pelvic exams or drawing blood) will further reduce this risk for both providers and patients.

There are probably several reasons why HIV infection in health care situations generates such a reaction. First, HIV and AIDS is a scary disease. Currently, it is fatal; its progression is slow and debilitating and emotionally taxing for everyone involved. Secondly, we do not handle incurable illness, chronic ill health, and death well in this culture. Much denial and discomfort exist in this culture about these issues. Thirdly, with a vaccine for polio and the development of highly effective antibiotics, we are a generation removed from the horror of not being able to "fix" a medical problem. That is a difficult situation for everyone to resolve. HIV infection humbles all of us.

Use of non-needle based drugs such as alcohol, amylnitrates ("poppers"), snorted cocaine or marijuana ("grass," "pot") does not directly transmit the virus. Their use, however, can put one at risk because they can impair judgment, making it more difficult to behave safely and healthily, and in the case of alcohol, cocaine, and amylnitrates, impair the immune system (IXth International AIDS Conference 1993). Nicotine also impairs the immune system. A compromised immune system makes one more susceptible to a variety of infections, including HIV.

Perinatal Transmission and Breastmilk

A third means of transmission is through perinatal and breastmilk infection. Perinatal and neonatal AIDS are not well understood. Breastmilk infection is less common than is transmission of the virus from an HIV+ mother to her child either in utero during the last trimester of pregnancy or during birth (NYSDH 1988; Vermund et al. 1991; Wilfert and McKinney, Jr. 1998). The virus may pass through the placental barrier during fetal development or through the blood in either a vaginal or cesarean birth.

The course of perinatal and pediatric AIDS (younger than 13 years old) is different than adult AIDS in part due to children's immature immune system (Wilfert and McKinney, Jr. 1998). As discussed, about 67–85% of babies born HIV+ to HIV+ mothers will spontaneously seroconvert to HIV-negative status within their first two years. At this time, the causes of seroconversion in neonates are not clearly understood (NYSDH 1989; Klass 1989; Vermund et al. 1991; IXth International AIDS Conference 1993; Wilfert and McKinney, Jr. 1998). Those who retain their HIV+ status will still have a high mortality rate by the age of nine, having had their life spans increased dramatically since 1987 due to earlier diagnosis of infection and intervention with new drug therapies (Wilfert and McKinney, Jr. 1998). These children fail to thrive, i.e., grow or gain weight well, are sick often, and have high fevers, thrush, and other ailments. They may be abandoned at birth, left as "boarder babies" to be raised by hospital staff. There are several care agencies in urban areas for HIV+ babies (NYSDH 1988, 1989; Hutching 1988; Woodruff and Sterzin 1988; Select Committee 1988).

A woman who knows she is HIV+ when she becomes pregnant has serious decisions to make. It is known that HIV+ women in general have a shorter life span from time of AIDS diagnoses to death than their male counterparts (WAN 1990; Nevid 1993). Whether this is due to female sex hormones, the fact that women tend to take care of their significant others' health first before their own and thus enter the health care system at a more progressed stage of the illness, often are misdiagnosed since many of their symptoms present gynecologically; or a combination of these variables is not yet clearly known (Klass 1989; WAN 1990; Altman 1989; SF AIDS Fdn 1998). This situation is changing now that the CDC has revised the AIDS diagnosis to include some gynecologic problems HIV+ women face. They now can be diagnosed with AIDS more readily, receive treatment and nonmedical support services only available to PWAS. Women are now being included in clinical trials and receiving the new combination anti-HIV drug therapies (Wilfert and McKinney, Jr. 1998; SF AIDS Fdn 1998).

The pregnant HIV+ woman's own health care is one set of decisions she needs to make. The other decisions relate to her fetus. There is about a 16-33% chance her baby will be born HIV+ and about the same chance that the child will either seroconvert to an HIV+ status or die within nine years of birth (Altman 1989; WAN 1990; *New York Times* 1993). New drug therapies administered after the first trimester and through labor and birth can reduce the risk even further (Wilfert and McKinney, Jr. 1998;

Beardsley 1998). Who gets these drugs? What effects will they have on the fetus? These issues are surrounded in controversy (Beardsley 1998). Does she terminate the pregnancy? Does she bear a child who may be healthy only not live to raise the child or be too ill herself to care for the baby? A new AIDS "crisis" is about to appear in the U.S.—surviving AIDS orphans who have no one to take care of them. While this situation has been problematic in parts of Africa for several years as a consequence of the epidemic, it is only recently being addressed in New York City (*New York Times* 1993; Conover 1994). Does she bear a child only to watch it waste away, suffer terribly (AIDS is not a pretty, quick disease), and die by the age of nine? These are questions that are not easily legislated and relate deeply to a person's culturally defined sense of what a woman, mother, and parent is; the meaning of life, and the role of children (Altman 1998; Beardsley 1998). These are societal and personal decisions about an illness that are very real and continuing to spread. As stated earlier, it is estimated that in the greater New York City area in the major city hospitals an HIV+ baby is born every eleven minutes (Altman 1989, Klass 1989). From the preceding discussions, it is obvious that women have different issues than men regarding HIV infection. Increasingly, more biomedical, preventive, treatment, and support resources are being directed to the specific needs of women and HIV infection.

The tragedy and hope in HIV transmission is that it is readily preventable. It is preventable through knowledge of HIV transmission and its course of infection, and through modification of one's behavior and attitudes so that the virus is not spread. The incidence and prevalence rates can be reduced by consistent sexual and needle use risk reduction behaviors.

Testing

The issue of HIV testing is controversial in U.S. culture. A test to detect antibodies to HIV has been available here since 1985. The test is not an AIDS test. It does not indicate whether or not one has AIDS. There is **no** AIDS test.

The **HIV test** detects infection with HIV through the appearance of **antibodies** to the virus. Antibodies are the body's specific reaction to the presence of a pathogen. A decision to take the test and receive the results is serious; one that necessitates careful thought and consideration, as well as competent pre- and post-test counseling. Currently, most cities in the U.S. have HIV test sites, referred to as ATS or alternative test sites. ATS's provide anonymous, free testing, with pre- and post-test counseling available at the site.

Anonymous testing means that there is no way of identifying the person being tested other than by a number code. Names, addresses, and phone numbers are not used. **Confidential testing**, common in clinics, hospitals, and doctors offices, allows for identification of the person being tested. Names, addresses, and phone numbers are on file with select access to the information. Given the economic, social, medical, and political discrimination presently occurring against people who are HIV infected, a decision whether to test anonymously or confidentially is critical. Individual states have their own disclosure laws which can be obtained from departments of health. It may be worthwhile to find out your state's medical disclosure and confidentiality laws.

Attitudes about testing are beginning to shift. Combinations of new drug therapies known as "the cocktail" can now be given in some situations as prophylactic treatment in the early stages of HIV infection to try to slow the progression of the virus, suppress viral load, i.e., how much of the virus is actively present, and possibly reduce the chances of contracting an OI (Bartlett and Moore 1998). Therefore, early diagnosis of infection may be important in the attempt to change HIV infection from a fatal disease to a chronic, but manageable illness. To achieve this, some people who engage in AIDS work are encouraging carefully conducted HIV testing for anyone who may be at risk of infection (AHCA Teleconference 1989; Stine 1993).

Since it may take anywhere from two weeks to four years, with an average of 2-6 weeks for the antibodies to the virus to appear in one's blood, it is important to follow testing guidelines. Ideally, one needs to have engaged in risk-free behavior for a minimum of thirty days before first being tested. If those results are negative, a 3–6 month period of risk-free behavior followed by a second test is recommended to be even more certain of one's HIV status. A negative test result, i.e., testing HIV-, means that on the day the blood was drawn, antibodies to the virus were not present. It is not a license to engage in risk taking behavior from that time.

The most commonly used initial screening test is called the ELISA test. A negative result is assumed to be accurate. A positive result necessitates a second ELISA test on the blood sample. If that result is negative, the person is assumed to be HIV negative. If that result is positive, the Western Blot test, an even more refined HIV antibody sensitive test, is performed. If the results are negative, seronegativity is assumed. However, a positive result from this test is interpreted as HIV+ status. These are the standard HIV tests generally performed at ATS and other test sites. Tests to actually detect the presence of the virus itself or to detect antibodies earlier are available, are expensive, and require a physician's or clinic's involvement. An oral HIV test is available in some parts of the U.S. (CDC 1998c. Nevid 1993; *Self* 1993).

The need for good pre- and post-test counseling becomes obvious. People need to understand what the test means, what their results indicate, and what behavioral and attitudinal changes need to occur to retain their HIV- status or to take care of themselves if they test HIV+. Most states have ATS's which can serve as a valuable source of information, counseling, and ethical testing.

Since 1988, there has been some promotion for at-home HIV test kits (San Francisco AIDS Foundation Meeting January 1988). This is a controversial topic. Proponents claim that people who otherwise would not be tested could do so in the privacy of their homes. Their HIV status can be self-monitored directly. The tests would be less expensive than those given by clinics and physicians, but are obviously more expensive than a free test at an ATS. Opponents refer to the total lack of direct one-to-one pre- and post-test counseling and problems with accuracy and reliability in obtaining the blood sample and receiving results. Confidentiality may be compromised in obtaining results; anonymity would not be guaranteed. At-home HIV tests currently (1998) are available. To obtain information about these tests, contact the CDC National AIDS Clearinghouse, at 1-800-458-5231 (CDC 1998c). It appears that every aspect of AIDS is rooted in controversy, risk, and widespread economic, personal, social, and medical consequences.

The Course of Infection

When a person's immune system is sufficiently compromised by HIV infection and one or more of the OI's appear, the person then generally receives a diagnosis of AIDS. As of January 1, 1993, the revised CDC definition of AIDS is:

1. an HIV+ antibody test and

2. a T cell count of 200 or less (normal range is 800–1200) and/or

3. one or more of over 20 opportunistic infections (OI's) to which have been added TB (tuberculosis), recurrent pneumonias of any type, and invasive cervical cancer, in order to reflect the "changing face of AIDS." Translated,

this means that the demographics of whom in this country is infected is changing from men who have sex with men to needle users and sharers (IDU's) and women. Infection with the virus is not AIDS; AIDS generally develops several years after infection, during which time the infected individual is contagious, usually asymptomatic (symptom free) and healthy, that is, able to function as if uninfected.

HIV Infection Relative to Acute, Chronic, and Terminal Illness

HIV disease, caused by the HIV virus' destruction of the immune system, is an acute, chronic, and terminal illness which is transmitted by behavior. The means of transmission, its acute, chronic, and fatal characteristics, the current lack of a cure or vaccine, its incidence and prevalence combine to trigger everything from fear, anxiety, and denial responses to compassion, acceptance, and caring among those people affected by it. Acute illnesses are those which appear suddenly, and have a relatively short duration, such as appendicitis. An HIV-infected person with PCP generally experiences an acute episode of the illness. Often they are hospitalized and given antibiotics to control the specific incidence of PCP. Chronic illnesses are long term, sometimes genetically predisposed or congenital (innate), and usually treatable but have a wearing effect on the body. Examples include such illnesses as arthritis or diabetes. HIV-infected people often experience chronic illnesses or health problems such as ongoing fevers, continuous abnormalities in PAP smears in women (Orenstein 1991), or thrush, a yeast infection usually found in the mouth. They are often chronically tired and have reduced levels of energy. They may live day-to-day not feeling ill or well, but on a plateau of not feeling as they used to. A terminal illness is one whose outcome usually is fatal, such as some forms of cancer (e.g., liver, pancreatic). At present, most people who are HIV infected develop AIDS. Most people diagnosed with AIDS die of one of the OI's associated with the illness within one to four years of diagnosis (CDC 1990; Bartlett and Moore 1998:86). As an acute, chronic, and terminal illness, AIDS has all the associated physical, psychological, socio-economic complications and complexities of these illnesses. These complexities affect both the PWA's and those who are part of their social network. HIV infection and AIDS take its toll on the individual, the kin group, and larger socio-economic, political, and health care structures.

Table 14.6
HIV Continuum as of January 1, 1993

HIV+, Symptomatic (formerly referred to as ARC-AIDs Related Complex)	AIDS diagnosis (Generally occurs 10.8 years after infection with the virus)
Symptoms generally begin several years after infection, with the range from several months to several years. Some or all of the following may appear:	HIV+ antibody test—ELISA confirmed by Western Blot, and T cell count at or below 200, and/or one or more opportunistic infections (OI's):

- HIV + antibody test, confirmed by ELISA and Western Blot

- T cell count above 200 (normal range is 800–1,200)

- Night sweats

- Swollen lymph glands, without other pathology

- Fatigue, malaise

- Chills

- Loss of appetite, nausea*

- >10% loss of body weight without dieting or other known causes

- Gynecologic problems such as *recurrent,* uncontrollable yeast infections, Pelvic Inflammatory Disease (PID), and oral or vaginal thrush*

- Body cavity infections, acute or chronic illness episodes, herpes virus infections such as CMV (cetamegolo virus)

Some people do not progress beyond this point, have chronic problems with these symptoms, or gradually become weakened and debilitated to the point of death. A few individuals go into remission.

- Kaposi's sarcoma (KS) decreasing in frequency in the U.S.

- Pneumocystis Carinii Pneumonia (PCP) most common OI in the U.S., recurrent viral or bacterial pneumonias

- Tuberculosis (TB) increasing incidence in U.S., most common in Latin America

- Invasive cervical cancer— common in HIV+ women. Added by CDC as of January 1, 1993. Will see an increase in number of women in the U.S. diagnosed with AIDS with invasive cervical cancer now included

- AIDS dementia—physical, psychological, and cognitive deterioration. May resemble some symptoms of Alzheimer's Disease

Most people with AIDS in the U.S. die within 4 years of diagnosis. Longevity varies by age, gender, and ethnicity. People do not die of AIDS per se, but due to one of the OI's, general debilitation, or heart or kidney failure.

In Africa, these constitute a wasting phenomenon referred to as "the slims."

Treatment

As the preceding chart indicates, HIV infection occurs along a continuum. Generally, it is a progressive illness whose stages are relatively demarcated by gradual deterioration and debilitation. From the time of infection to a diagnosis of AIDS averages about 10.8 years (Kloser 1989; IXth International AIDS Conference 1993). As of early 1990, the average life span from time of an AIDS diagnosis to death is slightly over 4 years for middle class, gay, white males in this culture. This is an increase since 1986, when the average life span was 18 months (Kloser 1989). Individual variations occur based on age, gender, substance abuse patterns, and ethnic identity (Page et al. 1990; Singer et al. 1990). For example, currently, women with AIDS usually die more quickly than do gay white males (Orenstein 1991; WAN Newsletter 1990; Nevid 1993).

One reason for the increase in average life span of PWA's may be the development of treatments which slow the progression of the disease, mitigate the impact of OI's on the individual, or reduce the risk of their occurrence (Kloser 1989; Bartlett and Moore 1998). There is hope that until either a cure or vaccine can be found, the use of effective new drug (chemotherapy) treatments, called the "cocktail," or combination anti-HIV drug therapy in infected individuals will both extend the person's life span and turn HIV infection and AIDS into a chronic, yet manageable illness rather than a terminal disease (Kloser 1989; AIDS Teleconference 1989; Bartlett and Moore 1998). The belief is that the earlier the person is diagnosed, the sooner overall health care and treatment can begin, thus improving both the quality and quantity of the individual's life, eventually transforming this disease from a fatal to incurable, but manageable one. At present, this position appears to be attainable for people who have access to the drugs—usually middle class people in industrialized countries who can tolerate their side effects and who can adhere to the strict scheduling and dietary restrictions required for these drugs to be effective (Mann and Tarantola 1998; Bartlett and Moore 1998). This perspective has implications for testing and alters some of the earlier views about being tested for HIV infection (see section on testing).

> **ASIDE**: There is ongoing research into all aspects of the biomedical management of HIV infection. This research includes cures, vaccines, both those to stop the progression of the virus in those infected and to prevent infection with the virus to people exposed to it (similar in effect to vaccines for polio, measles or mumps); as well as treatments to slow the progression of the virus, prevent the onset of OI's or to treat OI's which occur. This research is

unquestionably necessary, vital, worthwhile and receives splashy media attention. However, a note of caution is necessary. Much of this research currently is either preliminary, laboratory-based, or animal tested. "Findings" may be presented by the media or heard by the viewer as further along or more definitive than they actually are. In addition, there is a qualitative leap from "success" in the laboratory or with animals and "success" with humans (preventative or curative). Support and hopeful but cautious optimism may be the most apropriate response for some of these research reports.

Several drugs currently are administered to symptomatic and asymptomatic infected individuals as preventative and treatment measures. In addition to AZT and DDI to slow the progression of the virus, and aersolized pentamidine or TMP/SMX to prevent PCP, classes of drugs known as protease inhibitors and reverse transociptase inhibitors have been developed. They slow viral replication and decrease viral load (the amount of virus one has), and thus help to maintain the immune system (Bartlett and Moore 1998). These drugs are expensive and have their own possible side effects. Most notably, they include the toxic effects of AZT such as severe anemia and liver problems, or the overall intolerance to the drug which some PWAs and infected individuals experience (Kloser 1989; Bartlett and Moore 1998). Some people develop resistance to their drug treatments and the treatments stop working (Richman 1998). Other new drugs and treatments, including "alternative" medicines, i.e., acupuncture, herbal remedies, and homeopathy, are continually being researched and tried in the lab and by HIV+ individuals and PWA's.

A cure and vaccine remain futuristic. Treatment and prevention of OIs, while definitely extending the life of the person are after the fact, reactive responses to HIV infection. The best way to deal with AIDS is to understand the transmission of the disease, debunk oneself of its myths, and then, as consistently as possible, engage in safe and risk reduction behaviors. In short, prevention remains the primary means of coping with the epidemic.

The Cross-Cultural Perspective

It is clear that AIDS is a global, human phenomenon. As stated earlier in this chapter, dealing with AIDS challenges societies on all levels: economically, politically, socially, psychologically, and emotionally. It evokes from people their cultural responses to severe illness, disfigurement, death and dying, grief and loss, and sexuality. It challenges a society's responses to culturally defined normal and deviant behaviors relative to sexuality and drug usage. In sum, AIDS challenges human society on some of its deepest, survival-based concerns. How each culture or subculture responds to AIDS then, is mediated through its own specific means of addressing these concerns. Thus, any preventive and treatment measures used must be culturally relative, i.e., respectful of the values and behaviors of each group and adapting educative and treatment programs according to the culture of each group. These programs need to be culturally sensitive if they are to succeed. AIDS work among college students in a rural area of the northeast U.S.; within the behavioral cognitive concept of *machismo* in Latin America, and among urbanizing trade-route Africans provide three examples of culturally sensitive responses to this disease.

United States College Students

It is well documented that college students in the U.S. are genitally sexually active, are poor contraceptors, and are risk takers as they explore life-styles, push social boundaries, and have the highest rates of STDs of any age group in this culture. All these behaviors put them at risk for HIV infection (Edgar et al. 1988; Gray and Saracino 1989; CDC Annual Report 1990; Whelehan and Moynihan 1984; Turner 1994). About 1 in 500 college students is believed to be HIV+ (Keeling 1990; Nevid 1993). They also tend to have a high cognitive knowledge of AIDS, but have not necessarily internalized that knowledge into safer sex behavior and attitudes (Moore et al. 1988; Keeling 1990; Herting et al. 1988; Gray and Saracino 1989; Edgar et al. 1988; Inman 1989; Turner 1994). College students are also highly mobile as they move from campus to home or ritual retreats such as the annual pilgrimmage to Ft. Lauderdale on the East Coast and Baja, California on the West Coast, over breaks, vacations, and summer recess. Thus, if infected and engaging in unsafe behaviors, they can act as vectors of transmission through their mobility. College students tend to listen to respected, credible peers rather than adult authority figures at this point in their development (Erikson 1950, 1963; Atwood 1990; DiClemente 1990).

Therefore, effective AIDS education aims at prevention and risk reduction relative to sexual behaviors and drug usage, particularly alcohol, marijuana, and cocaine which are popular in this subculture (Engs and Hanson 1985, 1993), needle sharing through ear and body piercing, and use of steroids (Keeling 1990; McCammon, Knox, and Schacht 1993). Education tends to be more effective in small groups, with close, intense, ongoing one-on-one interaction; and with the use of peer educators (Ostrow 1989; Keeling 1989 and 1990). Education needs to be nonjudgmental, accepting, and geared to the present as opposed to long term future consequences of behavior. Reality confrontations are also effective through the use of speakers who have lost children to AIDS and through PLWAs (Persons Living with AIDS). These people provide concrete examples with which students can identify the reality of HIV infection. Given the incubation time for the disease, college is probably not an environment where many cases of full blown AIDS will appear within the student population. Therefore, prevention strategies on college campuses are very important. It is during this period that 17–25 year old students may get infected, with the disease appearing a decade or more later. At present on U.S. college campuses, prevention strategies will probably focus on reduction of fears surrounding casual contact modes of transmission; reinforcement of the need for confidentiality regarding testing and known HIV infection on campus; a discussion of HIV testing concerns; safer/safe sex practices as positive and life enhancing; communication and self-esteem concerns; and acceptance of life-styles and values different from one's own. The economic and political dimensions of AIDS probably are not as directly relevant to college students.

Machismo/Marianismo

Machismo is a symbolic behavioral pattern common in Latin America. Key values in *machismo* include active male heterosexual behavior which reinforces virility and fertility and defense of family and individual honor to avoid shame. The men are active, aggressive, and protective of their masculinity and the chastity of their women—mothers, sisters, daughters, and wives. Family responsibility and honor are important. Condoms generally are only used with prostitutes, and even then irregularly (Gonzalez del Valle personal communication 1989). They are not acceptable for use with one's wife. In a situation of Latino men having sex with men, the inserter is male. The insertee is "female," with all the attributes of being female in that culture. The ideal female counterpart is a sexually passive, submissive, and manipulative female. A sexual double standard prevails and can be enforced with physical violence. Women embrace *marianismo*; they are caretakers of the home and family health, which they will aggressively protect. Part of health maintenance for which *Latinas* are responsible can include injectable vitamins (Dela Cancela 1989; Polaris 1988).

Based on normative sexual behavior and routine health care, two possible sources of HIV transmission include sexual contact and needle sharing. Since cultural norms do not support open discussions of sexuality between adult men and women, but do support women as responsible for family health, and support the kin group as the basic unit of society, AIDS education needs to work within this framework. Respected people from the community— older women, grandmothers, status holding men—are probably the most effective educators. Same gender educators and audience are probably the safest, at least initially, to avoid suspicions of sexual advances. Since the behaviors involved in HIV transmission are personal and private, small groups, personal communication, and "safe" environments (home, church, small meeting places) are probably good sites (Polaris 1988; Dela Cancela 1989).

The focus of educative efforts can emphasize maintaining family health, as family is a shared strong value. Men can be encouraged to use condoms so that they do not infect their wives and future children. Condom use can be promoted as a means of maintaining their own health so that they will be there for their families. A well respected man who acknowledges his own condom use and enjoyment, or at least acceptance of them, is an effective role model (Dela Cancela 1989; Polaris 1988).

Encouraging women to subtly incorporate condom use with their husbands may work if carefully done so as to be sensitive to the possibility of spousal anger. If orally ingested vitamins are an unacceptable alternative to injectable vitamins, clean needle usage is an important step. Again, emphasis on maintenance of family health and integrity are important values with which to work. It is important to keep the educational efforts within the community as much as possible. This means that visual, aural, and written materials need to be generated from within the group by the group and linguistically and cognitively sensitive to the group's needs (Polaris 1988; Dela Cancela 1989).

Africa

Subsaharan Africa along both coasts and the trade belt across Central Africa have high rates of HIV infection and AIDS (Mann et al. 1988; Ingstad 1990; Schoepf et al. 1988; Barnes and Findlay 1988; Quinn et al. 1986; Mann and Tarantola 1998). Some sources have estimated that 20% of the population of Zaire has AIDS (Schoepf et al. 1988). Sexual behavior and shared needle use in health care situations are the primary routes of transmission (Mann et al. 1988; Ingstad 1990). In an article that is bound to be controversial,

Patton raises issues of racism and ethnocentrism concerning western responses to the incidence and modes of transmission of AIDS in Africa. Contrary to a widely held belief that AIDS originated in Africa and spread from there, she believes HIV may have been introduced in Africa through contaminated blood supplies from western countries. She challenges the way the sexual transmission of AIDS is portrayed in Africa by the West to perpetuate the "exotic sexualness and otherness of dark Africans" comparable to the male-to-male "otherness" of AIDS in the U.S. (Patton 1992:218-234).

As Third World countries already seriously affected by the disease, prevention and treatment are concurrent concerns. African countries in the AIDS belt, or areas of high incidence and prevalence rates, have populations which are geographically mobile. Rapid urbanization is occurring as people migrate between rural and urban homes in search of employment, to follow trade routes, and to fulfill extended kinship obligations. This mobility allows for easy sexual transmission of the disease. Polygyny, mistresses, girlfriends, and prostitutes are culturally accepted male-female sexual behavior in some parts of Africa. For some women in migration, their only source of employment in the urban areas is prostitution. Since it is culturally and traditionally acceptable for men to have multiple sexual partners in some of these societies, men serve as vectors of transmission as they move between urban and rural settings, having unprotected penetrative sex with the women in these settings. Infected women can pass the virus onto their unborn and nursing children, making the risk of transmission intergenerational (Pheterson, ed. 1989; Eckholm 1990; Eckholm and Tierney 1990; Tierney 1990; Wilfert and McKinney, Jr. 1998).

Western health care resources are scarce, operate on limited budgets and limited supplies. A number of current endemic health problems, in addition to the risk of HIV infection, exist which require much of the budget. Resources are in limited supply. It is not uncommon for needles and surgical equipment to be shared and reused without adequate sterilization (Wilson 1988; Select Committee on Hunger 1988; Mann and Tarantola 1998). As a Third World country, its economic resources for all needs—shelter, food, medical care—are finite and scarce. AIDS treatment is chronic, long term, and expensive. Frequently, it is outside the range of health care resources. Funds for prevention and treatment, particularly in isolated, rural populations which are being infected, are inadequate. At the same time, children, and young mobile adults, the future of any group's survival, are contracting the disease. Education and treatment take on qualitatively more problematic dimensions. Condoms, for example, are not part of traditional birth control practices. Multiple sex partners for men are culturally acceptable in many African groups (Ingstad 1990; Tierney 1990). A scarcity of needles and adequate sterilization equipment make clean needle use improbable. The numerous linguistic and cultural differences require highly specific educational programs. With high geographic mobility and few economic sources, where are people going to be educated—in the cities or their villages? Solutions to HIV infection in Africa are problematic and in an early stage. They are highly impacted by social and economic considerations and constraints.

The Human Perspective

AIDS is the most recent disease to challenge our concepts of humanity, to test the limits of human physical and social endurance. It is a human disease and problem. It requires a humanistic, holistic response. AIDS does not discriminate on the basis of age, gender, ethnicity, socio-economic status, life-style, or sexual orientation. While a pandemic phenomenon, AIDS occurs to individuals who are members of groups with their own culturally defined responses to issues of life, death, illness, disfigurement, sexuality, and drug use. Thus, a culturally relativistic approach is necessary when dealing with AIDS on any level. As such, AIDS is intellectually fascinating and an ideal phenomenon for anthropological investigation and approach. In finding a cure, in treating the illness, in educating people to reduce its spread, we need to pull on the best of what is our humanity—caring, acceptance, and reaching out within a culturally sensitive context.

Summary

Chapter 14: HIV Infection and AIDS

1. HIV infection and AIDS is a complex, multi-dimensional disease and illness that has spread globally.

2. HIV infection is currently known to be spread by infected blood and blood products, semen, vaginal fluids, and through perinatal transmission in utero, during birth, or through breastfeeding.

3. HIV slowly destroys the body's immune system, leaving it vulnerable to opportunistic infections (OI's).

4. Currently there is no cure or vaccine for the disease. Education and prevention through behavior and attitude change are the most effective means of not contracting or spreading the virus.

5. New treatments are being developed to try to slow the progression of the disease in those infected.

6. AIDS is a disease of behavior, not groups, although some populations have been more affected by the disease than others.

7. There is a continuum of HIV infection.

8. Transmission is through infected blood and blood products, semen, vaginal fluids, and breastmilk. Therefore, any needle use or sexual behavior which involves sharing these infected fluids can put one at risk for HIV infection.

9. HIV infection and AIDS are not spread by casual contact; it is not an airborne virus.

10. Not using or sharing needles reduces the risk of contracting HIV infection by this means.

11. Safe sex will protect one from HIV infection, pregnancy, and STDs.

12. Safer sex will reduce the risk of contracting HIV infection and STDs, impregnating, or getting pregnant.

13. Abstinence and guaranteed monogamy are acceptable, reasonable choices. They also will protect you from HIV and STD infection.

14. Perinatal and pediatric AIDS are much different than adult AIDS. Women whose behavior put them at risk for HIV infection need to make careful decisions about pregnancy based on their life-styles and values.

15. HIV infection and AIDS are worldwide phenomenon, affecting all segments of society and an infected individual's life. Responses to the infection and its prevention need to be culture specific and culturally sensitive.

16. To date, there are three major paths of the infection cross-culturally: Western countries, Africa, the Caribbean and parts of Latin America, Asia, and the South Pacific.

17. Testing for HIV infection continues to be controversial. Adequate pre- and post-test counseling needs to be given with a clear understanding of the difference between anonymous and confidential testing. Currently, people who practice risky behavior are encouraged to get careful testing and counseling so prophylactic treatment can be started if they are HIV+.

18. It takes an average of 2-6 weeks for HIV antibodies to appear in one's blood after being infected. From time of infection to diagnosis of AIDS averages 10.8 years. The average life span from time of an AIDS diagnosis to death currently is about 4 years for middle-class white men.

19. A combination of new and old anti-HIV drugs are being given to asymptomatic HIV+ individuals to try to slow the progression of the disease by slowing replication of the virus and suppressing viral load.

20. Three case studies from U.S. college students, Latin America, and Africa illustrate different culturally sensitive strategies possible to reduce the risk of HIV infection.

21. HIV infection and AIDS gives the species an opportunity to express the best of our humanness: compassion, support, acceptance, and nurturance.

OVERVIEW

Chapter 15

Conclusion and Summary

This chapter:

1. Restates the anthropological perspective and how it applies to an exploration of human sexual behavior.

2. Reiterates the distinction between universal human sexuality and that which is culture specific.

3. Puts sexual behavior in a socio-cultural context and emphasizes the effects of culture change on traditional sexual behavior and values.

4. Makes a concluding statement about the potential for changes in hominid sexuality based on late twentieth century sexual and reproductive technology.

5. Places AIDS in a global context relative to its threat and the potential for responding to it in a humane, hominid tradition.

Chapter 15

Conclusions and Summary

An anthropological perspective: evolutionary, holistic, cross-cultural or comparative, and relativistic has been used in this exploration of human sexuality. This perspective allows for an examination of sexuality through space and time as a human phenomenon. The evolutionary perspective provides a framework for understanding biobehavioral aspects of sexuality as it has developed and changed through time and adapted to particular environments. It also allows us to explore continuities in our sexuality through our evolution as primates. Holism allows for examining sexuality in the context of group behavior and institutions. It relates various aspects of sexuality such as marriage forms to other dimensions of society such as the economic and political spheres. Rather than viewed as separate from society, sexual attitudes and behaviors are discussed as integrated into the fabric of the culture.

By aiming for a culturally relativistic or nonjudgmental perspective on sexuality, the comparative approach can be utilized. Generally, the comparisons are cross-cultural, i.e., intersocietal, but can be extended to include the higher primates, our closest nonhuman relatives. Through a comparative perspective, a better understanding of that which is universal human sexual behavior (e.g., bonding) and what is culture-specific sexual behavior (e.g., marriage form) can be achieved. The striking similarities and rich diversities of human sexuality emerge. These perspectives provide a basic foundation for exploring continuity and change, our uniqueness as humans, and the potential impact that westernization and recent technological developments have on sexuality.

As primates and hominids, we have evolved certain biobehavioral characteristics that affect our sexuality. A crucial characteristic is bipedalism and the accompanying pelvic and brain changes which

379

affected pregnancy, birth, and survival of the young. Our young are born immature and have a prolonged infant dependency on adults. The major human survival strategy is to adapt through learned behavior. Certain behavioral patterns adapted to specific econiches have developed to promote reproductive success. These include culture-specific socialization patterns, definitions of acceptable sexual behavior, and birth practices as examples.

In late twentieth century U.S. and Western European cultures, technical developments are available which can drastically change our hominid sexuality. Since 1978, the availability of technological innovations such as in vitro fertilization, embryo transplants, chromosomal sex selection, sperm banks, and artificial insemination donor, fetal reduction, and surrogate mothers have the potential to qualitatively alter our means of reproduction, alter our definitions of parenthood, our concepts of kinship, and of relationships. Most simply, penile vaginal intercourse is no longer required for reproductive success. While societies have needed fewer adult males than adult females to survive for most of our evolution, this sex ratio can be altered further with the technical ability to store sperm indefinitely in sperm banks and to predetermine gender through chromosomal selection. Amniocentesis and chronic villi sampling (CVS) also allow for gender selection in utero. For example, while developed in the West, amniocentesis is used by some people in China to select for boys (Kelly 1994).

Paralleling the technological changes are socioeconomic shifts. Currently, fewer than 20% of all U.S. households model the earlier twentieth century traditional arrangement in which the female stays home full time and the male works full time outside the home. Over 25% of our children under 18 grow up in single-parent households. There are over 1 million single-parent fathers in U.S. society. Dual career couples are the norm (U.S. Bureau of the Census 1985; Hochschild 1989). The U.S. has the highest teenaged pregnancy rate in the western world as well as the highest maternal and infant mortality rates of western countries (WHO 1986, 1988; Wattleton 1990). Our marriage rate is high; but so is our divorce rate. In 1989 about 50% of marriages in the U.S. ended in divorce. There are numerous journals, books, and self-help groups which inform us about how to be attractive, successful, find a mate, and achieve intimacy. There are a comparable number of books which instruct us on parenting. Behaviors, values, and courting patterns which previously occurred in the context of extended kin groups or neighborhoods are now the arena of sex professionals, educators, counselors, and therapists. Simultaneously, Westerners are cognitively more knowledgeable about sexuality, more openly sexually active, and more openly engaged in life-styles which include singlehood, serial monogamy, cohabitation, bisexuality, and homosexuality.

At the same time that Westerners are more open about certain sexual behaviors and life-styles, we continue to hold rather rigid concepts concerning sexual orientation, gender, and gender role/identity. To balance some of that rigidity in our own culture, a greater interest in the study of orientation and gender identity is being established. An emerging body of literature on cross-cultural concepts of gender identity and gender role is becoming more available and increasing in scope. The combination of these developments may heighten our awareness of human sexual diversity and lead us to a more accurate, realistic, and accepting comprehension of identities and orientations.

Human sexuality is further explored from a life cycle developmental context. Using cross-cultural and western perspectives, attitudes, and behaviors regarding pregnancy and childbirth, early childhood and adolescent, adult, and aging sexuality are examined. The biobehavioral aspects of sexuality is reinforced through discussions of the universalistic and culture specific aspects of these topics as well as of birth control and sexual response.

Globally, we are presented with severe population pressures relative to available resources, conflicts between western and indigenous sexual behaviors and values in acculturating societies and among assimilating individuals, and AIDS. AIDS could easily be our most serious health and sexual problem we face as a species. The disease has spread worldwide, with devastating effects on the individuals and groups involved. It may also serve to unite us in our humanness. AIDS requires a cooperative, culturally relativistic and culturally sensitive approach. It can elicit from us hominid characteristics such as bonding, flexibility, and the ability to change and adapt to new surroundings and challenges. In the past generation, changes in our sexuality are probably as significant to us now as bipedalism was to our early ancestors.

Human sexuality has evolved over several million years. Its richness, diversity, complexity, and commonality are a reflection of us as a species. Fear, prejudice, and ethnocentrism can limit our appreciation of its depth and scope. Respect, cooperation, and an integration of the cognitive and affective dimensions of our sexual selves may help us to address the challenges we have created for ourselves as sexual beings.

Notes

Chapter 1

1. *Coming of Age in Samoa* is at the center of a current debate in anthropology launched by Australian anthropologist Derek Freeman in his book *Margaret Mead and Samoa: The Making and Unmaking of an Anthropological Myth*, 1983. Derek Freeman argued strongly for a very different Samoa from the one studied by Mead. Based on his own research in Samoa from 1940–1943 and extensive subsequent research, Freeman took issue with the picture of easy going family life and low affect, citing punitive family relationships, competition and aggression, and a stormy puberty. Freeman, however, did not disagree with Mead's depiction of casual attitudes toward adolescent sex (Freeman 1983:202, 244, 260 in Barnouw 1985:98–99).

For a review and critique, see James W. Cote's *Adolescent Storm and Stress: An Evaluation of the Mead-Freeman Controversy*. Lawrence Erlbaum Associates, Inc., 1994. Also Barnouw 1985 and Howard 1983.

2. I would like to thank Dr. Jane Granskog, Department of Anthropology, California State University at Bakersfield, for this model. Although I have modified it to meet the needs of our text and given it a new metaphor, Dr. Granskog was instrumental in providing the mainsprings for this approach.

Chapter 2

1. Sue-Ellen Jacobs states in "Native American Two-Spirits":

The term *"berdache"* [*sic*] as used by anthropologists is outdated, anachronistic, and does not reflect contemporary Native American conversations about gender diversity and sexualities. To use this term is to participate in and perpetuate colonial discourse, labeling Native American people by a term that has its origins in Western thought and languages.

The preferred term of Native Americans who are involved in refining understanding about gender diversity and sexualities among Native American peoples is "two spirit" . . . or terms specific to tribes (1994:7).

I have adopted this usage where it seems appropriate to refer to gender transformed/alternative genders throughout the nonwestern ethnographic record. *However, due to issues related to manuscript production, "Berdache" is used in this text with authors' apologies.*

Chapter 3

1. Reclassification has occurred. The **hominid** category is now known as **hominine** and includes hominids, gorillas, and chimpanzees. The reclassification has occurred based in biochemical evidence relative to chromosomal and hematological (blood) characteristics which indicate very strong similarities among these species (Weiss and Mann 1990). We will continue to use hominid in this text, since the new classification system is not yet widely used.

2. There are several theories as to the conditions that may have led to the development of the visual center of the brain as well as the grasping hand. Collins (1921) has proposed that binocular vision would be favored in species who must leap from branch to branch as in the conditions encountered by the earliest tree dwellers. Cartmill's (1974) visual predation theory suggests that diet may have selected for the grasping hand in tandem with binocular vision in situations where prey, such as insects, were found on slender vines. Sussman (1978) is of the opinion that grasping hands would be adaptive for an arboreal niche where early primates traveled on small branches. In this theory, reliance on vision occurred because these early primates were probably nocturnal and they had to be able to locate plant foods in the dark (Ember and Ember 1990:59).

3. Evidence for large game hunting appears relatively late in human history and may represent one of several possible strategies for hunting and survival. In fact, microscopic analysis of the earliest tools dated between two and two and a half million years ago reveals that these were not used in actual hunting. Wear patterns indicate use in modifying plant materials, scraping, and cutting up animal skins (Zihlman 1989).

Chapter 5

1. Baldness tendencies are a genetic trait in men carried by females.

2. This does not include taking steroids by some male and female athletes in order to increase muscle size.

3. Sex hormones can promote cancerous growth.

Chapter 6

1. Does this mean that self-identified, behaviorally consistent gays and lesbians are always virgins since they do not have P-V intercourse?

2. Aspirin dissipates prostglandins. It also is an anticlotting agent. If a woman has blood clotting disorders or is to undergo surgery, she should limit her aspirin intake and inform medical personnel as to how much and when she last took aspirin.

3. The craving for chocolate may be related to phenylethylamine. One of its chemical compounds is related to phenylalanine, an amino acid. These compounds may serve as mood elevators in humans.

4. Currently, regulated sperm banks test donations for HIV positivity, since the virus is carried by the semen.

Chapter 7

1. It takes about 24–30 hours to replenish the supply of sperm after ejaculation (Stewart et al. 1979).

2. Medical terminology for various sexual and reproductive conditions frequently have pejorative connotations. These connotations, while not consciously intended to harm clients, may inflict psychological and emotional discomfort or harm on a client. A distraught, infertile couple does not need to hear about "hostile" cervical mucous or "incompetent" cervices in their attempt to remedy their situation.

3. Since 1986, AI-D donations are supposed to be tested for the HIV virus, since the virus is carried in the semen of an infected person.

4. Later differentiation may occur in the female so that Wolffian duct development of the urinary tract can take place.

5. The role of H-Y antigen in male differentiation is controversial.

6. They have been referred to as "degenerate testicles." See note two.

7. Efforts have been made to adjust the language usage in reference to the terms transsexual, transvestite, and transgender so as not to suggest clinical syndromes but rather cultural and social identity, e.g., instead of transsexual, the terminology transsexual people is adopted. Unfortunately, not all instances could be eliminated due to stylistic requirements.

8. Relative to specific sexual activity, the major difference between heterosexual and homosexual behavior is that homosexual sexual behavior is nonreproductive, heterosexual behavior can be. In terms of its evolution, most of the heterosexual acts themselves are nonreproductive.

Chapter 8

1. This is starting to change since about 30% of the currently trained obstetricians and gynecologists are female. However, these women are socialized into a biomedical, interventionist model and tend to carry that into their practice (Arms 1975; Jordan 1988; Sargent and Stark 1989; Davis-Floyd 1992).

2. Even the term "coach" implies some form of external management akin to an athletic event.

Chapter 9

1. The interested reader is encouraged to explore the works of Cohen (1978); Fox (1980); Levi-Strauss (1969); Livingstone (1969); Murdock (1949); Phelan (1986); White (1948); among others too numerous to mention.

Chapter 11

1. For the reader interested in exploring the broad spectrum of ethnographic sexology, we recommend Edgar Gregersen's *The World of Human Sexuality* (1994) with its classification by culture area including Africa, the Middle East, India and Southern Asia, the Far East, Oceania, and the Americas.

Also, David N. Suggs and Andrew W. Miracle's "Culture and Human Sexuality" 1993 provides an excellent collection of anthropological contributions from the evolutionary as well as ethnographic record.

2. "Impotence," "premature," and "retarded" ejaculation are being phased out in the field of sex therapy and the literature because of their pejorative connotations. Replacing them are "erectile control" and "ejaculatory control" problems.

Chapter 12

1. Transgenderism has several meanings within (emic) and outside (etic) the gender variant communities. Your authors recognize the different connotations associated with transgenderism. See note 7, chapter 7.

Chapter 13

1. The sections of this chapter are based primarily on one of the author's (Whelehan's work since 1988 with individuals and groups in the gay and

bisexual communities in northern and southern California and northern New York. These include "white bread" Americans and Latinos. She is deeply appreciative of the openness and acceptance she has found. She acknowledges that any misinterpretations of the individual's and group's message to the straight world are solely her responsibility.

2. Additional information on homosexual support groups and referral sources for gays, lesbians, their friends, and loved ones can be obtained from the Lambda Legal Defense Group and Lambda Rising. Both of these groups have chapters in Washington, DC.

3. Faludi's (1991) research challenges an earlier statistic that women's disposable income after divorce drops almost 73% (AAUW 1989:5).

Chapter 14

1. The Centers for Disease Control (CDC) recently has been renamed the Centers for Disease Control and Prevention. As of July 1993, the commonly used acronym for this organization remained the "CDC" and will be used here. Statistics are updated every six months by the CDC.

2. ARC (AIDS Related Complex) is no longer a diagnostic category at the time this book went to press.

3. To distinguish between sexual orientation and behavior, please see chapter 13.

4. All statistics and demographics will be updated as revisions in the text are made.

5. While the virus has been found in other body fluids such as tears, saliva, and perspiration, it is not in sufficient doses to be infectious.

6. Lubricated condoms break less readily than unlubricated condoms. Lubricants must be non-oil based to be effective. Oil-based lubricants such as vaseline, baby oil, and hand lotions will dissolve the latex in the condoms. K-Y jelly and PROBE are effective lubricants.

7. Since her death, it has become known that Kimberly Bergalis may have lied about becoming HIV infected through dental procedures. We may never know what, if any, relation existed among HIV infection, Dr. Acer, and his patients (60 Minutes Broadcast 6/1994; Kolata 1994).

Glossary

AASECT—American Association of Sex Educators, Counselors, and Therapists.

abstinence—For this book's purpose, the practice of refraining from sexual intercourse.

actual effectiveness—The statistical figure used to calculate accuracy in birth control. Since all couples do not use birth control properly or consistently, this figure will be lower than the theoretical effectiveness rate.

adolescence—The period of life from puberty to maturity terminating legally at the age of majority.

adolescent fertility—The ability to impregnate or become pregnant and give birth to live young during adolescence.

adolescent sterility—The period that occurs between menarche and reproductive maturity when pregnancy is not likely to result from intercourse. It has been discussed among the Kalahari San people as a mechanism for controlling population.

adrenocorticotrophic hormone (ACTH)—A protein hormone of the anterior lobe of the pituitary gland that stimulates the adrenal cortex.

affinal kin—Those individuals who are related by marriage. For example, "in-laws" in U.S. culture.

age-grade—A grouping of individuals based on shared biological maturity. It includes responsibilities, rights, cultural practices, and obligations that change from age stage to age stage.

age of majority—A marker established by Euro-American society where an individual gains the legal status of adulthood. These rights include marriage, adult responsibilities for criminal actions, and the right to make personal choices.

age set—An organized group of persons of or near the same age who will interact throughout their lives.

alloparenting—Use of extended kin and non-kin as a means of socioeconomic support and socialization of children as a continuation of our hominid behavior.

altruism—Unselfish regard for, or devotion to, the welfare of others.

ambilineal descent—The tracing of one's family through either the male or female parent or both.

amniocentesis—The surgical insertion of a hollow needle through the abdominal wall and uterus of a pregnant female, especially to obtain amniotic fluid for the determination of sex or chromosomal abnormality.

ampulla—The far end of the fallopian tube.

analogues—Something that is similar in function to something else.

androcentric—Sexist, male biased.

androgen—The hormones such as testosterone and adrosterone that produce or stimulate the male characteristics.

androgen insensitivity syndrome—The most common hormonal error that occurs in chromosomal XY males. The testes secrete amounts of testosterone that are generally defective so that these cells are unresponsive to testosterone. The fetus develops with partially feminized sex hormones.

androgyny—Having the characteristics or nature of both male and female.

anorgasmia—The inability to have an orgasm or lack of sexual responsiveness.

anthropoids (anthropoidea)—A suborder of primates consisting of the Platyrrhine infraorder (New World Monkeys), and the Catarrhine suborder (Old World primates) that include Old World monkeys and the hominoids (apes and humans). Characteristics include a well-developed brain, stereoscopic vision, sociality, and grasping hands.

apocrine glands—Associated with hair growth, these glands are scattered over the body. Also known for producing odors.

arboreal—Adapted to life in the trees.

associated polyandry—The practice where a woman is allowed multiple unrelated husbands. These men are referred to as "visiting husbands."

asymetrical gender relations—Nonegalitarian relationships between males and females.

australopithecines—Of or relating to extinct southern African hominids with near-human dentition and a relatively small brain.

avunculocality—Of or relating to an uncle.

basal body temperature—The body temperature at rest, usually taken before arising in the morning. Used to determine when a woman is ovulating.

Bem Sex Role Inventory (BSRI)—An instrument measuring masculinity and femininity. A characterization of the person's sex role as perceived by the individual taking the BSRI.

Berdache (two spirit)—A term used by anthropologists to describe the nonwestern institution in which a male or female takes on all or some of the role parameters associated with the other gender. Among some Native American circles, these people are thought to possess supernatural powers. The preferred term is "two spirit" or the indigenous terminology.

bilateral descent (non-unilineal, double descent)—The practice of tracing one's descent through both parents with each of these in control of different areas of activity and property.

bilateral kindred (kinship)—The system of kinship structure in which an individual belongs equally to the kindred of both parents.

bilineal descent—Similar to ambilineal type in which descent is traced through both the patrilineage and matrilineage with each controlling different areas of activity and property.

bilocality—Practice where the newly married couple may live near the bride's or groom's parents and follow the particular set of rules established by either the parents or culture.

biobehaviorally—A behavior or response which is an interaction of biological and learned mechanisms.

biological paternity—The person who is the biological father.

bipedalism—The ability to maintain and walk in an upright stance.

birth control—Methods used to control population size at the beginning of the life cycle.

blended family—A household made up of two parents, their biological children, and any children from previous marriages and other relatives.

blood products—Include all components of whole blood separately or as a whole particularly relevant in transfusions and with hemophiliacs.

Boasian—A school of thought in anthropology that emphasizes data collection of data and culture specific approaches.

bonding—The ongoing and continuous socio-emotional link between people.

broad ligament—A band of connective tissue across the lower abdomen that supports the uterus, fallopian tubes, and ovaries.

caul—The amnion or "bag of waters" which envelops the developing embryo/fetus.

cervical mucous checks—Birth control based on avoiding sexual intercourse during the time when the cervical mucous thins in order to allow sperm to pass through the os.

chorionic villi sampling (CVS)—A technique for diagnosing medical problems in the fetus as early as the eighth week of pregnancy; a sample of the chorionic membrane is removed through the cervix and studied.

circumcision—In the female, surgical procedure that cuts the prepuce exposing the clitoral shaft; in the male, surgical removal of the foreskin from the penis.

clans—A group of people whose unilineal descent is established upon a belief that they have a common real or mythical ancestor even if the claim cannot be proven. While biological links cannot be traced due to the large number of people concerned, the group shares mutual economic security, social control, political and marriage relations, religious practices, and ceremonies.

climacteric—Sexual aging experienced by both men and women including physical and social changes.

clitoridectomy—Surgical removal of the clitoris; practiced routinely in some cultures.

cognatic descent—The practice of tracing relations through birth, usually emphasizes the mother's side.

cognitive complexity—The ability to process a variety of data through the cerebral cortex. Characteristic of humans.

cohabitation—Living together and sharing sex without marrying.

coital—The physical union of male and female genitalia accompanied by rhythmic movements leading to the ejaculation of semen from the penis into the female reproductive tract.

coitus interruptus—A method of birth control in which the penis is withdrawn from the vagina prior to ejaculation.

comparative (methodology)—See cross-cultural.

consanguineal—Kinship related by descent or filliation rather than through marriage; "blood relatives."

contraception—Any natural barrier, hormonal or surgical, that prevents conception and pregnancy.

controlled comparison—A research method that uses social and cultural features to find similarities among a small sample of cultural groups.

corpora cavernosa—Hollow, sponge-like cylinders of tissue within the penile shaft that become engorged with blood during sexual excitement.

corpus luteum—Reddish yellow endocrine tissue that forms within a ruptured ovarian follicle. It produces progesterone.

corpus spongiosum—A spongy cylinder of tissue running through the underneath part of the penile shaft that also becomes engorged with blood causing erection.

cortices—The outer layer of the brain.

cremasteric muscle—A muscle located in the spermatic cord that elevates the testicles when contracted.

cross-cousin—A relative who is the child of one's mother's brother or father's sister.

cross-cultural—Comparing two or more cultures along a variety of dimensions; often comparing "western" with traditional groups.

crura—The innermost tips of the cavernous bodies that connect to the pubic bones.

culture—The learned behavior, skills, attitudes, beliefs, and values of a particular society. These are learned by observation, imitation, and social learning.

cunninlingus—Oral stimulation of the clitoris, vaginal opening, or other parts of the vulva.

cyclic—Having a rhythmic or periodic pattern.

cystitis—A non-sexually transmitted infection of the urinary bladder.

defloration—The name given to the act where a woman loses her virginity through penile-vaginal intercourse.

deleterious genes—Those genes, often recessive, which can be expressed as harmful or fatal physical structures (phenotypes).

demographics—The study of human populations with reference to size and density, distribution, and vital statistics.

descent—The tracing of one's relationship based on the parent/child connection that defines the social relationship. Often claims are based on common ancestry that may or may not actually exist. Descent may be based on the mother's, father's, or both parentage.

descent groups—A set of relationships that cannot be changed by location or death.

dichotomous—Dividing into two parts.

disease model—A model that examines phenomenon from the perspective of pathology or disease.

distal cause—For use in our discussion, the basis or origin of behavior and not the immediate cause.

double descent—See bilineal descent.

duolocality—The living arrangement where a young married couple lives apart from one another so only blood relations make up a household.

dysmenorrhea—Painful menstruation.

dyspareunia—Painful intercourse.

econiche—A particular environment to which a species is adapted.

ectopic pregnancy—The implantation of a blastocyst somewhere other than in the uterus, usually in the fallopian tube.

effeminate—Having feminine qualities which are attributed to a man in western societies.

ego—The term used for a person or persons in determining and tracing kinships.

ejaculatory duct—Continuation of the vas deferens through the prostate to the urethra.

Electra complex—The female version of the Oedipal complex. The young female sexually desires her father and thus regards her mother as the competitor.

embryo transplants—Procedures in which women other than the prospective mother is impregnated with the husband's sperm. After several days, the fertilized egg is removed from her womb and placed within the mother's uterus.

emic—The perception of a phenomenon as seen and felt by a participant inside the system.

endemic—Continuing and widespread within a population; usually refers to a disease.

endogamy—The rule which requires a member of a community to marry within their group, tribe, caste, or class. This allows the group to retain and maintain control of the wealth, power and prestige.

endometrium—Interior lining of the uterus, innermost of three layers.

Eocene—The second period of the Cenozoic era, about 30–40 million years ago.

epididymis—Tubular structure on each testis in which sperm cells mature.

episiotomy—A surgical incision in the vaginal opening made by the clinician or obstetrician if it appears the perineum will tear in the process of birth.

Epstein-Barr Virus Syndrome—A disease with symptoms similar to early HIV infection.

erogenous zone—The parts of the body where touching or stroking results in sexual excitement. The areas usually include, but are not limited to, the breasts, lips, genital or anal regions, and buttocks.

erogenous zone sensitivity—Those parts of the body that produce sexual arousal when stimulated.

erotic symbol/sex signal—Of, devoted to, or tending to arouse sexual love or desire.

erotocentricity—The process by which our culture's sexual attitudes, values, and mores distort our understanding of sexual practices of other cultures.

estradiol—An estrogen that effects the development and maintenance of the female reproductive organs as well as all female secondary sex characteristics.

estrogen replacement therapy—Controversial treatment of the physical changes of menopause by administering doses of the hormone estrogen.

estrus—The reproductive cycle in female nonhuman mammals, which is accompanied by physiological, anatomical, and behavioral changes.

ethnocentrism—The view that the cultural values and practices of one's own society are superior to all others.

ethnography—The descriptive study of a culture.

ethnologist—An anthropologist who studies cultures from a comparative or historical point of view.

ethnology—The study of how and why cultures differ and/or are similar.

ethnomethodological—The study of methods people use to communicate in everyday, routine activities.

etic—The interpretation of customs in a specific culture as seen by the outside observer. The use of descriptions and analyses in terms of conceptual schemes and categories associated with western scientific perspective.

etiology—The origin of a disease or abnormal condition.

excision—To remove the clitoris.

exogamy—The marriage rule that requires its members to marry outside their community.

explicit culture—The knowledge about ourselves that can easily be communicated. Examples include genealogies and marriages.

extended family—A group of people consisting of parents, children, grandparents, aunts, and uncles. Relationships are consanguineal (blood) and/or affinal (marriage).

fecundity rate—A figure used in determining reproductive rate of capability.

fellatio—Stimulation of the male genitals with the mouth.

fertility—The ability of a person to produce an offspring.

fetal reduction—Selectively aborting a fetus in a multiple pregnancy so that the other fetuses have an increased chance of survival.

fetishism—Sexual arousal triggered by objects or materials not usually considered to be sexual.

fictive kin—People who are not related to you, but stand in a kinship relation to you.

follicular stimulating hormone (FSH)—A hormone produced or secreted by the anterior pituitary gland which stimulates sperm production in males and follicle and ovum development in females.

follicular phase—The first phase of the menstrual cycle during which egg maturation and development occur.

foragers—A group depending upon hunting of animals and gathering of food substance.

fraternal polyandry—The marriage of brothers to the same woman with whom they live patrilocally. All husbands take responsibility for the woman's children.

frenum/frenulum—Thin tightly-drawn fold of skin on the underside of the penile glans; it is highly sensitive.

fundus—The rounded top part of the uterus.

gatherers and hunters—See foragers.

G spot (Grafenberg)—An area of sensitive tissue located approximately one to two inches inside the opening of the vagina on the anterior wall under the pubis. This stimulation often leads to an orgasm.

gender—A designation given to sexes in regard to biobehavioral and psychosocial qualities.

gender dysphoria—Dissatisfaction with one's gender.

gender dysphoric—Term to describe gender-identity/role that does not conform to the norm considered appropriate for one's physical sex.

gender sex predetermination—The use of technology to increase the chances of having a boy or a girl child.

genetic fitness—Refers to reproductive success related to adaptation to the environment.

genotypes—Genetic traits that an individual inherits.

Gestalt—The concept used by Ruth Benedict by which the configuration of a culture is presented as a whole formation which connotates more than its component parts.

gonadotropins—Sex hormones.

gynecomastia—Breast enlargement in the male.

hermaphrodite—An individual having sexual tissues of both sexes.

heterogeneity—The quality of being unlike or different.

hogopans—The ancestors of humans, gorillas, and chimps that diverged somewhere between eight and five million years ago.

holistic—The view that all parts of a culture are interrelated; leads to the study of all aspects of a culture and how they are interrelated.

hominid—Humans and their direct ancestors.

hominines—The descriptive term for a member of the primate subfamily Homininae, composed of modern humans, chimpanzees, and gorillas.

homoeroticism—Sexual attraction to people of the same gender.

homologoues—Structures that develop from the same embryological tissue.

homophobia—Strongly held negative attitudes and irrational fears of homosexuals.

hormone replacement therapy (HRT)—The use of supplemental hormones during and after menopause.

H-P-G Axis—Composed of the hypothalamus, the pituitary, and the gonads. This is the hormonal basis of sex and reproduction. This axis forms a network for understanding puberty, sexuality, and reproduction.

HSR—Human Sexual Response.

Human Relations Area Files (HRAF)—Classification scheme developed by G.P. Murdock that serves as the basis for statistical comparison for anthropologists.

hymen—Membranous tissue that can cover part of the vaginal opening.

hypothalamus—A ductless gland located in the brain that contains neurosecretions used in controlling certain metabolic actions like the maintenance of the body's water balance, sugar and fat metabolism, the regulation of body temperature, and the secretion of FSH, LH, LTH.

ideal culture—The normative expectation or behavior of individuals in a given culture.

impotence—The inability to obtain, achieve, or sustain an erection. Value laden; generally referred to currently as erectile problems.

incest taboo—The universal prohibition against marrying or mating within the primary family. The prohibition of sexual relations exists between mother and son, father and daughter, and brother and sister.

independent households—(Also known as neolocality.) A newly married couple moves away from both set of parents to make their home in a different area.

indigenous sexual behaviors—Sexual behaviors specific to a particular group.

infibulation—Surgical procedure, performed in some cultures, that seals the opening of the vagina and removes the clitoris.

inhibited sexual desire (ISD)—Lack of interest in sexual behavior.

initiation ceremonies—Public recognition that facilitates the movement from childhood to adulthood. Usually stressful, these often include rigorous tests, hazing, isolation from younger friends, and painful ordeals.

insectivore—A order of small mammal which eats insects.

interfemoral intercourse—Position where the penis is placed between the partner's thighs.

intersex—Also referred to as a third sex. See hermaphrodite.

interstitial cell stimulating hormone (ICSH)—Pituitary hormone that stimulates the testes to secrete testosterone. Also known as LH in females and males.

introitus—The outer opening of the vagina.

in utero—In the uterus.

invitro fertilization (IVF)—A process whereby the union of the sperm and egg occurs outside the mother's body.

jerungdu—The belief among the Sambia that semen has power. The Sambia believe that semen must be acquired for a male to achieve true masculinity.

joint household residence—The practice of living with or near one of the couple's parents.

kin groups—Descent that is determined by social position and cultural meaning. These links are formed through marriages and reproduction.

Kinsey report—One of the earlier formal studies in U.S. sexual behavior that has generated controversy.

labia majora—Major or large lips of tissue on either side of the vaginal opening.

labia minora—Smaller lips of tissue on either side of the vagina.

leutotropic hormone (LTH)—A pituitary hormone found in males and females. In females, it is involved in maintaining uterine tone and lactation. In the male, its function is unknown.

levirate—A tradition practiced in some cultures where a man is required to marry a deceased brother's wife.

leydig cells—Cells located between the seminiferous tubules that are the major source of androgen in males.

liaisons—Unstable relationship based solely on sex.

libido—A desire for sex. Freud considered it natural, present at birth, and focused in various sections of the body.

life-course—Marked and recognized by social and physiological changes.

liminality—The period during initiaton rites when an individual is in transition from one stage in the life cycle to another.

lineage—A descent pattern where members trace their kinship/relationship to one another and a founding ancestor. This is usually based on unilineal descent rule and constitutes a segment of a clan.

luteal phase—The third phase of the menstrual cycle. Conception occurs during this phase.

lutenizing hormone (LH)—Hormone secreted by the pituitary gland that precipitates ovulation in females and maintains the leydig cells in males.

macrophages—Tissue cells that protect the body against infection and noxious substances.

Maithuna—A sexual union in which people feel they merge with the sacred.

male climacteric—Sometimes termed the "male menopause." At about the same age that women experience menopause, men have decreased testosterone production and, in western countries, often begin to question the direction in which their lives are headed.

manhood—The sense that a male embodies the cultural definition of what it means to be a man.

marriage—A public and social ceremony between two or more people which creates relationships, sexual and economic rights, and obligations within the union. It also provides a means for incorporating offspring into the group.

Masters and Johnson—Sex researchers who defined the four cycles (phases) of human sexual response: excitement, plateau, orgasm, and resolution.

matrescence—The sociological process of becoming a mother.

matrilineal descent—The practice of tracing descent through the female line.

matrilocal residence—The practice where a married couple moves to live near or with the bride's family.

menarche—Onset of menstruation.

menopause—The period of natural cessation of menstruation occuring usually between the ages of 45 and 55.

mid-Pleistocene—Geological time period that spans approximately 750,000 years ago to 125,000 years ago.

migration—The movement of people into a region where they seek permanent residence.

Miocene—The fourth epoch (period) of the Cenozoic era, beginning twenty-five million years ago and ending five million two hundred thousand years ago.

moiety system—Society divided into two large social groups.

monogamous—Sharing sexual relations with only one person.

monogamy—Marriage to only one partner (at a time).

mons, mons pubis, mons veneris—Cushion of fatty tissue located over the female's pubic bone.

morphological—The study of the form and structure of an organism or social manifestation.

mortality—The incidence of death rates.

Mullerian ducts—Found in both embryonic males and females. Prenatal structures that develop into the broad ligament uterus, fallopian tubes, and upper third of the vagina.

myometrium—The middle layer of the uterus; consists of smooth muscle, thereby aiding the pushing of the newborn through the cervix.

natural selection—The evolutionary process through which factors in the evironment exert pressure that favors some individuals over others to produce the next generation.

negative feedback loop—A regulatory system that coordinates the production of gonadal hormones through the complex interaction of the gonads, the hypothalmus, and the pituitary gland.

neocortex—The position of the cerebral hemisphere that comprises most of the convoluted cortex.

neolocality—The practice where the young married couple lives apart from their parents in their own home usually in a different area.

neonatal—Focus on the care of newborns.

neotony—A condition often exhibited in physical characteristics associated with immaturity and dependency and need to have care provided by adults.

neurophysiological—The study of the nervous system of the body.

nonconsensual—Forced.

non-unilineal—Also known as bilateral descent. The tracing of one's ancestors through both parents equally.

novitiates—Recent initiates.

nuclear families—A grouping of individuals of a household made up of a married couple and their biological children.

Oedipus complex—Named for the tragic mythical hero, Oedipus, who unknowingly fell in love with and married his mother. Oedipus blinded himself when he realized what he had done. The term is used to refer to a young boy who covets his mother while competing with his father.

omnivorous—A diet composed of both meats and vegetables.

oophrectomy—Surgical removal of the ovary or ovaries.

orchidectomy—The surgical removal of the testes.

orgasm—Sexual climax. The intensely pleasurable feeling that comes at the end of stimulation to the genital organs (clitoris, vagina, penis).

os—The opening of the cervix.

pairbonding—Forming unions between two individuals.

paleocortex—Located in the brain, this limbic system triggers emotions in primates. It is directly connected to the expression of sexual behaviors in addition to fear and aggression.

pan-human—Of or relating to all humanity.

pan-sexual—Lacking highly specific sexual orientations or preferences; open to a range of sexual activities.

paralanguage—Optional vocal effects (as tone of voice).

parallel cousins—Children of one's mother's sister, father's brother.

parametrium—Outer covering of the uterus.

paternity—Blood relationship traced through the male or father.

pathogens—Specific causative agent (e.g., bacterium or virus) of disease.

patrescence—A status change for the male after birth of his child.

patrilineal descent—Tracing kin through the male line.

patrilocality—Postmarital residence where the couple lives near or with the groom's family.

perinatally—Occuring at the time of birth.

perineum—Sensitive area of skin between the scrotum or vagina and the anus.

petting—A practice of experimentation in sexual relationships. This often includes both oral and manual sexual stimulation but does not include vaginal intercourse.

phallic stage—Found in the Freudian model occuring between the ages of three and five years. Often this is characterized by love, hate, envy, and guilt and fascination with one's genitals.

phenotypes—Observable physical expressions of genetic traits.

phenotypic sex—The physical expression of either the xx or xy genotype including structures such as ovaries, labia majora and minor, or the penis, testes, scrotum.

pheremones—Hormones produced by the body that either attract or repel the other sex.

phratries—A large grouping of people of unilineal descent composed of several clans.

Pleistocene—The epoch beginning one million six hundred thousand years ago that continues through today some argue. It is known for its periods of glaciation and for the proliferation of humankind.

Pliocene—The fifth epoch (period) of the Cenozoic era beginning five million two hundred thousand years ago and ending one million six hundred thousand years ago; a warm period during which early human types became differentiated from early apes.

polyandry—The practice where one woman has several husbands at the same time. A form of polygamy practiced in Tibet, Nepal, India, Sri Lanka, and the Marquesans of the Pacific.

polygamy—An individual married to multiple partners (spouses) at the same time.

polygyny—The practice where one man has several wives at the same time.

polymorphously preverse—A Freudian concept of sensuality in which sexual arousal is diffused over the body, not genitally focused sexuality.

population control—Any means a group uses throughout the life cycle to control its size.

prepuce—In the female, the upper part of the labia minora that covers the clitoral shaft, known as the "clitoral hood." In males, the foreskin, a loose tissue covering the glans.

preputial glands—Small lubricating glands located in the foreskin of the penis.

primary (immediate) family—An individual and his/her siblings, parents, and children.

primates—An order in the phylogenetic tree characterized by large-brained, highly social creatures of which prosimians, monkeys, apes, and humans are members.

progeny—Offspring; the descendants of two parents.

prolactin (PRL)—A hormone found in the pituitary gland that is involved with breast development and the production of milk during lactation (nursing or breast-feeding).

prophylactic asymptomatic drug treatment—The prescription of a drug to prevent symptoms. For example, AZT is given to people who are HIV infected to slow the progression of the virus.

prosimians—One of the two suborders of primates that include the lemurs, lorises, and tarsiers.

prostatic fluid—Secretions of the prostate gland that contribute to the semen.

proximal cause—Term used to describe the present and ongoing factors causing a behavior; the immediate cause.

psychoanalysis—Freudian technique used in treating personal and psychological disorders in which people explore their pasts to gain insights into their current behaviors and feelings.

psychocultural—Relating to the interaction of psychological and cultural factors in the individual's personality or in the characteristics of a group.

psychosexual maturation—The Freudian notion that early sexual experiences are important to personality development. The individual must successfully pass through a series of five sequential stages to reach psychosexual maturity.

puberty—The stage of life when a person develops secondary sexual characteristics and begins spermatogenesis or the menstrual cycle.

pubococcygeus muscles—Muscles encircling the vagina and the crura.

purdah—Seclusion of women from public observation among Muslims and some Hindus in India.

ramage—Ambilineal descent in which groups are formed by household and common ancestors, may be stratified internally.

relativism—The belief that cultures and groups are examined in relation to their own values and behaviors.

rhythm—A natural birth control method entailing abstinence during the ovulatory phase of the menstrual cycle.

rites of passage—The symbolic or ceremonial observance that occurs when an individual moves from one stage in the life cycle to the next. This includes three stages: separation, transition (liminality), and incorporation.

Romantic Era—The period that began around the year 1800. The artists of the period celebrated nature and the unspoiled.

sebaceous—An oil gland that is stimulated by testosterone and contributes to acne.

seminiferous tubules—Tightly coiled tubes in the testes where the sperm develop.

separation—First stage in the rite of passage where an individual is separated, either symbolically and/or actually, from their previous ways of life.

serial monogamy—Successive marriages or sexual unions.

seroconversion—After exposure to HIV, antibodies to the virus accumulate in one's blood and can be detected by a blood test.

serostatus—Refers to the presence or absence of antibodies to HIV.

sexology—The study of sexual attitudes and behaviors by scientific means.

sex positive—Generally attributed to an individual or group which expresses comfort with ones body and sexuality and views sexuality as a positive and normal behavior.

sexual dysfunctions—Problems individuals have with their sexuality or sexual relationships.

sexual orientation—The gender to which one is sexually attracted.

SIECUS—Sex Information and Education Council of the United States.

Skene's glans (Paraurethral glands)—Glands on either side of the Grafenberg spot that secrete a fluid into the urethra upon sexual stimulation.

smegma—White, lubricating, cheeselike substance secreted underneath the foreskin of the penis and prepuce of the clitoris.

social system—A set of patterned interactions incorporating various institutions, including the sex and gender system.

society—An organized system of statuses and roles.

sociobiology—Role played by genes in explaining both social and cultural behavior especially in regard to reproduction.

somatotype—Body type.

sororal polygyny—A situation where men marry their wives' sisters in the belief that this will contribute to household harmony and productivity.

sororate—The practice where a man marries his dead wife's sister. In some cultures this is required after the death of a spouse.

species—A group of individuals that are capable of interbreeding due to isolation from others.

spermatic economy—A widely held nineteenth century western belief that ejaculate exists in finite quantities in a man's life.

spermatogenesis—Process by which sperm are developed.

sperm banks—Repositories of donated ejaculate that have been medically and genetically screened and HIV tested, which are used for artificial insemination-donor.

sphincter—A small closure between the bladder and urethra that closes during arousal and ejaculation so that urine does not leak into the urethra and damage sperm.

stipulated descent—A group's claim to a common ancestor that cannot be traced or documented.

subculture—An ethnic, regional, economic, or social group exhibiting characteristic patterns of behavior sufficient to distinguish it from the larger culture or society of which it is a part.

superfamily—The laws and principles used for classification between an order and a famliy.

tacit culture—Shared rules that an individual learns unconsciously from those around them regarding behavior.

theoretical effectiveness rate—The effectiveness rate of a given method of birth control based on laboratory conditions.

threshold level—The amount of testosterone required to maintain the libido in men and women.

tonic—Ongoing releasing patterns.

transgender—A recent addition to western gender variant social identities. A community term that regards gender variant identities as a continuum that includes the traditional transsexual and transvestite and identities that lie in between and beyond.

transition—The middle level in the rite of passage where an individual is prepared to rejoin their community. See liminality.

transsexual—An individual who has the phenotype of one gender, but the gender identity of another. Two spirit—see Berdache.

transurethral resection of the prostate (TURP)—A surgical procedure to reduce benign (or noncancerous) enlargement of the prostate.

transvestite—An individual, usually male, who dresses in clothing of the opposite sex.

tubal patency—Having unblocked fallopian tubes.

unilineal descent—Tracing descent through either the male or female line.

urinary tract infection (UTI)—An infection of the urethra more often found in women which can be caused by an abrasion of the urethra or infection by a pathogen.

utero-penile erection—Fetal penile erections.

vaginismus—Vaginal muscle spasms causing penile penetration to be painful or difficult.

vasocongestion—A physical condition occuring during sexual intercourse when blood vessels in the genitals engorge.

vernix—A waxy, protective substance covering the fetus.

viability—The ability of the neonate to live.

Victorian Era—Historical rule of Queen Victoria of England (1837–1901). This era saw strict rules in regard to morality and the necessity for the unspoiled or the pristine state. Sexual impulses were associated with shame, and were suppressed or never publicly shared.

vulva—External sex organs of the female, including the mons, major and minor lips, clitoris, and opening of the vagina.

Western World—Generally refers to Western Europe and Anglo North America heavily influenced by a Judeo-Christian ideology.

woman marriage—The practice found in some cultures where two women marry. One becomes the "social" male and makes arrangements for the wife to become pregnant by another man.

Wolffian ducts—Embryonic structures that develop into male sexual and reproductive organs if male hormones are present and into part of the urinary tract system in both males and females.

xanith—A third gender identity found among the Omani in the Middle East.

Bibliography

AAUW Briefs. 1989. 83(3) (May/June).

"Abortion in American History." 1990. *Parade Magazine*. April 22:8.

Abramson, P., and Herdt, G. 1990. "The Assessment of Sexual Practices Relevant to the Transmission of AIDS: A Global Perspective." *Journal of Sex Research*. 27(2) (May):215–32.

ACHA. 1989. Conference on AIDS. Keynote address by Dr. R. Keeling.

ACHA. 1989 (Jan.). Statement on AIDS and International Education Issues. Rockville, MD: ACHA:1–4 with NAFSA.

ACHA Conference. April 1989. Syracuse, NY.

AIDS Teleconference. 1989.

AIDS Task Force Central NY Newsletter. 1993. April.

Alexander, P. 1987. "Prostitutes Are Being Scapegoated for Heterosexual AIDS." In *Sexwork: Writings by Women in the Sex Industry*. F. Delacoste and P. Alexander (eds.). Pittsburgh, PA: Cleis Press:248–63.

Alexander, P., and Delacoste, F. (eds.). 1987. *Sexwork: Writings by Women in the Sex Industry*. Pittsburgh, PA: Cleis Press.

Allgeier, E. 1989. "Review of *The Construction of Homosexuality*." D. Greenberg. Chicago, IL: The University of Chicago Press (1988), and "Review of *Gay, Straight, and In-between*." J. Money. New York, NY: Oxford University Press, (1988). *Journal of Sex Research*. 26(4) (November):552–54.

Allgeier, E., and Allgeier, A. 1991. *Sexual Interactions*. 3rd ed. Lexington, MA: D.C. Heath and Co.

Allen, B. 1988. "Aids 201." In the *AIDS Caregiver's Handbook*. T. Eidson (ed.). New York, NY: St. Martin's Press:13–45.

Altman, K. A. 1989. "Who's Stricken and How: AIDS Pattern is Shifting."
    New York Times (Sunday, Feb. 5):1, 28(col. 1).

Altman, L. K. 1998. "U.N. Breast-Feeding Warning for Moms with AIDS."
    San Francisco Examiner (Sunday, July 26): A19.

Alzate, H. and Londono, M. L. 1984. "Vaginal Erotic Sensitivity." *Journal
    of Sex and Marital Therapy.* 10(49).

American Health. 1994. (July/August):8.

American Health. 1994. (July/August):101.

American Health. 1989. "The Facts of Life for Infertile Couples." October
    (8) 8.

American Psychiatric Association. 1980. *Diagnostic and Statistical Manual
    of Mental Disorders.* 3rd edition. Washington, DC: American Psychi-
    atric Association.

American Psychiatric Association. 1973. *Diagnostic and Statistical Manual
    of Mental Disorders.* 2nd Edition. Washington, DC: American Psy-
    chiatric Association.

Angier, N. 1992. "A Male Menopause? Jury is Still Out." *New York Times:*20
    May 6.

Appignanesi, R. 1979. *Freud for Beginners.* New York, NY: Pantheon Books.

Aries, P. 1962. *Centuries of Childhood: A Social History of Family Life.*
    New York, NY: Random House.

Arms, S. 1975. *Immaculate Deception.* South Hadley, MA: Bergin and Garvey
    Publishers.

"Asia: Discarding Daughters." *Time.* 136(19):40.

"Attacking the Last Taboo." *Time.* 115(15):72. April 14.

Atwater. L. 1982. *The Extramarital Connection: Sex, Intimacy, and Iden-
    tity.* New York, NY: Irvington.

Atwood, J. D. 1990. "A Multi-Systematic Approach to AIDS and Adoles-
    cents." Paper presented at XXIII National AASECT Conference,
    Arlington, VA: February 17.

Ault, A. 1998. "First Pill for Male Impotence Approved in USA." *Lancet*
    351 (9108): 1037.

Austin, E. 1993. "Four New Improved Birth Control Pills." *Self.* 1993:57.

Avery, C. 1989. "How Do You Build Intimacy in an Age of Divorce?" *Psy-
    chology Today,* May:27–31. In *Annual Editions of Human Sexuality
    1990–91.* O. Pocs (ed.). Sluice Dock, Guilford, CT: Dushkin Publish-
    ers:131–34.

Avery, C. S. 1990. "Flirting with AIDS." In *Annual Reviews in Human Sexuality* 1990–91. O. Pocs (ed.). Sluice Dock, Guilford, CT: Dushkin Publishers: 232–34.

Baker, R. R. and Bellis, M. A. 1993. "Human Sperm Competition: Ejaculate Manipulation by Females and a Function for the Female Orgasm." *Animal Behavior.* 46:887–909.

Bakwin, H. 1974. "Erotic Feelings in Infants and Young Children." *Medical Aspects of Human Sexuality.* 8(10):200–215.

Baldwin, J., and Baldwin, J. 1988. "Factors Affecting AIDS-Related and Sexual Risk-Taking Behavior Among College Students." *Journal of Sex Research.* 25(2) (May):181–196.

Baldwin, J., Whiteley, S., and Baldwin, J. 1990. "Changing AIDS and Fertility-Related Behavior: The Effectiveness of Sexual Education." *Journal of Sex Research.* 27(2) (May):245–262.

Baltimore, D., and Heilman, C. 1998. "HIV Vaccines: Prospects and Challenges." *Scientific American.* 279(1) (July):98–103.

Barker-Benfield, G. J. 1975. *Horrors of the Half-Known Life.* New York, NY: Harper and Row.

Barkow, J. H. 1984. "The Distance Between Genes and Culture." *Journal of Anthropological Research.* 40(3):367–379.

Barnes, J. and Findlay, S. 1988. "AIDS in Africa: Ravager of the Nation Builders." *US News and World Report.* 104(6/27):32–36.

Barnouw, V. 1985. *An Introduction to the Field, Culture and Personality.* 4th Edition. Homewood, IL: The Dorsey Press.

Barth, F. 1969. *Ethnic Boundaries.* Boston, MA: Little, Brown and Company.

Bartlett, J. G., and Moore, R. D. 1998. "I'm Proving HIV Therapy." *Scientific American.* 279(1) (July):84–87, 89, 91–93.

Bateson, G. 1958. *Naven.* Stanford, CA: Stanford University Press.

Beach, F. 1973. "Human Sexuality and Evolution." In *Reproductive Behavior.* W. M. Montagna and W. A. Sadler (eds.). New York, NY: Plenum Press: 333–365.

Beardsley, T. 1998. "Coping with HIV's Ethical Dilemnas." *Scientific American.* 279 (1) (July): 106–107.

Becker, G. 1990. *Healing the Infertile Family. Strengthening Your Relationship in the Search for Parenthood.* New York, NY: Bantam Books.

Bell, A. and Weinberg, M. 1978. *Homosexualities: A Study of Diversity Among Men and Women.* New York, NY: Simon and Schuster.

Bell, A. and Weinberg, M., and Hammersmith, S. 1981. *Sexual Preference: Its Development in Men and Women*. Bloomington: Indiana University Press.

Bell, D. 1983. *Daughters of the Dreaming*. Sidney, Australia: McPhee Gribble/ George Allen and Unwin.

Belzer, E. G., Whipple, B., Moger, W. 1984. "On Female Ejaculation." *Journal of Sex Research*. 20(4):403–406.

Bem, S. 1977. "Psychological Androgyny." In *Beyond Sex Roles*. Alice G. Sargent (ed.). New York, NY: West Publishing Company: 319–324.

Bem, S. L. 1975a. "Probing and the Promise of Androgyny." Keynote Address for APA-NIMH Conference on the Research Needs of Women. Madison, WI. 31 May.

Bem, S. 1975b. "Androgyny vs. the Tight Little Lives of Fluffy Women and Chesty Men." *Psychology Today*. September: 58–62.

———. 1974. *Bem Sex-role Inventory*. Princeton, NJ: Educational Testing Service.

Benderly, B. L. 1987. "Rape Free or Rape Prone." In *Conformity and Conflict: Readings in Cultural Anthropology*. J. P. Spradley and D. W. McCurdy (eds.). Boston, MA: Little, Brown and Company: 184–188.

Bendet, P. 1989. "Hostile Eyes." In *Annual Editions Human Sexuality 1989/90*. O. Pocs (ed.). Sluice Dock, Guilford, CT: Dushkin Publishers: 193–96.

Benedict, R. 1959. *Patterns of Culture*. New York, NY: The New American Library.

———. 1934. "Anthropology and the Abnormal." *Journal of General Psychology*. 10:59–82.

Benjamin, H. 1966. *The Transsexual Phenomenon*. New York, NY: The Julian Press.

Berndt, C. 1981. "Interpretations and Facts in Aboriginal Australia." In *Woman the Gatherer*. F. Dahlberg (ed.). New Haven, CT: Yale University Press: 153–203.

Bernstein, R. J. 1983. *Beyond Objectivism and Relativism: Science, Hermeneutics and Praxis*. Philadelphia: University of Pennsylvania Press.

Beyene, Y. 1989. *From Menarche to Menopause. Reproductive Lives of Peasant Women in Two Cultures*. Albany: State University of New York Press.

"Birth-Control Popular Here." 1990. *Parade Magazine*, April 15:15.

Bjorklund, D. and Bjorklund, B. 1988. "Straight or Gay?" *Parents*, 98 (Oct.):93–96.

Blackwood, E. 1986. "Breaking the Mirror: The Construction of Lesbianism and the Anthropological Discourse on Homosexuality." In *The Many Faces of Homosexuality: Anthropological Approaches to Homosexual Behavior*. E. Blackwood (ed.). New York, NY: Harrington Park Press.

Blackwood, E. 1984. "Lesbian Behavior in Cross-Cultural Perspective." San Francisco, CA: San Francisco State University: MA Thesis.

Blaffer-Hrdy, S. 1981. *The Woman That Never Evolved*. Cambridge, MA: Harvard University Press.

————. 1986. "Empathy, Polyandry and the Myth of the Coy Female." In *Feminist Approaches to Science*. New York, NY: Perdamon Press: 119–145.

Blumenfeld, W. J., and Raymond, D. 1989. *Looking at Gay and Lesbian Life*. Boston, MA: Beacon Press.

Blumstein, P., and Schwartz, P. 1983. *American Couples*. New York, NY: Pocket Books.

Bock, P. K. 1988. *Rethinking Psychological Anthropology*. New York, NY: W. H. Freeman and Company.

Bogoros, W. 1907. "The Chuckchee Religion." Vol. II, *Memoirs of the American Museum of Natural History*. Leiden, Netherlands: E.S. Brill.

Bohlen, J.G., et al. 1982. "Development of a Woman's Multiorgasmic Pattern: A Research Case Report." *Journal of Sex Research*. 18(2):130–145.

Bolin, A. 1996. "Transversing Gender: Cultural Context and Gender Practices." In *Gender Reversals and Gender Cultures: Anthropological and Historical Perspectives*. Sabrina Petea Ramet (ed.). New York, NY: Routledge pp. 22–52.

————. 1992. *Review of Gender Blending: Confronting the Limits of Duality* by Holly Devor. *The Journal of the History of Sexuality* 2(3):497–498.

————. Forthcoming. "Vandalized Vanity: Feminine Physiques Betrayed and Portrayed." In *The Denaturalization of the Body in Culture and Text*. F. Mascia-Lees (ed.). Albany: State University of New York Press.

————. 1994. "Transcending and Transgendering. Male-to-Female Transsexuals, Dichotomy and Diversity." In *Third Sex, Third Gender: Beyond Sexual Dimorphism in Culture and History*. Gilbert Herdt, ed. New York, NY: Zone Books, 447–485.

————. 1992a. "A Transsexual Coming of Age: The Cultural Construction of Adolescence." In *Gender Constructs and Social Issues*, Tony L. Whitehead and Barbara V. Reid, eds. Urbana, IL: University of Illinois Press. Pp. 13–39.

————. 1992b. "Gender Subjectivism in the Construction of Transsexualism." *The Chrysalis Quarterly*, 1(3):23–26,39. Reprint of "Sexism in the Diagnosis and Treatment of Transsexuals" (1985).

————. 1988a. *In Search of Eve: Transsexual Rites of Passage.* South Hadley, MA: Bergin and Garvey Publishers.

————. 1988b. "Minds and Bodies: Gender Identity, Role and Body Rituals." Paper presented at the annual meetings of the American Anthropological Association, Phoenix, AZ: 17–20 November.

————. 1987a. "Transsexualism and the Limits of Traditional Gender Analysis." *American Behavioral Scientist* 31(1):41–65.

————. 1987b. "Transsexuals and Caretakers: A Study of Power and Deceit in Intergroup Relations." *City and Society* 1(2):64–79.

————. 1982. "Advocacy with a Stigmatized Minority." *Practicing Anthropology* 4(2):12–13.

————. 1974. "God Save the Queen: An Investigation of Homosexual Subculture." Boulder, CO: Norlin Library, University Microfiche. Ann Arbor, MI: MA Thesis.

Bolling, D., and Voeller, B. 1987. "AIDS and Heterosexual Anal Intercourse: Letter." *Journal of the American Medical Association.* 258:474.

Boston Women's Health Collective. 1992, 1984, 1976. *The New Our Bodies, Ourselves.* New York, NY: Simon and Schuster, Inc.

Boswell, H. 1991. "The Transgender Alternative." *Chrysalis Quarterly* 1(2):29–31.

Boswell, J. 1980. *Christianity, Social Tolerance, and Homosexuality.* Chicago: University of Aldine Press.

Boxer, A., and Cohler, B. 1989. "The Life Course of Gay and Lesbian Youth: An Immodest Proposal for the Study of Lives." G. Herdt (ed.). *Gay and Lesbian Youth.* New York, NY: Harrington:315–355.

Boyd, R., and Richerson, P. J. 1989. "The Evolution of Ethnic Markers." In *Ideas in Anthropology.* Santa Fe, NM: School of American Research.

Brandes, S. 1987. *Forty: The Age and the Symbol.* Knoxville, TN: University of Tennessee Press.

Brandt, G. S. 1985. "Running and Menstrual Dysfunction: Recent Medical Discoveries Provide New Insights Into the Human Division of Labor by Sex." *American Anthropologist.* 87(4):878–882.

Brisset, C. 1979. "Female Mutilation: Cautious Forum on Damaging Practices." *The Guardian.* (March 18):12–15.

Broderick, C. B. 1972. "Children's Romances." *Sexual Behavior* 2(5):16–21.

Brooke, J. 1993. "AIDS in Latin America." *New York Times* 1/25: 1, A6.

————. 1989. "Rapid Spread of AIDS Alarming the Ivory Coast." *New York Times.* (March 12):16.

Brooks, T. R. 1993. "Sexuality in the Aging Woman." *The Female Patient.* 18(11):27–28, 31–32, 34.

Broude, G. J. 1980. "Extramarital Sex Norms in Cross-Cultural Perspective." *Behavior Science Research.* 15:181–218.

Broude, G. J. and Greene, S.J. 1976. "Cross-Cultural Codes on Twenty Sexual Attitudes and Practices." *Ethnology.* 15:409–429.

Broverman, I. K., Broverman, D. M., Clarkson, F. E., Rosenkrantz, P., and Vogel, S. R. 1970. "Sex Role Stereotypes and Clinical Judgements of Mental Health." *Journal of Consulting and Clinical Psychology.* 34:(1)(Feb.):1–7.

Broverman, L., Vogel, S., Broverman, D., Clarkson, F., and Rosenkrantz, P. 1972. "Sex-role Stereotypes: A Current Appraisal." *Journal of Social Issues.* 28:59–78.

Brown, D. E. 1991. *Human Universals.* New York, NY: McGraw-Hill, Inc.

Brown, J. 1970. "A Note on the Division of Labor by Sex." *American Anthropologist.* 72:1073–78.

Brown, J. K. and Kerns, V. 1985. *In Her Prime, A New View of Middle-Aged Women.* So. Hadley, MA: Bergin and Garvey Publishers.

Browner, C. H. 1990. Review of *From Menarche to Menopause: Reproductive Lives of Peasant Women in Two Cultures* by Yewoubdar Beyene. American Ethnologist Book Review. 17(3):569–570.

Bryant, S. and Demian (eds.). 1990. "Partners' National Survey of Lesbian and Gay Couples." *Partners' Newsletter for Gay and Lesbian Couples.* (May/June):1–6.

Buchbinder, S. 1998. "Avoiding Infection after HIV Exposure." *Scientific American.* 279(1)(July):104–105.

Buckley, T. and Gottlieb, A. (eds.). 1988. *Blood Magic. The Anthropology of Menstruation.* Berkeley: University of California Press.

Buffum, J. 1982. "Pharmacosexology: The Effects of Drugs on Sexual Function." *Journal of Psychoactive Drugs.* 14:5–44.

Bullough, V. L. 1979. *Homosexuality: A History.* New York, NY: New American Library.

Bullough, V. 1976. *Sexual Variance in Society and History.* New York, NY: S. Wiley and Sons.

Burnham, D. 1978. "Biology and Gender." In *Genes and Gender:* I. Ethel Tobach and Betty Rosoff (eds.). New York, NY: Gordian Press: 51–59.

Burton, R., and Arbuthnot, F. 1984. *The Kaama Sutra of Vatsayana.* New York, NY: Berkeley Publishers.

Burton, R. V. and Whiting, J. W. M. 1961. "The Absent Father and Cross-Sex Identity." Reprinted from *Merrill-Palmer Quarterly of Behavior Development.* 7(2):86–95.

Burton, S. 1990. "Condolences, It's a Girl." *Time.* 136(19):36.

Butler, R. W. and Lewis, M. I.1988. *Love and Sex after Sixty.* New York, NY: Harper & Row.

Byer, C. O., and Shainberg, L. W. 1991. *Dimensions of Human Sexuality.* Dubuque, IA: Wm. C. Brown Pub.

Caffrey, M. M. 1989. *Ruth Benedict: Stranger in this Land.* Austin: University of Texas Press.

Calderone, M. 1985. "Adolescent Sexuality: Elements and Genesis." *Pediatrics Supplement*: 699–703.

Callender, C. and Kochems, L. 1983. "The North American Berdache." *Current Anthropology.* 24(4):443–456. See also comments and reply, pp. 456–70.

Callero, P. and Howard, J. 1989. "Biases of the Scientific Discourse on Human Sexuality: Toward a Sociology of Sexuality." In *Human Sexuality. The Societal and Interpersonal Context.* K. McKinney and S. Sprecher (eds.). Norwood, NJ: Ablex Publishing Corp.:425–38.

Campbell, B. 1988. *Humankind Emerging.* (5th Ed.). Scott, Foresman and Co.: Glenview, IL.

Canadian News Report on 6th International AIDS Conference. 1989. Montreal, Canada, 6/4/89.

Carlson, M. 1990. "It's Our Turn." *Time.* 136(19):16–18.

Carrier, J. 1989. "Sexual Behavior and the Spread of AIDS in Mexico." *Medical Anthropology,* 10:129–142.

Cassell, C. 1984. *Swept Away: Why Women Fear Their Own Sexuality.* New York, NY: Simon and Schuster.

Catania, J., Coates, T., Greenblatt, R., Dolcini, P., Kegeles, S., Puckett, S., Corman, M., and Miller, J. 1989. "Predictors of Condom Use and Multiple Partnered Sex Among Sexually-Active Adolescent Women: Implications for AIDS-Related Health Interventions." *Journal of Sex Research.* 26(4) (November):514–525.

Caulfield, M. D. 1985. "Sexuality in Human Evolution: What is 'Natural' in Sex?" *Feminist Studies.* 11(2):343–363.

CBS News Broadcast, 6/4/89. 6 pm.

CBS 1994. 60 Minutes. June.

CDC 1998(a). HIV/AIDS Statistical Surveillance Reports through December 1997.

CDC. 1998(b). "Testing the Blood Supply." http://www.cdcnac.org/ ques07.html.

CDC. 1998(c). "HIV Tests: Types, Accuracy, HIV Oral Fluid Tests, Home Collection Systems." http://www.cdcnac.org/ques01–04.html.

CDC 1994. AIDS Statistical Surveillance Reports, May.

CDC Health Resources and Services Administration. 1993. "Surgeon General's Report to the American Public on HIV Infection and AIDS." Atlanta, GA: National Institutes of Health.

CDC, 1990. AIDS Monthly Surveillance Report, January.

CDC, 1989. MMWR AIDS Update. 12/89. Atlanta, GA: CDC.

Center for Disease Control and Prevention AIDS Hotline. 1994. April.

Center for Disease Control and Prevention AIDS Hotline. 1993. July 28.

Chagnon, N. 1983. *Yanomamo: The Fierce People.* New York, NY: Holt, Rinehart and Winston.

Chapple, E. and Coon, C.S. 1942. *Principles of Anthropology.* New York, NY: Holt.

Chodorow, N. 1974. "Family Structure and Feminine Personality." In *Woman, Culture, and Society.* M. Z. Rosaldo and L. Lamphere (eds.). Stanford, CA: Stanford University Press.

Clanton, G. 1987. "A Historical Sociology of Sex and Love." *Teaching Sociology.* 15(3) (July):307–11.

Clark, M. and Anderson, B. G. 1975. "An Anthropological Approach to Aging." In *Introducing Anthropology.* James R. Hayes and James M. Henslin (eds.). Boston, MA: Holbrook Press, Inc.: 331–344.

Clement, U. 1990. "Surveys of Heterosexual Behavior." In the *Annual Review of Sex Research.* Vol. 1. John Bancroft (ed.). Lake Mills, IA: The Society for the Scientific Study of Sex:45–74.

Cobb, A. K. 1981. "Incorporation and Change: The Case of the Midwife in the United States." *Medical Anthropology.* 5(1)(winter):73–88.

Cohen, C. B. 1986. "Lasers in the Jungle. Human Sexuality: Biological and Social Reproduction?" Paper presented in the invited session, "Human Sexuality in Biocultural Perspective." 85th Annual Meeting, The American Anthropological Association. Philadelphia, PA. Dec. 2–7.

Cohen, C. B., and Mascia-Lees, F. E. 1989. "Lasers in the Jungle: Reconfiguring Questions of Human and Non-Human Primate Sexuality." *Medical Anthropology*. 11(4):351–366.

Cohen, E. N., and Eames, E. 1982. *Cultural Anthropology*. Boston, MA: Little, Brown and Company.

Cohen, Y. 1978. "The Disappearance of the Incest Taboo." *Human Nature*. 1(7):72–78.

Coleman, E., Gooren, L. and Ross, M. 1989. "Theories of Gender Transpositions: A Critique and Suggestions for Further Research." *Journal of Sex Research*. 26(4) (November):525–539.

Coleman, V. and Harris, G. 1989. "A Support Group for Individuals Recently Testing HIV Positive: A Psycho-Educational Group Model." *Journal of Sex Research*. 26(4) (November):539–549.

Comfort, A. 1972. *The Joy of Sex*. New York, NY: Crown Publishers.

Connell, R., and Kippax, S. 1990. "Sexuality in the AIDS Crisis: Patterns of Pleasure and Practice in an Australian Sample of Gay and Bisexual Men." *Journal of Sex Research*. 27(2) (May):167–198.

Conover, T. 1994. "The Hand-off." *New York Times Magazine*. Sunday, May 8: 28–36, 58, 61, 62, 71.

Consumer Reports. 1989. "Can You Rely on Condoms?" *Consumer Reports*. (March):135–141.

Cotes, J. E. 1994. *Adolescent Storm and Stress: An Evaluation of the Mead-Freeman Controversy*. Hillsdale, NJ: Lawrence Erlbaum Assoc., Inc.

Crandon, L. 1986. "Review of Reproductive Rituals: The Perception of Fertility in England from the Sixteenth Century to the Nineteenth Century. *American Anthropologist*. 88(2):470–471.

Crapo, R. H. 1987. *Cultural Anthropology*. Sluice Dock, Guilford, CT: Dushkin Publishers.

Crawford, A. et al. (ed.). 1989. *Our Lives in the Balance: U.S. Women of Color and the AIDS Epidemic*. Freedom Organizing Series #6. Latham, NY: Kitchen Table: Women of Color Press.

Critchlow-Leigh, B. 1990. "The Relationship of Substance Use During Sex to High-Risk Sexual Behavior." *Journal of Sex Research*. 27(2) (May):199–213.

———. 1989. "Reasons for Having and Avoiding Sex: Gender, Sexual Orientation, and Relationship to Sexual Behavior." *Journal of Sex Research*. 26(2) (May):199–210.

Cromwell, J. 1994. "Making the Visible Invisible: Female Gender Variance Cross-culturally." In *Gender Reversals and Gender Cultures: Anthropological and Historical Perspectives*. Sabrina Petra Ramet, ed. London and New York: Routledge.

———. 1993. "Not Female Berdache, Not Amazons, Not Cross-Gender Females, Not Manlike Women: Locating Female-to-Male Transgendered Persons Within Discourses on the Berdache Tradition." Paper presented at the American Anthropological Association Annual Meeting, Washington, D.C.

———. 1991. "Talking About Without Talking About: The Use of Protective Language Among Transvestites and Transsexuals." Paper presented at the American Anthropological Annual Meeting, Chicago, Illinois.

Crooks, R. and Baur, K. 1990. *Our Sexuality*. 4th edition. Redwood City, CA: Benjamin/Cummings.

———. 1987. *Our Sexuality*. 3rd edition. Menlo Park, CA: Benjamin/Cummings.

Cross, R. J. 1993. "What Doctors and Others Need to Know." *Siecus Report*. 21(5):7–9.

Crossette, B. 1989. "India Studying 'Accidental' Deaths of Hindu Wives." *The New York Times*. January 15:10.

Cutler, W. B., et al. 1985. "Sexual Behavior Frequency and Biphasic Ovulatory Type Menstrual Cycles." *Physiological Review*. 34:805–810.

Dahlberg, F. (ed.). 1981. *Woman the Gatherer*. New Haven, CT: Yale University Press.

Dale, E., Gerlach, D. H., and Wilhite, A. L. 1979. "Menstrual Dysfunction in Distance Runners." *Obstetrics and Gynecology*. 54(1):47–53.

Daly, M. and Wilson, M. 1978. *Sex, Evolution, and Behavior*. North Scituate, MA: Duxbury Press.

Danziger, S. and Wertz, D. 1989. "Sociological and Social Psychological Aspects of Reproduction." In *Human Sexuality. The Societal and Interpersonal Context*. K. McKinney and S. Sprecher (eds.). Norwood, NJ: Ablex Publishing Corp.:265–286.

Darling, C. J. and Hicks, M. 1982. "Parental Influence on Adolescent Sexuality: Implications for Parents as Educations." *Journal of Youth and Adolescence*. 11:231–245.

Darwin, C. 1874. (Original 1871). *The Descent of Man and Selection in Relation to Sex.* New York, NY. Thomas Y. Crowell.

Davenport, W. H. 1977. "Sex in Cross-Cultural Perspective." In *Human Sexuality in Four Perspectives.* Frank A. Beach (ed.). Baltimore, MD: The Johns Hopkins University Press:115–163.

———. 1976. "Sex in Cross-Cultural Perspective." *Human Sexuality in Four Perspectives.* Baltimore, MD: The Johns Hopkins University Press:115–163.

Davis, D. L. and Whitten, R. G. 1987. "The Cross-Cultural Study of Human Sexuality." *Annual Review in Anthropology.* 16:69–98.

Davis, D. 1994. "Cultural Sensitivity and the Sexual Disorders of DSM-IV: Review and Assessment." In J. E. Mezzich, A. Kleinman, H. Fabrega and D. Parron (eds.). *DSM-IV Sourcebook Culture and Psychiatric Diagnosis.* Washington, D.C.: American Psychiatric Press, Inc.

Davis, D. n.d. Draft of Article on Adolescence: 1–10.

Davis-Floyd, R. 1992. *Births as an American Rite of Passage.* Berkeley: University of California Press.

Davis-Floyd, R. E. 1988. "Birth As an American Rite of Passage." In *Childbirth in America.* K. Michaelson (ed.). Granby, MA: Bergin and Garvey Publishers: 153–172.

Davis-Floyd, R. 1989/90. "The Technological Model of Birth." In *Annual Editions Anthropology.* E. Angeloni (ed.). Sluice Dock, Guilford, CT: Dushkin Publishers: 163–171.

Dela Cancela, V. 1989. "Minority AIDS Prevention: Moving Beyond Cultural Perspectives Towards Sociopolitical Empowerment." *AIDS Education and Prevention.* 1(2):141–153.

Delameter, J. 1989. "The Social Control of Human Sexuality." In *Human Sexuality: The Societal and Interpersonal Context.* K. McKinney and S. Sprecher (eds.). Norwood, NJ: Ablex Publishing Co.:30–62.

Demac, D. 1988. *Liberty Denied.* New York, NY: PEN American Center.

Denny, D. 1991. "Dealing with Your Feelings." AEGIS Transition Series. Decatur, GA: American Educational Gender Information Service, Inc. p. 6.

Devereux, G. 1955. *A Study of Abortion in Primitive Societies: A Typological, Distributional, and Dynamic Analysis of the Prevention of Birth in 400 Pre-Industrial Societies.* New York, NY: The Julian Press.

———. 1967. "A Typological Study of Abortion in 350 Primitive, Ancient and Pre-Industrial Societies." In *Abortion in America.* H. Rosen (ed.). Boston, MA: Beacon Press.

———. 1937. "Institutionalized Homosexuality of the Mohave Indians." *Human Biology.* 9(4):498–527.

Devi, K. 1977. *The Eastern Way of Love: Tantric Sex and Erotic Mysticism.* New York, NY: Simon and Schuster.

Devor, H. 1989. *Gender Blending: Confronting the Limits of Duality.* Bloomington: Indiana University Press.

DHEW. 1983.

Diamond, J. 1989/90. "Everything Else You Always Wanted to Know About Sex . . . but That We Were Afraid You'd Never Ask." In *Annual Editions Anthropology.* E. Angeloni (ed.). Sluice Dock, Guilford, CT: Dushkin Publishers: 66–72.

Dickermann, M. 1990. "A Sister Reclaimed. A Review of Ruth Benedict: *Stranger in This Land.*" In *Solga Newsletter.* 12(2):5–9.

DiClemente, R. 1990. "Predictors of Risky Adolescent Sexual Behavior." Medical Aspects of Human Sexuality. April: 39–40. *In Annual Editions of Human Sexuality.* 1991–2. O. Pocs (ed.). Sluice Dock, Guilford, CT: Dushkin Publishers:186–7.

Dirks, N. B. 1989. Review of *The Construction of Homosexuality* by D.F. Greenberg. *New York Times* Book Review. 1/15:9–16.

Divale, W. and Harris, M. 1976. "Population, Warfare and the Male Supremacist Complex." *American Anthropologist.* 78:521–538.

Donnerstein, E. I. and Linz, D. G. 1986. "The Question of Pornography." *Psychology Today.* 20(12):56–59.

Douglas, C. 1990. "Lesbian Child-Custody Cases Redefine Family Law." *New York Times.* 7/8 Opinion, 1.

Dowdle, W. R. 1987. Quoted in the *New York Times.* 16 October: 11.

Dowie, M. 1991. "Reluctant Crusader." *In Annual Editions of Human Sexuality 91/92.* O. Pocs (ed.). Sluice Dock, Guilford, CT: Dushkin Publishers: 147–140.

Dowling, C. 1981. *The Cinderella Complex: Women's Hidden Fear of Independence.* New York, NY: Summit Books.

Downie, D. C. and Hally, D. J. 1961. "A Cross-Cultural Study of Male Transvestism and Sex-Role Differentiation." Unpublished manuscript Dartmouth College.

Doyle, J. A. and Paludi, M. A. 1991. *Sex and Gender: The Human Experience.* Dubuque, IA: Wm. C. Brown Pub.

Draper, P. 1975. "!Kung Women: Contrasts in Sexual Egalitarianism in Foraging and Sedentary Contexts." In *Toward the Anthropology of Women.* R.R. Reiter (ed.). New York, NY: Monthly Review Press: 77–109.

Dugan, A. 1988. "Compadorazgo As a Protective Mechanism in Depression." In *Women and Health, Cross-Cultural Perspective.* P. Whelehan (ed.). Granby, MA: Bergin and Garvey Publishers:143–153.

Duggan, L. 1990. "From Instincts to Politics: Writing the History of Sexuality in the U.S." *The Journal of Sex Research.* 27(1):95–109.

Dyk, P. H. 1994. "Review of *Adolescence: An Anthropological Inquiry.* Alice Schlegel and Herbert Barry, III." *Sociological Inquiry* 64(2):251–254.

Eckholm, E. 1990. "Confronting the Cruel Reality of Africa's AIDS Epidemic." *New York Times.* 9/19:1, 14, 15.

Eckholm, E. and Tierney, J. 1990. "AIDS in Africa: A Killer Rages On." *New York Times.* 9/16:1, 14, 15.

Edelman, L. 1992. "Tearooms and Sympathy, or, the Epistemology of the Water Closet." In *Nationalisms and Sexualities.* A. Parker, M. Russo, D. Sommer, and P. Yaeger (eds.). New York, NY: Routledge:263–285.

Edgar, T. et al. 1988. "Communicating the AIDS Risk to College Students: The Problem of Motivating Change." *Health Education Research: Theory and Practice.* 3(1) (Aug.–Sept.):59–65.

Ehrenreich, B. 1973. *Complaints and Disorders: The Sexual Politics of Sickness.* Old Westbury, NY: The Feminist Press.

Ehrenreich, B. and English D. 1978. *For Her Own Good: 150 Years of the Experts Advice to Women.* Garden City, NY: Anchor Press.

Ehrenreich, B., Hess, E., and Jacobs, G. 1986. *Re-making Love: The Feminization of Sex.* New York, NY: Anchor Press, Doubleday.

Eicher, W. et al. 1981. "Transsexuality and H-Y Antigen." Paper presented at the 7th International Gender Dysphoria Symposium. The Harry Benjamin International Gender Dysphoria Association. Lake Tahoe, Nevada, 4–8 March.

Eidson, T. (ed.). 1988. *The AIDS Caregivers Handbook.* New York, NY: St. Martin's Press.

Elias, J. and Gebhard, P. 1969. "Sexuality and Sexual Learning in Childhood." Phi Delta Kappan. 50:401–405.

Ellis, A. 1976. *Sex and the Liberated Man.* Secaucus, NJ: Lyle Stuart.

Elshtain, J. B. 1989. "Why We Need Limits." In *Annual Editions Human Sexuality,* 1989–90. O. Pocs (ed.). Sluice Dock, Guilford, CT: Dushkin Publishers: 6–8.

Ember, C. and Ember, M. 1996. *Anthropology.* 8th edition. Englewood Cliffs, NJ: Prentice Hall.

———. 1990. *Anthropology.* 6th edition. Englewood Cliffs, NJ: Prentice Hall.

———. 1988. *Anthropology.* 5th edition. Englewood Cliffs, NJ: Prentice Hall. 31: 63–65.

Engs, R. C. and Hanson, D. J. 1993. "Drinking Games and Problems Related to Drinking Among Moderate and Heavy Drinkers." *Psychological Reports.* 73:115–120.

———. 1985. "The Drinking Patterns and Problems of College Students: 1983." *Journal of Alcohol and Drug Education*. 31(1)(fall):65–83.

Erhardt, A. A. 1979. "Interactional Model of Hormones and Behavior." In *Human Sexuality: A Comparative and Developmental Perspective.* H.A. Katchadourian (ed.). Berkeley: University of California Press: 150–160.

Erick, M. 1994. "Morning, Noon, and Night Sickness." *Parenting.* February: 72–78.

Erikson, E. 1963. *Childhood and Society.* 2nd edition. New York, NY: W.W. Norton.

Erikson, E. H. 1950. *Childhood and Society.* New York, NY: W.W. Norton.

Eshleman, Cashion & Basirico. 1988. *Sociology: An Introduction.* Boston, MA: Scott, Foresmans and Company.

Estioko-Griffin, A. 1993. "Daughters of the Forest." In *The Other Fifty Percent.* M. Womack and J. Marti. Prospect Heights, IL: Waveland Press, Inc. Pp. 225–232.

Estioko-Griffin, A. and Griffin, P. G. 1981. "Woman the Hunter: The Agta." In *Woman the Gatherer.* F. Dahlberg, ed. New Haven, CT: Yale University Press. Pp. 121–151.

Evans, H. 1993. "The Morning After Pill is Already Available." *Self:* 52, February.

Evans-Pritchard, E. E. 1970. "Sexual Inversion Among the Azande." *American Anthropologist.* 72(6):1428–34.

Fabrega, H., Jr. 1974. *Disease and Social Behavior: An Interdisciplinary Perspective.* Cambridge, MA: MIT Press.

Faludi, S. 1991. *Backlash: The Undeclared War Against American Women.* Crown Publishers: New York.

Fan, D. and McAvoy, G. 1989. "Predictions of Public Opinion on the Spread of AIDS: Introduction of New Computer Methodologies." *Journal of Sex Research.* 26(2) (May):159–187.

Farrell, W. 1986. *Why Men Are the Way They Are.* New York, NY: McGraw-Hill.

———. 1974. *The Liberated Male.* New York, NY: Bantam Books.

Faust, B. 1988. "When Is a Midwife a Witch? A Case Study from a Modernizing Maya Village." *Women and Health.* P. Whelehan (ed.). Granby, MA: Bergin and Garvey Publishers.

Fausto-Sterling, A. 1985. *Myths of Gender: Biological Theories About Women and Men.* New York, NY: Basic Books.

Feinbloom, D. H. 1976. *Transvestites and Transsexuals: Mixed Views*. New York, NY: Delacorte Press/Seymour Lawrence Books.

Fernea, E. W. 1965. *Guests of the Sheik*. New York, NY: Doubleday.

Ficher, I.V. and Eisenstein, T. 1984. "Current Approaches in Sex Therapy." In *Sexual Arousal*. M. Ficher, R. E. Fishkin, and J. A. Jacobs (eds.). Springfield, IL: Charles C. Thomas:142–156.

Ficher, M., Fishkin, R. E., and Jacobs, J. A. (eds.). 1984. *Sexual Arousal*. Springfield, IL: Charles C. Thomas.

Findlay, S. "Birth Control." In *Annual Editions of Human Sexuality 91/92*. O. Pocs (ed.). Sluice Dock, Guilford, CT: Dushkin Publishers Group:126–129.

Finkelhor, D. 1984. *Child Sexual Abuse: New Theory and Research*. New York, NY: Free Press.

Fisher, H. 1992. *Anatomy of Love*. New York, NY: W.W. Norton and Company.

Fisher, H. 1989. "Evolution of Human Serial Pair Bonding." *American Journal of Physical Anthropology*. 78: 331–354.

Fisher, S. 1973. *The Female Orgasm: Psychology, Physiology, Fantasy*. Second Title *Understanding the Female Orgasm*. New York, NY: Basic.

Fisher, T. D. 1987. "Family Communication and the Sexual Behavior and Attitudes of College Students." *Journal of Youth and Adolescence*, 16(5):481–495.

Fiske, A. P. and Mason, K. F. 1990. "Introduction." (Issue devoted to Moral Relativism). *Ethos*. 18(2):131–139.

FOCUS. "A Guide to AIDS Research and Counseling." 1993. 8(6) (May).

Ford, C., and Beach, F. 1951. *Patterns of Sexual Behavior*. New York, NY: Harper Torchbooks.

Fox, J. R. 1962. "Sibling Incest." *British Journal of Sociology*, 13:128–150.

Fox, R. 1987. "In the Beginning: Aspects of Hominid Behavioral Evolution." *Man*, 2:415–33.

———. 1980. *The Red Lamp of Incest*. New York, NY: E. P. Dutton.

Francoeur, R. T. 1993. *The Gender Rainbow: An Ancient Myth of Sexuality Reinterpreted for a New Millenium*. Rockaway, NJ: Robert T. Francoeur.

———. 1992. "Sexuality and Spirituality: The Relevance of Eastern Traditions." For *The Siecus Report*. 20(4):1–8.

———. 1991a. *Becoming a Sexual Person*. New York, NY: Macmillan Publishing Co.

——. 1991b. *Instructors Manual for Becoming a Sexual Person.* New York, NY: Macmillan Publishing Co.

——. 1991c. "Sex and Spirituality: The Relevance of Tantric and Taoist Experiences." For *The Siecus Report.* No Publisher: 1–12

——. 1991d. "Should Prenatal Testing for Sex Selection Be Permitted?" Issue Summary. In *Taking Sides, Clashing Views on Controversial Issues in Human Sexuality.* R. Francoeur (ed). Guilford, CT: Dushkin Publishers:159.

——. (ed.). 1989. *Taking Sides: Clashing Views on Controversial Issues in Human Sexuality.* Guilford, CT: Dushkin Publishers.

Frayser, S. 1989. "Sexual and Reproductive Relations: Cross-Cultural Evidence and Biosocial Implications." *Medical Anthropology.* 11(4):385–407.

——. 1985. *Varieties of Sexual Experience: An Anthropological Perspective on Human Sexuality.* New Haven, CT: HRAF.

Frayser, S. G. and Whitby, T.J. 1987. *Studies in Human Sexuality: A Selected Guide.* Littleton, CO: Libraries Unlimited, Inc.

Freeman, D. 1983. "Margaret Mead and Samoa." *The Making and Unmaking of an Anthropological Myth.* Cambridge, MA: Harvard University Press.

Freud, S. 1950. [1913]. *Totem and Taboo.* New York, NY: W. W. Norton & Company.

——. 1929 [1959]. *Group Psychology and the Analysis of the Ego.* J. Strachey (ed.), and trans. London: Hogarth Press.

——. 1920a. "Three Essays on the Theory of Sexuality." Standard original edition of *The Complete Psychological Works of Sigmund Freud,* V VII. London: Hogarth Press.

——. 1920b. *Selected Papers on Hysteria and Other Psychoneurosis.* 3rd edition. New York, NY: Nervous and Mental Disease Publishing Company.

Freund, K., Steiner, B.W., and Chang, S. 1982. "Two Types of Cross-gender Identity." *Archives of Sexual Behavior.* 11(1):49–63.

Friday, N. 1974. *My Secret Garden: Women's Sexual Fantasies.* New York, NY: Pocket Books.

Friedan, B. 1963. *The Feminine Mystique.* New York, NY: W. W. Norton.

Frisch, R. 1980. "Fatness, Puberty, and Fertility." *Natural History.* 89:16–27.

——. 1978. "Population, Food Intake, and Fertility." *Science.* 199:22–30.

Frisch, R. E. and McArthur, J. W. 1974. "Menstrual Cycles: Fatness as a Determinant of Minimum Weight for Height Necessary for Their Maintenance or Onset." *Science*. 185:949–51.

Fuller, N. and Jordan, B. 1981. "Maya Women and the End of the Birthing Period: Postpartum Massage-and-Binding in Yucatan, Mexico." *Medical Anthropology*. 5(1)(winter):35–51.

Fullilove, M. T., Fullilove, R. E., Haynes, K., and Gross, S. 1990. "Black Women and AIDS Prevention: Understanding the Gender Rules." *Journal of Sex Research*. 27(1) (February):47–65.

Gagnon, J. H. 1979. "The Interaction of Gender Roles and Sexual Conduct." In *Human Sexuality: A Comparative and Developmental Perspective*: 225–245. H.A. Katchadourian (ed.). Berkeley: University of California Press.

———. 1978. "Reconsiderations." *Human Nature*. 1(10):92–96.

Gagnon, J. and Simon, W. 1973. *Sexual Conduct, The Social Sources of Human Sexuality*. Chicago, IL: Aldine.

Gagnon, J. H., Simon, W., and Berger, A. J. 1970. "Some Aspects of Sexual Adjustment in Early and Late Adolescence." In *Psychopathology of Adolescence*. J. Zubin, A. N. Freedman (eds.). New York, NY: Greene and Stratton, 278.

Gallup, G. 1989. *Gallup Opinion Poll 1989*. Wilmington, DE: Scholarly Resources Inc.

Gallup, G. 1992. *Gallup Opinion Poll 1992*. Wilmington, DE: Scholarly Resources Inc.

Garfinkel, H. and Stoller, R. J. 1967. "Passing and the Managed Achievement of Sex Status in an 'Intersexed' Person." In H. Garfinkel, *Studies in Ethnomethodology*. Englewood Cliffs, NJ: Prentice-Hall. Pp. 116–85.

Garrison, O. 1983. *Tantra: The Yoga of Sex*. New York, NY: Julian Press.

Geertz, C. 1984, 1973. *The Interpretation of Cultures*. New York, NY: Basic Books.

Gelber, M. G. 1986. *Gender and Society in the New Guinea Highlands*. Boulder, CO: Westview Press.

Gerlman, D. and Hager, M. 1988. "Body and Soul." *Newsweek*, November: 7.

Gilligan, C. 1982. *In a Different Voice*. Cambridge, MA: Harvard University Press.

Gladue, B., Green, R., and Hellman, R. 1984. "Neuroendocrine Response to Estrogen and Sexual Orientation." *Science*. 225:1496–1498.

Gleitman, H. 1987. *Basic Psychology*. 2nd ed. New York, NY: W. W. Norton.

Glenmullen, J. 1993. *The Pornographer's Grief and Other Tales of Human Sexuality*. New York, NY: Harper Perennial.

Godziehen-Shedlin, M. 1981. "Notes From a Field Log: Doña Bernarda at Work." *Medical Anthropology.* 5(1):(winter):13–15.

Goettsch, S. L. 1987. "Textbook Sexual Inadequacy? A Review of Sexuality Texts." *Teaching Sociology.* 15(3)(July):324–38.

Goldberg, D. et al. 1983. "The Grafenberg Spot and Female Ejaculation: A Review of Initial Hypotheses." *Journal of Sex and Marital Therapy.* 9(1):19–22.

Goldberg, D. 1984. *The New Male-Female Relationship.* New York, NY: New American Library.

Goldberg, H. 1985. "Sex at Cross-Purposes." In *Human Sexuality 85/86.* O. Pocs (ed.). Guilford, CT: Dushkin Publishers: 147–149.

———. 1980. *The New Male: From Self-destruction to Self-care.* New York, NY: New American Library.

———. 1979. *The New Male: From Macho to Sensitive but Still All Male.* New York, NY: New American Library.

———. 1976. *The Hazards of Being Male: Surviving the Myth of Masculinity.* New York, NY: Nash Publishers.

Goldberg, J. 1989. "'60s Sexual Revolution Didn't Occur, Kinsey Study Says." *Los Angeles Times.* (June 30):18.

Goldman, R. and Goldman, J. 1982. *Children's Sexual Thinking.* Boston, MA: Routledge and Kegan Paul.

Goldstein, B. 1990. "Genetic and Biochemical Functioning in Human Sexual Behavior." Paper presented at AASECT XXII Annual Conference 16, Feburary, Arlington, VA.

Goldstein, M. 1987. "When Brothers Share a Wife." *Natural History.* 96(3):39–48.

Goleman, D. 1990. "Study Defines Major Sources of Conflict Between Sexes." In *Annual Editions of Human Sexuality, 1990–91.* O. Pocs (ed.). Sluice Dock, Guilford, CT: Dushkin Publishers: 129–130.

Goodale, J. C. 1971. *Tiwi Wives; A Study of the Women of Melville Island, North Australia.* Seattle, WA: University of Washington Press.

Goodman, R. E. 1983. "Biology of Sexuality: Inborn Determinants of Human Sexual Response." *British Journal of Psychiatry.* 143(September):216–220.

Gordon, M. 1988. "College Coaches' Attitudes Toward Pregame Sex." *Journal of Sex Research:* 24:256–62.

———. 1978. "Demography: Births, Deaths and Migration." In *The American Family in Social-Historical Perspective.* M. Gordon (ed.). New York, NY: St. Martin's Press: 13–15.

Gordon, S. and Gilgun, J. F. 1987. "Adolescent Sexuality." In *Handbook of Adolescent Psychology.* V. B. Van Hasselt and M. Hersen (eds.). Elmsford, NY: Pergamon Press.

Gordon, S. and Snyder, C. W. 1986. *Personal Issues in Human Sexuality.* Boston, MA: Allyn and Bacon, Inc.

Graham, C. A. and McGrew, W. C. 1980. "Menstrual Synchrony in Female Undergraduates Living on a Coeducational Campus." *Psychoneuroendocrinology.* 5:245–52.

Gray, L. A. and Saracino, M. 1989. "AIDS on Campus: A Preliminary Study of College Students' Knowledge and Behaviors." *Journal of Counseling and Development.* 68(Nov./Dec.):199–202.

Green, F. 1987. *The "Sissy Boy" Syndrome and the Development of Homosexuality.* New Haven, CT: Yale University Press.

Green, R. 1979a. "Biological Influences on Sexual Identity." In H.A. Katchadourian (ed.). *Human Sexuality: A Comparative and Developmental Perspective.* Berkeley: University of California Press.

———. 1979b. "Childhood Cross-gender Behavior and Subsequent Sexual Preference." *American Journal of Psychiatry,* 135:692–97.

———. 1978. "Sexual Identity of 37 Children Raised by Homosexual or Transsexual Parents." *American Journal of Psychiatry,* 135(6):692–697.

———. 1974a. *Sexual Identity Conflict in Children and Adults.* New York, NY: Basic Books.

———. 1974b. "Children's Quest for Sexual Identity." *Psychology Today.* 7(9):45–51.

Greenberg, D. F. 1988. *The Construction of Homosexuality.* Chicago: University of Chicago Press.

Greenburg, J., Bruess, C., and Mullen, K. 1993. *Sexuality Insights and Issues.* Madison, WI: W. C. Brown.

Greer, G. 1992. *The Change: Women, Aging and the Menopause.* New York, New York: Alfred Knopf.

———. 1971. *The Female Eunuch.* New York, NY: McGraw-Hill.

Gregersen, E. 1994. *The World of Human Sexuality: Behaviors, Customs and Beliefs.* New York, NY: Irvington Publishers, Inc.

Gregersen, E. 1992. (Second Ed.). *Sexual Practices: The Story of Human Sexuality.* New York: Franklin Watts.

Gregersen, E. 1983. (First Ed.). *Sexual Practices: The Story of Human Sexuality.* New York, NY: Franklin Watts.

Grimes, D. A. and Cates W. 1980. "Abortions: Methods and Complications." In *Human Reproduction: Conception and Contraception*. E. S. Hofiz (ed.). Hagerstown, MD: Harper and Row.

Gutman, D., Seeley, D., Barry, D., and Chesanow, N. 1990. "What Do Men Want?" Special Section. *Woman*. 11(6) (June):55–60.

Haas, K. and Haas, A. 1987. *Understanding Sexuality*. St. Louis, MO: Times-Merron.

Haley, C. 1988. "AIDS 101." In *The AIDS Caregiver's Handbook*. T. Eidson (ed.). New York, NY: St. Martin's Press: 3–13.

Hall, E. T. 1966. *The Hidden Dimension*. Garden City, NY: Doubleday.

Hall, E. 1989. "A Conversation with Clifford Grolo Stein and When Does Life Begin?" *Psychology Today*. 1989:42–46.

Hamilton, J. A. and Gallent, S. J. 1990. Professional Psychology Research and Practice. 2(1): 60–68.

Hammer-Burns, L. 1987. "Infertility and the Sexual Health of the Family." *Journal of Sex Education and Therapy*. 13(2)(fall-winter):30–34.

Handy, B. 1998. "The Viagra Craze." *Time Magazine* 15(17):50–57.

Harden, B. 1985a. "Female Circumcision: Painful, Risky, and Little Girls Beg for It." *Washington Post National Weekly Edition*. July 29:15–16.

Harlow, H. F., Harlow, M. K. and Suomi, S.S. 1971. "From Thought to Therapy: Lessons from a Primate Laboratory." *American Scientist*. 59:538–49.

Harris, M. 1993. *Culture People and Nature*. New York, NY: Harper Collins.

———. 1989. *Our Kind*. New York, NY: Harper Perennial.

———. 1974. *Cows, Pigs, Wars, and Witches: The Riddles of Culture*. New York, NY: Random House.

Harrison, G. A., Tanner, J. M., Pilbeam, D. R., and Baker, P. T. 1988. *Human Biology: An Introduction to Human Evolution, Variation, Growth, and Adaptability*. 3rd edition. Oxford, England: Oxford Science Publications.

Hart, C. W. M., and Pilling, A. 1960. *The Tiwi of North Australia*. New York, NY: Holt, Rinehart, and Winston.

Hart, C. W. M., Pilling, A., and Goodale, J. C. 1988. *The Tiwi of North Australia*. Fort Worth, TX: Holt, Rinehart and Winston.

Hatcher, R. A. et al. 1986. *Contraceptive Technology 1986–1987*. 13th edition. New York, NY: Irvington.

Haviland, W. A. 1989. *Anthropology*. New York, NY: Holt, Rinehart, and Winston.

Haynes, J. D. 1994. "Pheremones." In *Human Sexuality: An Encyclopedia*. New York, NY: Garland Publishing, Inc.:441–442.

Helman, C. 1990. *Culture, Health and Illness*. 2nd edition. London: Wright.

Henderson, W. J. 1969. *Early History of Singing*. New York: AMS Press.

Hendrick, S. and Hendrick, C. 1987. "Multidimensionality of Sexual Attitudes." *Journal of Sex Research*. 23(4) (Nov.):502–527.

Henshaw, S. 1987. "Characteristics of U.S. Women Having Abortions, 1982–1983." *Family Planning Perspectives*. 19(1):5–9.

Henshaw, S. K., Kenny, A. M., Somberg, D., and Van Vort, J. 1989. *Teenage Pregnancy in the United States: The Scope of the Problem and State Responses*. New York: Center for Population Options, Alan Guttmather Institute.

Herdt, G. 1993. "Semen Transactions in Sambia Culture." In *Culture and Human Sexuality*. D. Suggs and A. Miracle (eds.). Belmont, CA: Brooks/Cole Publishing Company:298–327.

Herdt, G., and Stoller, R.J. 1989. "Commentary to the Socialization of Homosexuality and Heterosexuality in a Non-Western Society." *Archives of Sexual Behavior*. 18(1):31–34.

Herdt, G. 1988. "Cross-Cultural Forms of Homosexuality and the Concept 'Gay'." *Psychiatric Annal*. 18(1):37–39.

———. 1987. *The Sambia: Ritual and Gender in New Guinea*. New York, NY: Holt, Rinehart, and Winston.

———. (ed.). 1984a. *Ritualized Homosexuality in Melanesia*. Los Angeles and Berkeley: University of California Press.

———. 1984b. "Ritualized Homosexual Behavior in the Male Cults of Melanesia, 1862–1983: An Introduction in Ritualized Homosexuality in Melanesia." G. Herdt (ed.). Los Angeles and Berkeley: University of California Press: 1–82.

———. (ed.). 1982. *Rituals of Manhood: Male Initiation in Papua New Guinea*. Berkeley: University of California Press.

Herdt, G. 1981. *Guardians of the Flute: Idioms of Masculinity*. New York, NY: McGraw-Hill.

Herek, G. M. 1988. "Heterosexuals' Attitudes Toward Lesbians and Gay Men: Correlates and Gender Differences." *Journal of Sex Research*. 25(4) (Nov.):451–478.

Herman, J. and Hirschman, L. 1981. "Father-Daughter Incest." In *Female Psychology.* S. Cox (ed.) New York, NY: St. Martin's Press.

Herting, D. L., et al. 1988. "AIDS and Student Sexual Behavior: Who's Concerned, Who Isn't, and Why." Paper presented at the Annual Meeting of the Southeastern Psychological Assoc., 34th. New Orleans, LA, 3/31–4/2.

Hevesi, D. 1990. "Parents Vow to Protest Brooklyn AIDS Student." *New York Times.* Sunday 9/10:34.

Heyl, B. 1989. "Homosexuality: A Social Phenomenon." In *Human Sexuality. The Societal and Interpersonal Context.* K. McKinney, and S. Specher (eds.). Norwood, NJ: Ablex Publishing Corp.: 321–350.

Heyward, W. and Curran, J. 1988. "The Epidemiology of AIDS in the U.S." *Scientific American.* 259(Oct.)(4):72–82.

Hill, W. W. 1938. "Note on the Pima Berdache." *American Anthropologist.* 40:338–40.

———. 1935. "The Status of the Hermaphrodite and Transvestite in Navajo Culture." *American Anthropologist.* 37:273–9.

Hino, S. 1987. "Breaking the Cycle of HTLV-I Transmission via Carrier Mother's Milk [letter]." *Lancet,* July 18; 2(8551):158–9.

Hirschorn, M. 1987a. "Persuading Students to Use Safer Sex Practices Proves Difficult Even with the Danger of AIDS." *The Chronicle of Higher Education.* June, 10:30.

———. 1987b. "AIDS Is Not Seen as a Major Threat by Many Heterosexuals on Campuses." *The Chronicle of Higher Education.* April, 29:1, 32, 33.

Hite, S. 1987. *Women and Love.* New York, NY: Knopf

———. 1981. *The Hite Report on Male Sexuality.* New York, NY: Knopf.

———. 1976. *The Hite Report on Female Sexuality.* New York, NY: Knopf.

"HIV-Prevention Strategy Highlights Women." 1994. *The Female Patient.* 19(1):26.

Hochschild, A., with Machung, A. 1989. *Second Shift: Working Parents and the Revolution at Home.* New York, NY: Viking Press.

Hoebel, E. A. 1949. *Man in the Primitive World.* New York, NY: McGraw Hill.

Holmes, L. P. and Schneider, K. 1987. *Anthropology.* Prospect Heights, IL: Waveland Press.

Hostetler, J. and Huntington, G. 1980. *The Hutterites in North America.* New York, NY: Holt, Rinehart, and Winston.

Howard, J. 1993. "Angry Storm Over the South Seas of Margaret Mead." Smithsonian, 14(1):67–74.

Huffman, S. L. et al. 1978. "Postpartum Amenorrhea: How Is It Affected by Maternal Nutritional Status?" *Science.* 200:1555–56.

Hunt, M. 1975. *Sexual Behavior in the 1970's.* New York, NY: Dell.

Hutching, J. 1988. "Pediatric AIDS: An Overview." *Children Today.* (May–June):8–9.

Huxley, A. 1946. *Brave New World.* New York, NY: Harper and Bros.

Hyde, J. S. 1994. *Understanding Human Sexuality.* (5th edition). New York, NY: McGraw-Hill.

———. 1985. *Half The Human Experience.* Lexington, MA: D. C. Heath and Co.

———. 1982. *Understanding Human Sexuality.* New York, NY: McGraw-Hill.

IASHS. 1987. "Guide to Safer Sex." San Francisco, CA: Multifocus Publishers.

Ingstad, B. 1990. "The Cultural Construction of AIDS and Its Consequences for Prevention in Botswana." *Medical Anthropology Quarterly.* 4(1) (March):28–41.

"Injection Vasectomy." 1992. Hotline. *Muscle and Fitness.* 53(4):28.

Inman, J. 1989. "A Model for Effective AIDS Education on North Country College Campuses." Senior paper.

"In the Public Eye." 1998 *Urology Times* 26(6):35–36.

Irvine, J. 1990. *Disorders of Desire: Sex and Gender in Modern American Sexology.* Philadelphia, PA: Temple University Press.

Irvine, J. M. 1990. "From Difference To Sameness: Gender Ideology in Sexual Science." *The Journal of Sex Research.* 27(1):7–24.

IXth International AIDS Conference. June 7–11, 1993. Berlin, Germany.

J. 1969. *The Sensuous Woman.* New York, NY: Dell.

Jackson, D. 1990. "Our Sexual Past, Our Sexual Future." *New Woman.* New York Magazine, vol. 20, issue 10: 180–182, 184, 186.

Jacobs, S. E. 1994. "Native American Two Spirits." *Anthropology Newsletter* 35(8):7.

Jacobs, S. E. and Roberts, C. 1989. "Sex, Sexuality, Gender and Gender Variance." In *Gender and Anthropology.* Sandra Morgen, ed. Washington, DC, American Anthropological Association. Pp. 438–462.

JAMA. 1998. "Prevention: Do Condoms Work? Does HIV Prevention Work?" (JAMA. HIV/AIDS Information Center) http://www.ama_assn.org/special/hiv/prevent/htm.

Jarrett, L. 1984. "Psychosocial and Biological Influences on Menstruation: Synchrony, Cycle, Length, Regularity." *Psychoneuroendocrinology.* 9:21–28.

Johanson, D. C. and White, T.D. 1979. "A Systematic Assessment of Early African Hominids." *Science.* 203:321–30.

Johanson, D. C. and Edey, M. 1981. *Lucy: The Beginnings of Humankind.* New York, NY: Simon and Schuster.

Johnson, M. 1990. "Silent Debate Over Abortion." *Syracuse Herald American.* March 11, B9.

Jolly, A. 1985. *The Evolution of Primate Behavior.* 2nd edition. New York, NY: MacMillan Publishing Co.

Jones, H. W., and Jones, G. S. 1981. *Novak's Textbook of Gynecology.* Baltimore, MD: Williams and Wilkins.

Jordan, B. 1993. 4th edition. *Birth in Four Cultures.* Prospect Heights, IL: Waveland Press

———. 1983. *Birth in Four Cultures: A Cross-cultural Investigation of Childbirth in Yucatan, Holland, Sweden and the United States.* London: Eden Press.

Jordan, B. and Irwin, S. 1987. "Knowledge, Practice, and Power: Court-ordered Caesarean Section." *Medical Anthropology Quarterly.* 1(3) (Sept.):319–334.

Josephs, L. 1989. "Fighting AIDS All the Way." *New York Times Magazine.* Sunday, 10/9:42–46.

Jost, A. 1972. "A New Look at the Mechanisms Controlling Sex Differentiation in Mammals." *Johns Hopkins Medical Journal*: 130:38.

———. 1961. "Role of Hormones in Prenatal Development." *Harvey Lect.* 55:201.

Kaas, M. U. 1978. "Sexual Expression of the Elderly in Nursing Homes." *Gerontologist* 18(4):372–378.

Kaplan, H. S. and Owett, T. 1993. "The Female Androgen Deficiency Syndrome." *Journal of Sex and Marital Therapy.* 19(1):3–24.

Kaplan, H. S. 1989. PE: *How to Overcome Premature Ejaculation.* New York, NY: Brunner/Mazel.

Kaplan, H. 1983. *The Evaluation of Sexual Desires.* New York, NY: Brunner/Mazel.

———. 1979. *Disorders of Sexual Desire and Other New Concepts and Techniques in Sex Therapy.* New York, NY: Brunner/Mazel.

———. 1974. *The New Sex Therapy; Active Treatment of Sexual Dysfunctions.* New York, NY: Brunner/Mazel.

Katchadourian, H. A. 1990. *Biological Aspects of Human Sexuality.* 4th edition. Fort Worth, TX: Holt, Rinehart, and Winston, Inc.

———. 1985. *Fundamentals of Human Sexuality.* New York, NY: Holt, Rinehart, and Winston.

———. (ed.). 1979. *Human Sexuality: A Comparative and Developmental Perspective.* Berkeley: University of California Press.

Katchadourian, H. A. and Lunde, D. T. 1975. *Fundamentals of Human Sexuality.* New York, NY: Holt, Rinehart, and Winston.

Keeling, R. 1990. "AIDS Prevention Challenges for Colleges and Universities." Focus: *A Guide to AIDS Research and Counseling.* 5(4) (March):1–2.

———. 1989. "We Are Not Socially, Politically, or Economically Prepared to Cope with the Spread of AIDS Among Young Americans." *The Chronicle of Higher Education.* September 20:B1–B3.

Kelly, G. 1994. *Sexuality Today: The Human Perspective.* 4th edition. Sluice Dock, Guilford, CT: Dushkin Publishers.

———. 1990. *Sexuality Today: The Human Perspective.* 2nd edition. Sluice Dock, Guilford, CT: Dushkin Publishers.

———. 1988. *Sexuality Today: The Human Perspective.* Sluice Dock, Guilford, CT: Dushkin Publishers.

Kelso, A. J. 1980. "The Evolution of Human Genders." In *The Turbulent Years.* John Scanlon (ed.). New York, NY: Academy for Educational Development: 81–89.

———. ND. "Female Sexuality: Basic Changes in Human Evolution." Unpublished manuscript. Pg. 1–12.

Kennedy, K., Fortney, J., and Sokal, D. 1989. "Breastfeeding and HIV." *Lancet.* 1:333.

Kessler, S. J. and McKenna, W. 1978. *Gender: An Ethnomethodological Approach.* New York, NY: John Wiley and Sons.

Kimball, L. and Craig, S. 1988. "Women and Stress in Brunei." In *Women and Health.* P. Whelehan, (ed.). Granby, MA: Bergin and Garvey Publishers: 170–183.

King, B. M., Camp, C. J., and Downey, A. M. 1991. *Human Sexuality Today.* Englewood Cliffs, NJ: Prentice Hall.

Kinsey, A. C., Pomeroy, W. B., and Martin, C. E. 1948. *Sexual Behavior in the Human Male*. Philadelphia, PA: Saunders.

Kinsey, A. C. et al. 1953. *Sexual Behavior in the Human Female*. Philadelphia, PA: Saunders.

Kirby, D. 1984. *Sexuality Education: An Evaluation of Programs and Their Effects*. Santa Cruz, CA: Network Publications.

Kirk, M. and Madsen, H. 1989. *After the Ball*. New York, NY: Doubleday.

Klass. P. 1989. "AIDS." *The New York Times Magazine*. Sunday (June 18):34.

Klee, L. 1988. "The Social Significance of Elective Hysterectomy." *Women and Health*. P. Whelehan (ed.). Granby, MA: Bergin and Garvey Publishers.

Klein, D. 1981. "Interview: Helen Singer Kaplan." *Omni*. 3/911/0:73–75, 77, 92.

Klein, F. 1978. *The Bisexual Option*. New York, NY: Arbor House, Berkeley Books.

Kloser, P. 1989. Highlights: Fifth International AIDS Conference, Montreal, June 4–9. "Medical Aspects of Human Sexuality," August: 57, 59, 63–64. Reprinted in *Annual Editions Human Sexuality, 1990–91*. O. Pocs (ed.). Sluice Dock, Guilford, CT: Dushkin Publishers.

Kluckhohn, C. and Leighton, D. 1962. *The Navaho*. New York, NY: Doubleday.

Klugman, L. 1993. "The New Contraceptives." *New Body*. August: 58–59, 70.

Kolata, G. 1994. "The Face That Haunts." *New York Times*. July 10:5.

————. 1992. "After Five Years of Use, Doubt Still Clouds Leading AIDS Drug." *New York Times*. 6/2.

Komarovsky, M. 1962. *Blue Collar Marriage*. New York, NY: Vintage Books.

Konker, C. 1992. "Rethinking Child Sexual Abuse: An Anthropological Perspective." *American Journal of Orthopsychiatry*. 62(1):147–153.

Konner, M. and Worthman, C. 1980. "Nursing Frequency, Gonadal Function and Birth Spacing Among !Kung Hunter-Gatherers." *Science*. 207:788–91.

Konner, M. 1989. "Homosexuality: Who and Why?" *New York Times Magazine*. (April 2):60–62.

————. 1990. "Mutilated in the Name of Tradition." New York Times Book Review of *Prisoners of Ritual*, H. Lightfoot-Klein. Binghamton, NY: The Haworth Press. *New York Times*. Sunday, 4/15.

———. 1982. *The Tangled Wing: Biological Constraints on the Human Spirit.* New York, NY: Harper & Row Publishers.

Korbin, J. (ed.). 1981. *Child Abuse and Neglect: Cross-cultural Perspectives.* Berkeley: University of California Press.

Kottak, C. P. 1991. *Anthropology.* New York, NY: McGraw Hill.

———. 1982. *Researching American Culture.* Ann Arbor, MI: The University of Michigan Press.

Kroeber, A. L. 1953. *Handbook of the Indians of California.* Berkeley: University of California Press.

Kuttner, R. 1989. "She Minds the Child, He Minds the Dog." *The New York Times Book Review.* (June 25):3–4.

Lacayo, R. 1990. "Do the Unborn Have Rights? *Time.* 13(19):22–23.

Ladas, A., Whipple, B. and Perry, J. D. 1982. *The G-spot and Other Recent Discoveries About Human Sexuality.* New York, NY: Dell.

Lambda Legal Defense Fund Newsletter. January 1990.

Lambert, B. 1989. "Koch's Record on AIDS: Fighting a Battle Without a Precedent." *The New York Times.* (August 27):30.

Lancaster, J. B. 1985. "Evolutionary Perspectives on Sex Differences in the Higher Primates." In *Gender and the Life Course.* Alice S. Rossi (ed.). Hawthorne, NY: Aldine:3–27.

Lander, L. 1988. *Images of Bleeding: Menstruation as Ideology.* New York, NY: Orlando Press.

Laws, J. L. and Schwartz, P. 1977. *Sexual Scripts: The Social Construction of Female Sexuality.* Hinsdale, IL: Dryden Press.

Leacock, E. 1978. "Society and Gender." In *Genes and Gender: I.* Ethel Tobach and Betty Rosoff (eds.). New York, NY: Gordian Press: 75–85.

Leavitt, D. 1989. "The Way I Live Now." *The New York Times Magazine.* (July 9):28–32, 80–83.

Lebolt, S. A., Grimes, D.A., and Cates, W. 1982. "Mortality from Abortion and Childbirth." *Journal of the American Medical Association.* 248:188–191.

Lee, R. 1985. "Work, Sexuality, and Aging Among !Kung Women." *In Her Prime: A New View of Middle Aged Women.* K. Brown and V. Kerns, eds. South Hadley, MA: Bergin and Garvey Publishers.

Leibowitz, L. 1978. *Females, Males, Families: A Biosocial Approach.* Belmont, CA: Wadsworth Publishing Co.

Lemonick, M. D. 1994. "How Man Began." *Time* 143(11):80–87.

Leo, J. 1984. "The Revolution Is Over." *Time*. 123(5):74–78, 83.

Leonard, T. L. 1990. "Male Clients of Female Street Prostitutes: Unseen Partners in Sexual Disease Transmission." *Medical Anthropology Quarterly*. 4(1) (March):41–55.

Lett, J. 1987. *The Human Experience: A Critical Introduction to Anthropological Theory*. Boulder, CO: Westview Press.

LeVay, S. 1991. "A Difference in Hypothalamic Structure Between Heterosexual and Homosexual Men." *Science*. August 30:1034–37.

Levine, M. P. and Troiden, R.R. 1988. "The Myth of Sexual Compulsivity." *The Journal of Sex Research*. 25(3):347–363.

Levine, N. 1988. *The Dynamics of Polyandry: Kinship, Domesticity, and Population in the Tibetan Border*. Chicago, IL: University of Chicago Press.

Levine, N. and Sangree, W. 1980. "Women with Many Husbands." *Journal of Comparative Family Studies*. 11(3) (special issue).

Levinson, D. J. 1978. *The Seasons of a Man's Life*. New York, NY: Alfred A. Knopf.

Levi-Strauss, C. 1969. *The Elementary Structures of Kinship*. Boston, MA: Beacon Press.

Levy, R. E.. 1973. *Tahitians: Mind and Experience in the Society Islands*. Chicago, IL: University of Chicago Press.

Lewin, R. 1989. *Human Evolution*. Boston, MA: Blackwell Scientific Publications.

Lewin, T. 1993. "The Strain on the Bonds of Adoption." *New York Times*. 8/8:1, 5.

Lewis, H. B. 1978. "Psychology and Gender." In *Genes and Gender: I*. Ethel Tobach and Betty Rosoff (eds.). New York, NY: Gordian Press: 63–73.

Lightfoot-Klein, H. 1990. *Prisoners of Ritual*. Binghamton, NY: The Haworth Press.

———. 1989. "The Sexual Experience and Marital Adjustment of Genitally Circumcised and Infibulated Females in the Sudan." *Journal of Sex Research*. 26(3) (Aug.):375–393.

Lips, H. M. 1993. *Sex and Gender. An Introduction*. (2nd ed.) Mountain View, CA: Mayfield Publishing Co.

Livingstone, F. 1969. "Genetics, Ecology, and the Origin of Incest and Exogamy." *Current Anthropology*. 10:45–62.

LoPiccolo, J., and LoPiccolo, L. (eds.). 1978. *Handbook of Sex Therapy*. New York, NY: Plenum.

Lowie, R. H. 1935. *The Crow Indians*. New York, NY: Farrar and Rinehart.

Lubbock, J. 1873. (Original 1870). *The Origin of Civilization and the Primitive Condition of Man: Mental and Social Condition of Savages*. New York, NY: D. Appleton.

Lurie, N. (ed.) 1973. *Mountain Wolf Woman. Sister of Crashing Thunder. The Autobiography of a Winnebago Indian*. Ann Arbor, MI: University of Michigan Press.

MacCorquodale, P. 1989. "Gender and Sexual Behavior." In *Human Sexuality. The Societal and Interpersonal Context*. K. McKinney, and S. Sprecher (eds.). Norwood, NJ: Ablex Publishing Corp.:91–115.

MacMahon, B. et al. 1974. "Urine Estrogen Profiles of Asian and North American Women." *Lancet*. (October 23):900–902.

Mahoney, E. R. 1983. *Human Sexuality*. New York, NY: McGraw-Hill.

Males, M. 1992. "Adult Liaison in the Epidemic of 'Teenage' Birth, Pregnancy and Vernereal Disease." *Journal of Sex Research*. 29(4) (Nov.): 525–47.

Malinowski, B. (original 1927) 1961. *Sex and Repression in Savage Society*. Cleveland, OH: World.

———. 1929. *The Sexual Life of Savages in North-Western Melanesia: An Ethnographic Account of Courtship, Marriage and Family Life Among the Natives of the Trobriand Islands, British New Guinea*. London: George Routledge.

Maltz, W. 1989. "Counter Points: Intergenerational Sexual Experience or Child Sexual Abuse?" *Journal of Sex Education and Therapy*. 15(1) (spring):13–16.

Mange, A. and Mange, E. J. 1980. *Genetics: Human Aspects*. Philadelphia, PA: Saunders College.

Mann, J. M. 1988. "The Global AIDS Situation." *World Health*. (June):6–8.

Mann, J., Chin, J., Piot, P., and Quinn, T. 1988. "The International Epidemiology of AIDS." *Scientific American*. 259(October)(4):82–90.

Mann, J. and Tarantola, D. J. M. 1998. "HIV 1998: The Global Picture." *Scientific American* 279(1) (July):82–83.

Marino, E. 1993. "The T Word." In *Democrat and Chronicle*. 7/25.

Mariposa Newsletter. December 1992.

Mariposa Newsletter. September 1992.

Marmor, J. (ed.). 1980. *Homosexual Behavior*. New York, NY: Basic Books.

Marmor, J. A. 1988. "Health and Health-Seeking Behavior of Turkish Women in Berlin." In *Women and Health Cross-Cultural Perspectives*. P. Whelehan (ed). Granby, MA: Bergin and Garvey Publishers: 84–97.

Marrone, J., Dimitroff, L. J., and Van Veckten, K. 4/17/90. Workshop 10: AIDS and Sexually Transmitted Diseases. In *Enriching the College Experience: Addressing the Needs of Women Students. Conference Proceedings*, M.F. Stuck (ed.). Albany, NY: Office of Student Affairs and Special Programs, SUNY: 25–26.

Marshall, D. S. 1971. "Sexual Behavior on Mangaia." In *Human Sexual Behavior*. D. S. Marshall and R. C. Suggs, eds. New York, NY: Basic.

Martin, E. 1987. *The Woman in the Body*. Boston, MA: Beacon Press.

Martin, M. K. and Voorhies, B. 1975. *Female of the Species*. Irvington, NY: Columbia University Press.

Marx, J. 1989. "Circumcision May Protect Against the AIDS Virus." *Science*. 245:470–471.

Masters, W. H. and Johnson, V. E. 1979. *Homosexuality in Perspective*. Boston, MA: Little, Brown and Company.

———. 1974. *The Pleasure Bond: A New Look at Sexuality and Commitment*. Boston, MA: Little, Brown and Company.

———. 1970. *Human Sexual Inadequacy*. Boston, MA: Little, Brown and Company.

Masters, W. H., and Johnson, V. E. 1966. *Human Sexual Response*. Boston, MA: Little, Brown and Company.

Masters, W. H., Johnson, V. E., and Kolodny, R. C. 1988. *Crisis: Heterosexual Behavior in the Age of AIDS*. New York, NY: Grove Press.

Masters, W. H., Johnson, V. E., and Kolodny, R. C. 1985. *Human Sexuality*. 2nd edition. Boston, MA: Little Brown and Company.

Masters, W. H., Johnson, V. E., and Kolodny, R. C. 1982. *Human Sexuality*. Boston, MA: Little, Brown and Company.

Maticka-Tyndale, E., et al. 1994. "Knowledge, Attitudes and Beliefs about HIV/AIDS Among Women in Northeastern Thailand." *AIDS Education and Prevention*. 6(3):205–219.

Matthews, T. and Bolognesi, D. 1988. "AIDS Vaccines." *Scientific American*. 259(October) (4):120–128.

Meyer-Bahlburg, H. F. L. 1979. "Sex Hormones and Female Homosexuality: A Critical Examination." *Archives of Sexual Behavior*. 8(2): 101–120.

————. 1977. "Sex Hormones and Male Homosexuality in Comparative Perspective." *Archives of Sexual Behavior.* 6(4):297–35.

Mayo Foundation for Medical Education and Research. 1993. "Sexuality and Aging." *Mayo Clinic Health Letter.* February.

Mays, V. and Cochran, S. 1988. "Issues in the Perception of AIDS Risk and Risk Reduction Activities by Black and Hispanic/Latina Women." *American Psychologist.* 43(11) (November):949–957.

McCabe, M. 1987. "Desired and Experienced Levels of Premarital Affection and Sexual Intercourse During Dating." *Journal of Sex Research.* 23(1) (February):23–34.

McCammon, S., Knox, D. and Schacht, C. 1993. *Choices in Sexuality.* St. Paul, MN: West Publishers.

McCarthy, P. 1990. "The Romance Factor: Fertility Clinic Sends Couples to the Bedroom." *American Health.* IX(1) (Jan./Feb.):14.

McClintock, M. K. 1971. "Menstrual Synchrony and Suppression." *Nature.* 229:244.

McGill, M. E. 1987. *The McGill Report on Male Intimacy.* New York: Harper & Row.

McGraw, W. C. 1981. "The Female Chimpanzee as Evolutionary Prototype." In *Woman the Gatherer.* F. Dahlberg (ed.). New Haven, CT: Yale University Press: 35–73.

McKinney, K. and Sprecher, S. (eds.). 1989. *Human Sexuality. The Societal and Interpersonal Context.* Norwood, NJ: Ablex Publishing Corp.

McKusick, L. et al. 1985. Reported Changes in the Sexual Behavior of Men at Risk for AIDS. San Francisco, 1982–1984: The Aids Behavioral Research Project. *Public Health Reports.* 100:622–629.

McLennan, J. 1865. *Primitive Marriage.* Edinburgh: Adam and Charles Black.

McMillen, L. 1990. "An Anthropologists's Disturbing Picture of Gang Rape on Campus." *The Chronicle of Higher Education.* September 19, A3.

McNeilly, A. S. 1979. "Effects of Lactation on Fertility." *British Medical Bulletin.* 35:151–54.

McWhirter, D. P. and Mattison, A. M. 1984. *The Male Couple: How Relationships Develop.* Englewood Cliffs, NJ: Prentice-Hall.

Mead, M. 1959. "A New Preface." In *Patterns of Culture.* Ruth Benedict (ed.). New York, NY: The New American Library.

————. 1949. *Male and Female.* New York, NY: Morrow.

————. 1961. (Original 1928). *Coming of Age in Samoa.* New York, NY: Morrow Quill Paperbacks.

———. 1963. (Original 1935). *Sex and Temperament in Three Primitive Societies.* New York, NY: Morrow.

Mead, M. and Newton, N. 1967. "Cultural Patterning of Perinatal Behavior." In *Childbearing: Its Social and Psychological Aspects.* S.A. Richardson and A. Guttmacher (eds.). Baltimore, MD: Williams and Wilkins.

Medical Alert. 1994. "On the Pulse." 2(3) (May/June):8.

Medical Anthropology. 1981, 5(1) (winter). Special Issue: "Midwives and Modernization." L. F. Newman, guest editor.

Medical Aspects of Human Sexuality. 1991. (July):14.

"Men's Shots Used as Contraceptive." 1990. *Syracuse Herald American,* October 21: 1.

Mercer, R.T. and Stainton, M.C. 1984. "Perceptions of the Birth Experience: A Cross-cultural Comparison." *Health Care of Women International.* 5:29–47.

Mernissi, F. 1975. *Beyond the Veil.* Cambridge, MA: Schenkman Publishing Company, Inc.

Messenger, J. C. "Sex and Repression in an Irish Folk Community." In *Human Sexual Behavior.* New York, N.Y. D. S. Marshall and R. C. Suggs, eds. Pp 3–38.

Metropolitan Life Insurance Co. 1987. Life Insurance Tables.

Metts, S. and Cupach, W. 1989. "The Role of Communication in Human Sexuality." In *Human Sexuality. The Societal and Interpersonal Context.* K. McKinney and S. Sprecher (eds.) Norwood, NJ: Ablex Publishing Corp.: 139–162.

Michael, R. P. and Zumpe, D. 1970. "Aggression and Gonadal Hormones in Captive Rhesus Monkeys." *Animal Behavior.* 18:1–19.

Michaelson, K. (ed.). 1988. *Childbirth in America.* Granby, MA: Bergin and Garvey Publishers.

Michaelson, K. L. 1988. "Childbirth in America: A Brief History and Contemporary Issues." In *Childbirth in America.* K. Michaelson (ed.). Granby, MA: Bergin and Garvey Publishers: 1–32.

Migeon, C. V., Rivarola, M. A., and Forest, M. G. 1969. *Studies of Androgens in Male Transsexual Subjects: Effects of Estrogen Therapy in Transsexualism and Sex Reassignment.* R. Green and J. Money (eds.). Baltimore, MD: The Johns Hopkins University Press.

Miller, B. 1988. (Personal Communication).

Miller, E. S. 1979. *Introduction to Cultural Anthropology.* Englewood Cliffs, NJ: Prentice-Hall.

Miller, P.Y. and Simon, W. 1980. "The Development of Sexuality in Adolescence." In *Handbook of Adolescence*. Joseph Adelson (ed.). New York, NY: John Wiley and Sons.

Miller, P. Y. and Fowlkes, M. R. 1987. "Social and Behavioral Constructions of Female Sexuality." In *Sex and Scientific Inquiry*. Sandra Harding and Jean F. O'Barr (eds.)., Chicago, IL: University of Chicago: 147–164.

Miracle, A. W. and Suggs, D. N. 1993. "On the Anthropology of Human Sexuality." In *Culture and Human Sexuality: A Reader*. Pacific Grove, CA: Brooks/Cole Publishing Co. Pp. 2–6.

Mitford, J. 1992. *The American Way of Birth*. Dutton: New York, NY.

MN and TBA, District Court, Stearn City, MN: AIDS Litigation Reports 8/12.

Modell, J. 1989. "Last Chance Babies: Interpretations of Parenthood in an In-vitro Fertilization Program." *Medical Anthropology Quarterly*. 3(2):124–139.

Moghadam, V. 1992. "Revolution, Islam and Women: Sexual Politics in Iran and Afghanistan." In *Nationalisms and Sexualities*. A. Parker, M. Russo, D. Sommer and P. Yaeger (eds.). Routledge: New York: 424–447.

Mohr, J. C. 1978. *Abortion in America: The Origins and Evolution of National Policy*. New York, NY: Oxford University Press.

Money, J. 1988a. *Gay, Straight and In-between*. New York, NY: Oxford University Press.

———. 1988b. Lecture on Gender Dysphoria and Paraphilias. Institute for the Advanced Study of Human Sexuality, February 13 (IASHS): San Francisco, CA.

———. 1987. "Propaedeutics of Diecious G-1/R: Theoretical Foundations for Understanding Dimorphic Gender-identity/Role." In *Masculinity/Femininity: Basic Perspectives*. J. Reinisch, L.A. Rosenblum, and S.A. Sanders (eds.). New York, NY: Oxford University: 13–28.

———. 1986. *Lovemaps: Clinical Concepts of Sexual Erotic Health and Pathology, Paraphilia, and Gender Transposition in Childhood, Adolescence, and Maturity*. New York, NY: Irvington.

———. 1977. "Determinants of Human Gender Identity/Role." In *Handbook of Sexology*. J. Money, and H. Musaph (eds.). New York, NY: Excerpta Medica: 57–79.

Money, J. and Ehrhardt, A. 1972. *Man and Woman, Boy and Girl*. Baltimore, MD: The Johns Hopkins Press.

Money, J. and Russo, A. H. 1981. "Homosexual vs. Transvestite or Transsexual Gender-identity/Role: Outcome Study in Boys." *International Journal of Family Psychiatry*. 2(1–2):139–45.

Money, J. and Wiedeking, C. 1980. "Gender Identity Role: Normal Differentiation and its Transpositions." In *Handbook of Human Sexuality*. B.B. Wolman, and J. Money (eds.). Englewood Cliffs, NJ: Prentice-Hall; 270–284.

Moore, L. G., VanArsdale, P. W., Glittenberg, J. E., and Aldrich, R. A. 1980. *The Biocultural Basis of Health*. Prospect Heights, IL: The Waveland Press.

Moore, C. H. et al. 1988. "Sex Differences on Attitudes Towards AIDS Among College Students." Paper presented at the the Annual Meeting of the Southeastern Psychological Assoc., 34th, New Orleans, LA:3/31–4/2.

Morbidity and Mortality Weekly Report. 1998. "Public Health Service Guidelines for the Management of Health-Care Worker Exposure to HIV and Recommendations for Postexposure Prophylaxis." 47(7)(Mg).

Morbidity & Mortality Weekly Report 1994.

"More on Koop's Study of Abortion." 1990. *Family Planning Perspectives*. 22:36.

Morgan, L. H. 1870. *Systems of Consanquinity and Affinity of the Human Family*. Washington, DC: Smithsonian Institution.

———. 1877. *Ancient Society*. New York, NY: World Publishing.

Moss, A. M. 1978. "Men's Mid-Life Crisis and the Marital-Sexual Relationship." *Medical Aspects of Human Sexuality*. 13(2):109–110.

Muecke, E. C. 1979. "The Embryology of the Urinary System," in *Campbell's Urology*. M. Campbell (ed.), 4th edition, vol. 2, Philadelphia: Saunders: (Chpt. 37):1286–1307.

Munroe, R., Whiting, J. M. and Hally, D. J. 1969. "Institutionalized Male Transvestism and Sex Distinctions." *American Anthropologist*. 71(1):87–90.

Murdock, G. P. et al. 1964. *Outline of Cultural Materials*. 4th edition. New Haven, CT: HRAF Press.

Murdock, G. P. 1949. *Social Structure*. New York, NY: The Free Press.

Murphy, Y. and Murphy, R. 1974. *Women of the Forest*. New York, NY: Columbia University Press.

Musaph, H. 1978. "Sexology: A Multidisciplinary Science." *The Handbook of Sexology: History and Ideology*. Vol. 1. John Money and Herman Musaph, eds. New York, NY: Elsevier. Pp. 81–84.

Nag, M. 1962. "Factors Affecting Human Fertility in Non-Industrial Societies: A Cross-Cultural Study." In *Yale University Publications in Anthropology 66*. New Haven, CT: HRAF Press.

Nanda, S. 1987. *Cultural Anthropology*. Belmont, CA: Wadsworth Publishing Co.

National Public Radio (NPR). June 1994. "Morning Edition: Lesbian Mothers in Custody."

National Public Radio (NPR). January 1990. "Evening Edition: AIDS in Africa."

Navarro, M. 1993. "Growing Up Condemned with the AIDS Virus." *New York Times*. Sunday 3/21/:33,36.

NBC. 1994. Nightly News. July 1.

NBC. 1994. Special program on Fertility. April 6. 10p.

Nelson, J. 1989a. "Intergenerational Sexual Contact: A Continuum Model of Participants and Experiences." *Journal of Sex Education and Therapy*. 15(1) (spring):3–13.

Nelson, J. 1989b. "Joan Nelson's Rebuttal." *Journal of Sex Education and Therapy*. 15(1) (spring):16.

Netter, F. H. 1974 (fifth printing). *The Ciba Collection of Medical Illustrations*, vol. 2, The Reproductive System. Summit, NY: CIBA.

Nevid, J. 1993. *A Student's Guide to AIDS and Other Sexuality Transmitted Diseases*. Boston, MA: Allyn and Bacon.

"New Woman Report, The." 1986. In *Dimensions of Human Sexuality*. C. Byer and Louis W. Shainberg (eds.). Dubuque, IA: Wm. C. Brown Publishers: 379.

New York Times. 1993. "Incidence of HIV Infected Babies." March/April.

New York Times. 1990. "The Courts Are Again Asked to Redefine Family." 9/23:E7.

New York Times. 1989. "AIDS Spreading in Teens," 8 October: 28.

Newcomer, S. F. and Udry, J.R. 1985. "Oral Sex in an Adolescent Population." *Archives of Sexual Behavior*. 14:41–46.

Newman, L. F. 1981. "Introduction: Midwives and Modernization." *Medical Anthropology*. 5(1) (winter):1–12.

Newsweek. 1990. "Aids, The Next Ten Years." 6/25:20–27.

Newton, N. 1981. Foreword to *Anthropology of Human Birth*. M. Kay (ed.). Philadelphia, PA: F.A. Davis: ix–xi.

NYSDH. 1988. *A Physician's Guide to AIDS: Issues in the Medical Office*. Albany, NY: NYSDH.

Obituary Column. *New York Times*. 1992. Alison Gertz. 17 August.

Oesterling, J. 1991. "The Role of PSA in Detecting Prostate Cancer." *Medical Aspects of Human Sexuality.* (July):22–27.

Offir, C. W. 1982. *Human Sexuality.* New York, NY: Harcourt Brace Jovanovich, Inc.

Ohm, W. 1980. "Female Circumcision." *Sexology Today.* (June):21–25.

Ohno, S. 1979. *Major Sex-Determining Genes.* Berlin, Germany:Springer-Verlag.

O'Kelly, C.G. and Carney, L.S. 1986. *Women and Men in Society: Cross-Cultural Perspectives on Gender Stratification.* Belmont, CA: Wadsworth Publishing Co.

Olesen, V. L. and Woods, N. F. (eds.). 1986. *Culture, Society and Menstruation.* New York, NY: Hemisphere Publishing Corp.

Olsen, C. 1976. "The Demography of New Colony Formation in a Human Isolate: Analysis and History". Ph.D. dissertation. University of Michigan.

O'Neill, N. and O'Neill, G. 1972. *Open Marriage: A New Lifestyle for Couples.* New York, NY: M. Evans and Company.

Op Ed. 1990. *New York Times.* 9/30:7.

Orenstein, P. 1991. "Women At Risk." Condom Sense, 1991 edition. Oakland, CA: The Condom Resource Center: 1, 4, 5.

Orlando (Pseud.). 1978. "Bisexuality: A Choice, Not an Echo." *Ms.* October: 60.

Orwell, G. 1950. *Nineteen Eighty-Four.* San Diego, CA: Harcourt, Brace, Jovanovich, Inc.

Ostrow, D. 1989. "AIDS Prevention Through Effective Education." *Daedalus,* Living With AIDS Part II(118)(3)(summer):229–254.

Oswalt, W. 1986. *Lifecycles and Lifeways.* Palo Alto, CA: Mayfield Publishing Company.

Padian, N. 1987. "Heterosexual Transmission of Acquired Immunodeficiency Syndrome: International Perspectives and National Projections." *Reviews of Infectious Diseases.* 9(5) (Sept./Oct.):947–960.

Padian, N. et al. 1987. "Male-to-Female Transmission of Human Immunodeficiency Virus." *Journal of the American Medical Association.* 258(6) (Aug. 14):788–790.

Page, J. B. et al. 1990. "Intravenous Drug Use and HIV Infection in Miami." *Medical Anthropology Quarterly.* 4(1) (March):56–71.

Pagels, E. 1988. *Adam, Eve and the Serpent.* New York, NY: Random House.

Paplau, L. 1987. "What Homosexuals Want in Relationships." *Psychology Today*. (March):28–38.

Parade Magazine. 1989. "Report on Test Tube Babies," from *Syracuse Herald American*. 2 July.

Parker, R. 1987. "Acquired Immunodeficiency Syndrome in Urban Brazil." *Medical Anthropology Quarterly*. 1:155–75.

Parker, R., Herdt, G., and Carballo, M. 1991. "Sexual Culture, HIV Transmission, and AIDS Research." *Journal of Sex Research*. 28(1) (February):77–99.

Patton, C. 1992. "From Nation to Family: Containing 'African AIDS'." In *Nationalisms and Sexualities*. A. Parker, J. Russo, D. Sommer, and P. Yaeger (eds.). New York: Routledge, 218–234.

Paul, J. 1984. "The Bisexual Identity: An Idea Without Social Recognition." *Journal of Homosexuality*. 9:45–63.

Pela, A. O. and Platt, J. J. 1989. "AIDS in Africa: Emerging Trends." *Social Science and Medicine*. 28(1):1–9.

Percival-Smith, R. K. L. and Abercrombie, B. 1987. "Postcoital Contraception with Dl-Norgestrel/Ethinyl Estradiol Combination: Six Years in a Student Medical Clinic." *Contraception:* 35–287.

Perper, T. 1985. *Sex Signals: The Biology of Love*. Philadelphia, PA: ISI Press.

Perry, J. D. and Whipple, B. 1981. "Pelvic Muscle Strength of Female Ejaculators: Evidence in Support of a New Theory of Orgasm." *Journal of Sex Research* 17(1):22–39.

Pfafflin, F. 1981. "H-Y Antigen in Transsexualism." Paper presented at the 7th International Gender Dysphoria Association. Lake Tahoe, Nevada, 4–8 March.

Phelan, P. 1986. "The Process of Incest: Biologic Father and Stepfather Families." *Child Abuse and Neglect: The International Journal*. 10(4):531–539.

Pheterson, G. 1989. "Update on HIV Infection and Prostitute Women." In *A Vindication of the Rights of Whores*. G. Pheterson (ed.). Seattle, WA: Seal Press: 132–141.

Planned Parenthood. 1988–89. Planned Parenthood of Northern New York, Inc., 1988–89 Newsletter.

Plenary Sessions. IXth International AIDS Conference. June 7–11, 1993. Berlin, Germany.

Poirier, F. E., Stini, William, A. and Wrenden, K. B. 1990. *In Search of Ourselves*. Englewood Cliffs, NJ: Prentice Hall.

Polaris-Ursa Research Institute. 1988. Cal. State Evaluation of AIDS Agency Programs. Polaris Group San Francisco, CA.

Powers, M. N. 1986. *Oglala Women: Myth, Ritual and Reality*. Chicago, IL: University of Chicago Press.

Prendergast, A. 1990. "Beyond the Pill." *American Health*, 9(8):37–45.

Pritchard, E. E. 1977. *The Nuer*. New York, NY: Oxford University Press.

Pritchard, J., MacDonald, D., and Gant, N. 1985. *Williams' Obstetrics*. Norwalk, CT: Appleton-Century-Crafts.

"Putting the Pill (for Men) in Perspective." 1998 *Harvard Health Letter* 23(8): 4–5.

Quinn, T.C. et al. 1986. "AIDS in Africa: An Epidemiological Paradigm." *Science*. 234(11/12):955–962.

Radin, P. 1926. *Cashing Thunder: The Autobiography of a Winnebago Indian*. New York, New York: D. Appleton.

Ragone, H. 1994. *Surrogate Motherhood. Conception in the Heart*. Boulder, Colorado: Westview Press.

Ramey, E. 1973. "Sex Hormones and Executive Ability." *Annals of the NY Academy of Sciences*. 28:237.

Raphael, D. 1988. "The Need for a Supportive *Doula* in an Increasingly Urban World." In *Women and Health*. P. Whelehan (ed.). Granby, MA: Bergin and Garvey Publishers.

Rathus, S., Nevid, J., and Fichner-Rathus, L. 1993. *Human Sexuality In a World of Diversity*. Boston, MA: Allyn and Bacon.

Reiss, I. 1989. "Society and Sexuality: A Sociological Theory." In *Human Sexuality. The Societal and Interpersonal Context*. K. McKinney and S. Sprecher (eds.). Norwood, NJ: Ablex Publishing Corp.: 3–30.

⸻. 1986. *Journey into Sexuality: An Exploratory Voyage*. Englewood Cliffs, NJ: Prentice-Hall.

⸻. 1967. *The Social Context of Permissiveness*. New York, NY: Holt, Rinehart, and Winston.

Reiss, I. and Leik, R. 1989. "Evaluating Strategies to Avoid AIDS: Number of Partners vs. Use of Condoms." *Journal of Sex Research*. 26(4) (November):411–434.

Remafedi, G. 1989. "The Healthy Sexual Development of Gay and Lesbian Adolescents." *Siecus Report*. 17(5) (May/July):7–8.

Rice, A. 1982. *Cry to Heaven*. New York: Pinnacle Books.

Richardson, L. 1988. *The Dynamics of Sex and Gender: A Sociological Perspective.* New York, NY: Harper Collins Publishers.

Richman, D. D. 1998. "How Drug Resistance Arises." *Scientific American* 279(1) (July):88.

Richwald, G. et al. 1988. "Sexual Activities in Bathhouses in Los Angeles County: Implications for AIDS Prevention Education." *Journal of Sex Research.* 25(2) (May):169–180.

Riddle, J. M., Estes, J. W., and Russell, J. C. 1994. "Birth Control in the Ancient World." *Archaeology.* 47(2):29–35.

Riportella-Muller, R. 1989. "Sexuality in the Elderly: A Review." In K. McKinney and S. Sprecher, eds. *Human Sexuality: The Societal and Interpersonal Context.* Norwood, NJ: Ablex. Publishing Corp.

Robertson, J. A. and Plant, M. A. 1988. "Alcohol, Sex and Risks of HIV Infection." *Drug and Alcohol Dependence.* 22(1–2):75–78.

Robinson, J. 1987. "Senators Told of Family's Plight with AIDS." *Boston Globe.* 1(9/12).

Ross, E. and Rapp, R. 1983. "Sex and Society: A Research Note from Social History and Anthropology." In *Powers of Desire: The Politics of Sexuality.* Christine Stansell and Sharon Thompson (eds.). New York, NY: Monthly Review Press: 49–72.

Rosser, S. 1993. "Review of Women AIDS and Communities by G. Pearlberg." *AIDS Education and Prevention.* 5(1) (spring):92–93.

Rossi, A. S. 1977. "A Biosocial Perspective in Parenting." *Daedalus.* 106:1–31.

Rothman, D. 1971. *The Discovery of the Asylum.* Boston, MA: Little, Brown, and Co.

Rozenbaum, W. et al. 1988a. "HIV and Oragenital Transmission." *Lancet.* 2:1023–24.

———. 1988b. "HIV Transmission by Oral Sex." *Lancet.* 2:1395.

Rubin, L. B. 1976. *Worlds of Pain.* New York, NY: Basic Books.

Salisbury, J. E. 1988. Review of "Courtly Lovers and the Physiology of Pleasure" by D. Jaguart and C. Thomasset, translated by Matthew Adamson. *Journal of Sex Research.* 27(1):141–144.

San Francisco Dept. Health. 1993. "HIV Seroprevalence and Risk Behaviors Among Lesbians and Bisexual Women: The 1993 San Francisco/ Berkeley Women's Survey." San Francisco, CA: Surveillance Branch, AIDS Office. SFDH.

Sargent, C. and Stark, N. 1989. "Childbirth Education and Childbirth Models: Parental Perspectives on Control, Anesthesia, and Technological Intervention in the Birth Process." *Medical Anthropology Quarterly.* 3(1) (March):36–51.

Sarvis, B. and Rodmon, H. 1974. *The Abortion Controversy.* 2nd ed. Irvington, NY: Columbia University Press.

Sayers, J. 1986. *Sexual Contradictions.* New York, NY: Tavistock Publications.

Schapera, I. 1993. "Some Kgatla Theories of Procreation." In *Culture and Human Sexuality.* D. Suggs and A. Miracle (eds.). Brooks Cole Publishing Company: Pacific Grove, CA: 171–182.

Schenker, J. and Evron, S. 1983. "New Concepts in the Surgical Management of Tubal Pregnancy and the Consequent Postoperative Results." *Fertility and Sterility.* 40:709–23.

Scheper-Hughes, N. 1987. "Culture, Scarcity, and Maternal Thinking: Mother Love and Child Death in Northeast Brazil." In *Child Survival.* Nancy Scheper-Hughes (ed.). Boston, MA: D. Reidel: 187–208.

——. 1992. *Death Without Weeping: The Violence of Everyday Life in Brazil."* Berkeley: University of California Press.

Schlegel, A. (ed.). 1977. *Sexual Stratification A Cross Cultural View.* New York, NY: Columbia University Press.

Schlegel, A. and Barry, H. 1991. *Adolescence: An Anthropological Inquiry.* New York, New York: The Free Press.

Schlegel, A. and Barry, H. III. 1980. "The Evolutionary Significance of Adolescent Initiation Ceremonies. *American Ethnologist.* 7(4):696–715.

Schmidt, G. 1977. "Working-Class and Middle-Class Adolescents." In *The Handbook of Sexology.* J. Money and H. Musaph, (eds.). New York, NY: Elsevier/North-Holland Biomedical Press:283–295.

Schneider, D. M. 1968. "Abortion and Depopulation on a Pacific Island." In *People and Cultures of the Pacific.* Andrew P. Vayda (ed.). Garden City, NY: The Natural History Press: 383–406.

Schoepf, B. G. et al. 1988. "Aids in Africa: Death is the Only Certainty." *Christianity Today.* 32(4/8):36–40.

Schultz, E. A. and Lavenda, R.H. 1990. *Cultural Anthropology.* St. Paul, MN: West Publishing Company.

Schwarz, R. 1981. "The Midwife in Contemporary Latin America." *Medical Anthropology.* 5(1) (winter):51–71.

Science. 1991. "Spreading Strain of HIV Infection in the Western U.S." 1251:1022.

Scientific American. 1988. 259(4)(October). "What Science Knows About AIDS."

Scupin, R. and DeCorse, C.R. 1992. *Anthropology: A Global Perspective.* Englewood Cliffs, NJ: Prentice Hall, Inc.

Seaman, B. and Seaman, G. 1977. *Women and the Crisis in Sex Hormones.* New York, NY: Rawson Associates Publ., Inc.

Sears, R. A. 1979. "Sex Typing, Object Choice, Child Rearing." In *Human Sexuality: A Comparative and Developmental Perspective.* H.A. Katchadourian (ed.). Berkeley: University of California Press, 204–23.

Segal, J. 1984. *The Sex Lives of College Students.* Wayne, PA: Miles Standish Press.

Select Committee on Hunger, U.S. House of Representatives 100th Congress, second session. 1988. AIDS and the Third World: The Impact on Development. Washington, DC: U.S. Government Printing Office.

SELF. October 1993.

Seyler, L. E., Canalis, E., Spare, S., and Reichlin, S. 1978. "Abnormal Gonadotropin Secretory Responses to Leutinizing Releasing Hormone in Transsexual Women After Diethylstilbestrol Priming." *Journal of Clinical Endocrinology and Metabolism.* 47:176–83.

Shah, D. K., Walters, L., and Clifton, T. 1979. "Lesbian Mothers." *Newsweek.* (February 12):61.

SF AIDS Fdn. 1998. "Summary Sheets on HIV Treatment Strategies: Special Concerns of Women and Children." *Treatment Education and Advocacy.* http://www.sfaf.Org. 1–5.

Shannon, S. and Pyle, G. 1989. "The Origin and Diffusion of AIDS: A View from Medical Geography." *Annuals of the Association of American Geographers.* 79(1) (March):35–55.

Shapiro, H. L. 1958. *Man, Culture and Society.* New York, NY: Oxford University Press.

Shapiro, J. 1979. "Cross-Cultural Perspectives on Sexual Differentiation." In *Human Sexuality: A Comparative and Developmental Perspective.* Herant Katchadourian (ed.). Berkeley: University of California Press: 269–308.

Sharp, H. 1981. "The Null Case: The Chipeweyan." In *Woman the Gatherer.* F. Dahlberg (ed.). New Haven, CT: Yale University Press: 221–44.

Shreeve, James. "Sunset on the Savannah." *Discover* 17 (7): 116–125, July, 1996.

Sherfey, M. J. 1972. *The Nature and Evolution of Female Sexuality.* New York, NY: Vintage Books.

Shilts, R. 1987. *And the Band Played On: People, Politics, and the AIDS Epidemic*. New York, NY: St. Martin's Press.

Short, R. V. 1978. "Healthy Fertility." *Upsala Journal of Medical Science*. Suppl. 22:23–26.

Silber, S. 1981. *The Male*. New York, NY: Charles Scribner's Sons.

Simpson-Herbert, M. and Huffman, S. L. 1981. "The Contraceptive Effect of Breastfeeding." *Studies in Family Planning*. 12:125–33.

Singer, J. and Singer, I. 1972. "Types of Female Orgasm." *Journal of Sex Research*. 8(4):255–267.

Singer, M. et al. 1990. "SIDA: The Economic, Social, and Cultural Context of AIDS Among Latinos." *Medical Anthropology Quarterly*. 4(1) (March):72–114.

Singer, M. 1961. "A Survey of Culture and Personality Theory and Research." In *Studying Personality Cross-Culturally*. Bert Kaplan (ed.). New York, NY: Harper and Row, Pub.: 9–90.

Sittitrai, W. 1990. "Research on Human Sexuality in Pattern III Countries." In *Human Sexuality: Research Perspectives in a World Facing AIDS*. A. Chouinard and J. Albert (eds.). Ottawa, Ontario, Canada: IDRC: 173–190.

Sivin, I. et al. 1983. "Three Year Experience with Norplant Subdermal Contraception." *Fertility and Sterility*. 39:799–808.

Small, M. F. 1993. "The Gay Debate/Lesbianism: Less is Known." *American Health*. XII (2) (March):70–77.

Smith, A. 1990 "Nurse-Midwives a Growing Birth Alternative." *Syracuse Herald American*. Sunday, 4/8:D-1, D-6.

Smythers, R. 1989. "Instruction and Advice for the Young Bride on the Conduct and Procedures of the Intimate and Personal Relationship of the Marriage State for the Greater Spiritual Sanctity of This Blessed Sacrament and the Glory of God." *Transgender Views*. 3(9):5–7. (Original 1894).

Sommerville, C. J. 1990. *The Rise and Fall of Childhood*. New York, NY: Vintage Books.

SOLGAN (Society of Lesbian and Gay Anthropology Newsletter). 1992.14(2).

Sorensen, R. C. 1973. *Adolescent Sexuality in Contemporary America*. New York, NY: World.

Speroff, L., Glass, R. H., and Kane, N. 1978. *Clinical Gynecologic Endocrinology and Infertility*. 2nd edition. Baltimore, MD: Williams and Wilkins.

Spitzer, P. and Weiner, N. 1989. "Transmission of HIV Infection From a Woman to a Man by Oral Sex." *New England Journal of Medicine.* 320:251.

Spock, B. and Rothenberg, M.B. 1985. *Dr. Spock's Baby and Child Care.* New York, NY: Pocket Books.

Sponaugle, G. 1989. "Attitudes Towards Extramarital Relations." In *Human Sexuality. The Societal and Interpersonal Context.* K. McKinney and S. Sprecher (eds.). Norwood, NJ: Ablex Publishing Corp.: 187–210.

Spradley, J. P. 1987. "Ethnography and Culture." In *Conformity and Conflict.* Boston, MA: Little, Brown and Co.: 17–25.

Spradley, J. P. and McCurdy, D. W. 1989. *Anthropology: The Cultural Perspective.* Prospect Heights, IL: Waveland Press.

Sprecher, S. 1989. "Influences on Choice of a Partner and on Sexual Decision Making in the Relationship." In *Human Sexuality. The Societal and Interpersonal Context.* K. McKinney and S. Sprecher (eds.). Norwood, NJ: Ablex Publishing Corp.: 115–139.

Sprecher, S., McKinney, K., and Orbuch, T. L. 1991. "Has the Double Standard Disappeared? An Experimental Test." In *Human Sexuality 91/92.* O. Pocs (ed.). Sluice Dock, Guilford, CT: Dushkin Publishers: 87–95.

Stack, C. 1974. *All Our Kin.* New York, NY: Harper & Row.

Stall, R., Coates, T. J., and Hoff, C. 1988. "Behavioral Risk Reduction for HIV Infection Among Gay and Bisexual Men: A Review of Results from the United States." *American Psychologist.* 43(11):878–885.

Stall, R. et al. 1990. "Sexual Risk for HIV Transmission Among Singles-Bar Patrons in San Francisco." *Medical Anthropology Quarterly.* 4(1) (March):115–128.

Starka, L., Sipova, I., and Hynie, J. 1975. "Plasma Testosterone in Male Transsexuals." *The Journal of Sex Research.* 11(2):134–38.

Stearns, C. Z., and Stearns, P. N. 1985. "Victorian Sexuality: Can Historians Do It Better." *Journal of Social History.* 18:625–634.

Stern, P. N. (ed.). 1986. *Women, Health, and Culture.* New York, NY: Hemisphere Publishing Co.

Stewart, F., Guest, F., Stewart, G., and Hatcher, R. 1979. *My Body, My Health: The Concerned Woman's Guide to Gynecology.* New York, NY: John Wiley & Sons.

Stine, G. 1998. *Acquired Immune Deficiency Syndrome: Biological, Medical, Social, and Legal Issues,* 3rd. Ed. Upper Saddle River, NJ: Prentice Hall.

Stine, G. 1993. *Acquired Immune Deficiency Syndrome. Biological, Medical, Social, and Legal Issues.* Englewood Cliffs, NJ: Prentice Hall.

———. 1977. *Biosocial Genetics.* New York, NY: MacMillian Publishing Co. Stockholm Conference. 1988.

Stoller, R. J. 1968. *Sex and Gender: On the Development of Masculinity and Femininity.* New York, NY: Science House.

Stone, L. 1977. *The Family, Sex and Marriage in England 1500–1800.* New York, NY: Harper & Row.

Storms, M. 1980. "Theories of Sexual Orientation." *Journal of Personality and Social Psychology.* 38:783–792.

Strum, S. 1975. "New Insights into Baboon Behavior: Life With the Pumphouse Gang." *National Geographic.* 147:627–691.

Strunin, L. and Hingson, R. 1987. "Acquired Immunodeficiency Syndrome and Adolescents: Knowledge, Beliefs, Attitudes, and Behaviors." *Pediatrics.* 79(5):825–828.

Suggs, D. and Miracle, A. 1993. *Culture and Human Sexuality. A Reader.* Pacific Grove, CA: Brooks/Cole Publishing Company.

Suggs, R. 1966. *Marquesan Sexual Behavior.* New York, NY: Harcourt, Brace & Jovanovich.

Sukkary, S. 1981. "She Is No Stranger: The Traditional Midwife in Egypt." *Medical Anthropology.* 5(1) (winter):27–34.

Sullivan, K. 1998. "Gay Youths Struggle in Personal Hell." *San Francisco Examiner* (Sunday, July 26):D1, 4.

Symons, D. 1979. *The Evolution of Human Sexuality.* Oxford, England: Oxford University Press.

Tanner, N. 1981. *On Becoming Human.* New York, NY: Cambridge University Press.

Tanner, N. and Zihlman, A. 1976. "Women in Evolution. Part I: Innovation and Selection in Human Origins." *Signs.* 1(3):585–608.

Tavris, C. 1992. *The Mismeasure of Woman.* New York: Simon & Schuster.

Taylor, J. M. and Ward, V. 1991. "Culture, Sexuality and School: Perspectives from Focus Groups in Six Different Cultural Communities." *Women's Studies Quarterly* 1 and 2: 121–137.

"Teen Sexual Activity Rises." 1991. *Parade Magazine.* February: 24, 25.

Terenzi, L. 1992. "Abortion: A World View." *Self.* 54, November.

Terry, J. 1990. "Lesbians Under the Medical Gaze: Scientists Search for Remarkable Differences." *Journal of Sex Research.* 27(3) (August). (Special issue) (Part 2):317–341.

The Association of Reproductive Health Professionals. 1993. "Women in Transition: Defining the Perimenopause." *ARHP Clinical Procedings.* July.

The North American Menopause Society. 1993. "Managing Menopause: Our Readers Share Their Views." *Menopause Management.* 11(7):18, 21–23.

Tierney, J. 1990. "With 'Social Marketing' Condoms Combat AIDS." *New York Times.* 9/18:1, 10.

———. 1990. "AIDS Tears Lives of the African Family." *New York Times.* 9/17:1, 14.

Tiger, L. 1970. *Men in Groups.* New York, NY: Vintage Press.

Tiger, L. and Fox, R. 1971. *The Imperial Animal.* New York, NY: Dell Publishing Co.

Time. 1993. "Born Gay." 26 July: 36–39.

"Tragedy of Parental Involvement Laws, The." 1990. *The National Now Times.* January/February:5.

Trevathan, W.R. 1987. *Human Birth: An Evolutionary Perspective.* Hawthorne, NY: Aldine De Gruyter.

Troiano, L. 1990. "Verbal Birth Control." *American Health.* 9(4):101.

Turnbull, C. 1988. *The Forest People.* New York, NY: Simon and Schuster.

———. 1981. "Mbuti Womanhood." In *Woman the Gatherer.* F. Dahlberg (ed.). New Haven, CT: Yale University Press: 205–219.

———. 1972. *The Mountain People.* New York, NY: Simon and Schuster.

———. 1961. *The Forest People.* New York, NY: Simon and Schuster.

Turner, J. 1994. "Promoting Responsible Sexual Behavior Through a College Freshman Seminar." *AIDS Education and Prevention.* 6(3) (June): 266–278.

Turner, V. 1969. *The Ritual Process: Structure and Anti-Structure.* Chicago, IL: Aldine Press.

Tyler, S. L. and Woodall, G. M. 1982. *Female Health and Gynecology: Across the Lifespan.* Bowie, MD: Robt. J. Brady Co.

U.S. Bureau of the Census. 1980 and 1990.

U.S. Bureau of the Census. 1985. *Current Population Reports: Household, Families, Marital Status and Living Arrangements.* Series P–20, no. 382. Washington, DC: U.S. Government Printing Office.

U.S. Bureau of Labor Statistics. 1985. Washington, D.C.: U.S. Dept. of Labor.

Utian, W. H., Cohen, C., Isacs, J. H., Mikuta, J. J., and Plouffe, L. 1993. "Doctor, Will HRT Cause Cancer?" *Menopause Management.* 11(7):11–12, 14, 16.

Vance, C. S. 1983. "Gender Systems, Ideology and Sex Research." In *Powers of Desire: The Politics of Sexuality.* A. Snitow, C. Stansell, and S. Thompson (eds.). New York, NY: Monthly Review Press: 371–384.

Vance, C. S. and Pollis, C. A. 1990. "Introduction: A Special Issue on Feminist Perspectives on Sexuality." *Journal of Sex Research.* 27(1):1–5.

Van Gennep, A. 1960. *The Rites of Passage.* Translated by M.B. Vizedom and G.I. Caffee. London, England: Routledge and Kegan Paul. (Orig. ed. 1909).

Vatuk, S. 1985. "South Asian Cultural Conceptions of Sexuality." In *In Her Prime: A New View of Middle-Aged Women.* J. K. Brown and V. Kerns, eds. South Hadley, MA: Bergin and Garvey Publishers, Inc. Pp. 137–154.

Vermund, S., Sheon, A., Ebner, S., and Fischer, R. 1991. "Transmission of the Human Immunodeficiency Virus." *AIDS Research Reviews.* 1:81–136.

Vliet, E. L. 1993. "New Insights on Hormones and Mood." *Menopause Management.* 11(6):14–16.

Voeller, B. 1983. "Heterosexual Anal Sex." Mariposa Occasional Papers, 1.B:1–8.

vos Savant, M. 1991. "Ask Marilyn." *Parade Magazine.* May 26, 26.

Wade, C. and Cirese, S. 1991. *Human Sexuality.* 2nd ed. San Diego, CA: Harcourt Brace Jovanovich, Publ.

Wagenaar, T. C. (ed.). 1987. *Teaching Sociology,* 15(3) (July). "Special Issue on Teaching Human Sexuality." (also 241–263.)

Walter, A. 1990. "Putting Freud and Westermarck in Their Places: A Critique of Spain." *Ethos.* 18(4):439–446.

Walters, W. 1986. *The Spirit and the Flesh, Sexual Diversity in American Indian Culture.* Boston, MA: Beacon Press.

WAN. 1990. Newsletter update on Women and AIDS (1/90).

Warren, M. P. 1983. "Effects of Undernutrition on Reproductive Function in the Human." *Endocrine Reviews.* 4(4):363–377.

Washburn, S. L. and DeVore, I. 1961. "The Social Life of Baboons." *Scientific American.* 204:62–71.

Watter, D. N. 1987. "Teaching About Homosexuality: A Review of the Literature." *Journal of Sex Education and Therapy.* 13(2) (fall/winter):63–66.

Wattleton, F. 1990. "Teen-age Pregnancy. The Case for National Action." *The Nation.* (July 24/31):138–141.

Weil, M. 1990. *Sex and Sexuality. From Repression to Expression.* Lanham, MD: University Press of America.

Weiner, A. B. 1987. "Introduction." In *The Sexual Life of Savages.* Bronislaw Malinowski. Boston, MA: Beacon Press, xiii–xix.

Weinrich, J. 1987. *Sexual Landscapes.* New York, NY: Charles Scribner's Sons.

Weiss, M. L., and Mann, A. E. 1990. *Human Biology and Behavior/An Anthropological Perspective.* 5th edition. Glenview, IL: Scott, Foresman.

Welch, D. 1992. "The Birth-Control Shot." *Self.* 48 (November).

Wellness Letter. 1991. "Sex Quiz." Berkeley, CA: UCB:4.

Wells, R.V. 1978. "Family History and Demographic Transitions." In *The Family in Social-Historical Perspective.* N. Gordon (ed). New York, NY: St. Martin's Press: 516–532.

Wertheimer, D.M. 1989. "Victims of Violence: A Rising Tide of Anti-gay Sentiment." In *Annual Editions Human Sexuality.* O. Pocs (ed.). Sluice Dock, Guilford, CT: Dushkin Publishers: 201–203.

Westermarck, E. 1956. "Homosexual Love." In B.W. Corey, ed. *Homosexuality: A Cross-Cultural Approach.* New York, NY: The Julian Press. Pp. 101–138.

———. 1922. *The History of Human Marriage.* 5th edition. Vol. II. New York, NY: Allerton.

Wheeler, G. D., Wall, S. R., Belcastro, A. N., and Cumming, D.C. 1984. "Reduced Serum Testosterone and Prolactin in Male Distance Runners." *Journal of the American Medical Association.* 252(4):514–16.

Whelehan, P. 1987. Review of "Labor More Than Once." *American Anthropologist.* 89(1) (March).

Whelehan, P. and Moynihan, J. 1984. "Survey of Sexual Attitudes and Behaviors of Potsdam College Students." *College Student Personnel Association Journal.* (1).

Whelehan, P. N.D. Fieldnotes.

Whipple, B. and Perry, J.D. 1981. "Pelvic Muscle Strength of Female Ejaculators: Evidence in Support of a New Theory of Orgasm." *The Journal of Sex Research.* 17(1) (February):22–39.

Whitaker, R. 1990. "Intolerance the Rule for Gay Teenagers." *Times Union.* Albany. (May 27):G1–G2.

Whitam, F. (ed). 1980. "The Prehomosexual Male Child in Three Societies: The United States, Guatemala, Brazil." *Archives of Sexual Behavior.* 9:87–99.

White, L. 1948. "The Definition and Prohibition of Incest." *American Anthropologist.* 50:416–435.

"Who Gets Abortions." 1990. *Syracuse Herald American.* October 21, G1.

Wikan, U. 1977. "Man Becomes Woman: Transsexualism in Oman as a Key to Gender Roles." *Man.* 12(2):304–319.

Wilfert, C. M., and McKinney, Jr., R. E. "When Children Harbor HIV." *Scientific American.* 279(1)(July):94–95.

Williams, J. W. 1989. (18th ed.). *Williams' Obstetrics.* E. Norwalk, CT: Appleton and Lang.

Williams, L. 1989. "Inner City Under Siege: Fighting AIDS in Newark." *The New York Times.* (February 6):1, B–8.

Williams, W.L. 1986. *The Spirit and the Flesh.* Boston, MA: Beacon Press.

Wilson, C. 1988. "Uganda: An Open Approach to AIDS." *Africa Report.* 33(6):32–34.

Wilson, E. O. 1978. *On Human Nature.* Cambridge, MA: Harvard University Press.

_____. 1975. *Sociobiology, the New Synthesis.* Cambridge, MA: Harvard University Press.

Wilson, J. D. 1979. "Embryology of the Genital Tract." In *Campbell's Urology.* M. Campbell (ed.). 4th edition. Vol. 2. Philadelphia: Saunders (Chpt. 41):1469–1483.

Winick, C. 1970. *Dictionary of Anthropology.* Totowa, NJ: Littlefield, Adams and Co.

Winkelstein, W. et al. 1987. "The San Francisco Men's Health Study: III. Reduction in Human Immunodeficiency Virus Transmission Among Homosexual/Bisexual Men, 1982–1986." *American Journal of Public Health.* 76(9):685–89.

Winton, M. A. 1989. "Editorial: The Social Construction of the G-Spot and Female Ejaculation." *Journal of Sex Education and Therapy.* 15(3) (fall):151–163.

Wolf, A. P. 1970. "Childhood Association and Sexual Attraction: A Further Test of Westermark Hypothesis." *American Anthropologist.* 72:503–515.

————. 1968. "Adopt a Daughter-in-Law, Marry a Sister: A Chinese Solution to the Problem of the Incest Taboo." *American Anthropologist.* 70:864–874.

Wolf, A. P. 1966. "Childhood Association, Sexual Attraction and the Incest Tabu." *American Anthropologist.* 68:883–898.

Woman. 1990. "Are You Having an Office Affair?" VII(4) (April):56–59.

Woodruff, G. and Sterzin, E. 1988. "The Transagency Approach: A Model for Serving Children with HIV Infection and Their Families." *Children Today.* (May–June):9–14.

World Health Organization (WHO). 1990. Special issue. "Women, Mothers, Children and Global AIDS." Global AIDS Fact File. *GPA Digest.* Geneva, Switzerland: WHO.

World Health Organization (WHO). 1989. *Weekly Epidemiological Record.* No. 31. Geneva, Switzerland.

World Health Organization (WHO). 1988 (continuous 3 years). *World Health Statistics Annual.* Geneva, Switzerland: WHO.

World Health Organization (WHO). 1977. "The Ninth Revision of the International Classification of Diseases." London: HMSO.

World Health Statistics Annual. 1986. World Health Organization. Geneva, Switzerland: WHO: 73–75.

Young, F. W. 1965. *Initiation Rites: A Cross-Cultural Dramatization.* Indianapolis, IN: Bobbs-Merrill.

Zelman, E. C. 1977. "Reproduction, Ritual and Power." *American Ethnologist.* 4(4):714–733.

Zihlman, A. 1989. "Woman the Gatherer: The Role of Women in Early Hominid Evolution." In *Gender and Anthropology: Critical Reviews for Research and Teaching.* Sandra Morgen (ed.) Washington, DC: American Anthropological Association.

Zihlman, A. 1985. "Gathering Stories for Hunting Human Nature. A Review Essay." *Feminist Studies.* 11(2):365–377.

Zihlman, A.H. 1981. "Women as Shapers of the Human Adaptation." In *Woman the Gatherer.* F. Dahlberg (ed.). New Haven, CT: Yale University Press, 75–120.

Zihlman, A.L. 1978. "Women in Evolution, Part II: Subsistence and Social Organization Among Early Hominids." *Signs.* 4(1):4–20.

Zilbergeld, B. 1992. *The New Male Sexuality.* (2nd ed.). Boston, MA: Little, Brown Company.

————. 1978. *Male Sexuality.* Boston, MA: Little, Brown and Company.

Zelnik, M. and Kanter, J.F. 1980. "Sexual Activity, Contraceptive Use, and Pregnancy Among Methopolitan-Area Teenagers 1971–1979." *Family Planning Perspectives.* 12:230–327.

Zimmerman, R. 1989. "AIDS: Social Causes, Patterns, Cures, and Problems." In *Human Sexuality: The Societal and Interpersonal Context.* K. McKinney and S. Sprecher (eds.). Norwood, NJ: Ablex Publishing: 286–321.

Index

A

A-200 pyrinate, 351
abdomen, 89, 123, 168, 170;
  hernias, 115; organs, 120; pain,
  349, 351
abnormality(ies), 8, 131, 136,
  153, 366
aboreal adaptations, 43–44
aborigine, 202; peoples, 207
abortion, 131, 160, 231, 328;
  choice, 155; issue, 28; spontaneous, 131, 328
absence of competition, 36
absolute levels, 155
abstinence, 353, 355, 375;
  models, 227
abstract, 62; argument, 139
abuse(ive), 43, 86, 87, 140, 159,
  195, 200, 344, 368; abusive-
  addictive drug usage, 139
abuser, 194
acceptable practices, 344
accessory organs, 145
acculturation, 111, 381
Acer, Dr., 361, 387
achieved status, 220
acidic, 95, 114
ACLU AIDS Project, xvi
acne, 85, 105

Acquired Immuno-Deficiency
  Syndrome (AIDS). See AIDS,
  HIV
acupuncture, 345, 360, 369
acute, 342, 367; characteristics,
  366; episode, 366; illness, 366;
  illnesses, 335
adapt, adaptation, xiii, xv, 14,
  15, 16, 17, 25, 41, 42, 43, 44,
  51, 62, 64, 106, 122, 178, 179,
  181, 231, 318, 320, 324, 379,
  380, 381; advantage, 61;
  pattern, 327; strategies, 49
Adolescence: An Anthropological
  Inquiry, 11, 219–220
adolescent(s), 3, 68, 89, 111, 150,
  193, 198, 201, 207–231, 338,
  340; boy's sexuality, 221; boys,
  93, 216, 315; concerns, 321;
  experience, 231; female, 222;
  females, 107, 221, 224; girl, 9;
  girls, 221, 222, 225; groups,
  321; males, 221; period, 200;
  permissiveness, 219; pregnancy, 225, 227; pregnancy
  rate, 224, 225; sex, 383; sex
  activity, 224; sexual behavior,
  205, 216–228, 229; sexual
  experiences, 317; sexual
  experimentation, 222; sexual

adolescent(s) *(continued)*
practices, 137; sexuality, 203,
216–228, 231, 344, 381; stage,
207; sterility, 17, 205, 218,
224–225, 232; stress, 9;
subfertility, 224; trends, 220;
unrest, 9; women, 217. *See*
puberty, teenage
Adolescent Family Life Act
(AFLA), 227
*Adolescent Storm and Stress: An
Evaluation of the Mead-
Freeman Controversy*, 383
adopt, 26, 27, 138, 139, 142, 177,
193, 343, 357, 361; children,
326. *See* infertile
adrenal(s), 74; cortices, 148;
glands, 68, 75, 81, 85, 104
adreno genital syndrome (AGS),
148, 149
adult(s), adulthood, 111, 124,
129, 141, 143, 150, 159, 172,
187, 192, 193, 195, 205, 207,
208, 209–211, 214, 216, 218,
219, 226–231, 232, 321–323,
328, 336, 344, 357, 372, 373,
380; adultery, 220; AIDS, 342,
346, 362, 375; Americans, 3,
227; authority figures, 321,
370; caregivers, 48, 52; characteristics, 83; copulation, 203;
female(s), 194, 196; female-
child, 64; female-child bonds,
53; heterosexual relations, 93;
homosexuality, 217; issues,
203; male, man, men, 49, 54,
131, 196, 229; male-child
bonds, 64; male-female, 315;
parents, 325; perspective, 201;
population, 322; relationship,
325; sexual attraction, 197;
sexual contact, 200; sexual
expression, 185; sexual relations, 197; sexual standard,
322; sexuality, xv, xvi, 199,
202, 224, 381; size, 52; status,
226, 230; support, 321; woman,
229; women, 49, 54; world, 209
adult-child, 195, 196, 200, 201,
202; interaction, 200; sex, 196;
sexual contact, 195; sexual
interaction, 201, 202
adult-male child bonds, 54
adultery, 220
aerobic: exercise, 128; fitness,
107
affection, 46, 198, 199, 200, 231
affective, 22, 23, 33; definitions,
32; dimensions, 32, 37, 381;
foundation, 76; states, 5
affiliate(s), 13, 184
affinal, 53, 159; relationships,
185; relatives, 179, 180
Africa, Africans, 44, 63, 83, 111,
138, 139, 327, 333, 335, 337,
341, 363, 367, 370, 372–374,
375, 376, 386; apes, 62;
countries, 373; dark, 373;
females, 225; groups, 374;
societies, 107, 219
age grade(s), 207, 210
age group, 225, 370
age: of AIDS, 95; of majority,
230; of marriage, 217, 228; of
menarche, 224
aggression, aggressive, 35, 59,
61, 86, 148, 371, 383; violent
tendencies, 147
aged, aging, 85, 223; men, 73;
process, 84; sexuality, 381
agriculture, 182; adaptations, 15;
societies, 54, 177; systems, 17
Agta, 49
AIDS (acquired immuno-
deficiency syndrome), xiv, 54,
79, 87, 115, 155, 199, 200, 220,
224, 227, 322, 333, 335–376,
377, 381; belt, 373; cases, 340;
coordinator, 335; crisis, 363;
dementia, 367; dementia
complex, 338–339; diagnosis,
338, 339, 342, 362, 367; drugs,

xvi; education, 223, 344, 370,
372; orphans, 363; statistics,
339; test, 363; treatment, xvi,
373; virus, 352; work, 364,
370. *See* HIV
Ainu, 224
albumen, 92
alcohol, 56, 79, 86, 95, 105, 127,
128, 168, 171, 361, 370; abuse,
95; families 194
alkaline, 92, 95
all boys' schools, 313, 314
all girls' schools, 313, 314
all male clubhouses, 214
allergic reaction(s), 140, 350
Allgeier, E. and Allgeier, A., xiii
alliance(s), 55, 181, 192, 193,
217; theory, 179, 182
alloparenting, 327
Alorese, 202
alternate test sites (ATS), 364–
365
alternative, 129, 138, 180, 195,
354, 360, 372; gender, 32;
genders, 32, 384; life-styles,
326; medicines, 369; relationship styles, 323; view, 315
alyha, 32
Alzheimer's Disease, 367
ambiguous, 60, 199, 231; female,
149; male, 148, 149
ambilineal descent, 183
ambisexual, 312, 329
America(cans), 7, 9, 22, 178, 183,
186, 187, 190, 194, 195, 198,
199, 200, 207, 223, 225, 227,
228, 229, 231, 386, 387;
adolescence, 9, 228; adolescents, 223, 225, 229; adulthood, 231; anthropology, 7;
children, 199; descent system,
183; girls' adolscence, 9;
society, 9, 195, 199; standard,
231; teenager(s), teens, 9, 229,
231; women, 194; youths, 223,
228

American Association of Sex
Educators, Counselors and
Theapists (AASECT), 226
American Foundation for AIDS
Research, xvi
American Psychiatric Association
(APA), 317
American Psychiatric Association,
126
amino acid, 385
Amish, 139, 324
amniocentesis, 129, 131, 380
amnion, 131, 161, 163
amoxicilin, 348
ampicillin, 348
ampulla, 116
amylnitrates (poppers), 361
anal, 349, 351, 356; interaction,
352; penetration, 316; region,
353; sexual intercourse, 215,
354, 351, 355, 357; sphincter,
186; stage of development, 186,
187
analogous, 7, 42, 77, 101, 103,
109, 115, 116, 121, 129, 132,
145, 217, 329; processes, 104;
structures, 65, 70–71
anatomy, 33, 37, 67, 68, 91, 101,
129, 132, 136, 208; changes,
159; development, 69, 145;
differences, 60; external, 79,
109; gross, 250; level, 70;
structure 111; terms, 136
ancestor(s) xvi, 38, 44, 182, 183,
184, 381; females, 57–58;
forms, 58; males, 57–58;
models, 42; primates, 47;
relations, 39. *See* descent,
family, kin
Andaman Islanders, 213
androcentric, 36, 113, 155
androgen(s), 69, 74, 75, 81, 87,
98, 103, 104, 122, 148, 149,
317; therapy, 84
androgen insensitivity (AIS), 31,
149

androgen overdose (AO), 148
androgyny, androgynous, 311,
   330, 331; gender roles, 329
andrological health: men, 90
androsterone, 74
androstriol, 122
anecdotal reporting, 142
anemia, 369
anesthesia, 89
animal(s), 5, 11, 41, 42, 43, 189,
   331, 369; husbandry, 182;
   instincts, 14; protein, 105;
   skins, 384; domesticated, 272
animistim, 120, 143
anonymity, 141, 339, 365
anonymous testing, 364, 376
anorexia nervosa, 107
Antarctica, 335
anterior: lobe, 103; wall, 114
anthropology, 52, 383;
   approach(es), 3, 21, 37, 374;
   concepts, 19, 24; definition, 4;
   explanations, 151; investigation, 374; kinship, 183; literature, 25, 125; model, 151; of
   sex and gender, 4; perspective,
   xiii, xiv, 1, 3–13, 18, 19–37, 41,
   186, 190, 192, 193, 199, 336,
   377, 379; report(s), 181, 187;
   research method, 6; term(s),
   24, 52; theories, 193
anthropologist(s), xiii, 1, 4, 5, 6,
   13, 15, 17, 22, 23, 24, 26, 27,
   31, 42, 47, 151, 178, 182, 186,
   192, 219, 315, 383
anti-female ideology, 125
anti-Freudian, 190
anti-HIV, 368; drug therapies,
   362; drugs, 360, 376. *See* HIV
antibiotics, 94, 114, 361, 366
antibodies, 172, 342, 345, 346,
   363, 364, 365, 367, 376
anticlotting agent, 385
antigen, 145
antilibido hormone, 86
antiviral drugs, 352
anus, 90, 95, 113, 187, 351

ape(s), 10, 42, 46, 52, 190, 353;
   ape-like ancestor, 47
aphrodisiacs, 56, 142
apocrine glands, 56
appendicitis, 366
Arab, 89, 105; societies, 105
Arapesh, 35
arboreal, 42; adaptation, 44, 46,
   56; niche, 43, 64, 384; period,
   51
ARC (AIDS Related Complex),
   367, 387. *See* AIDS, HIV
archaeology, 42
architecture appropriate to
   culture, 15–17
areola, 106
aristocrats, 209
arousal, arouse(d), 4, 34, 57, 89,
   95, 97, 112, 136, 150
arthritis, 366; rheumatoid, 262
artifact, 59
artificial insemination, 50,
   326; AI-D (artificial insemination donor), 129, 132, 140,
   141, 143, 326, 385; children,
   141; donations, 385; donor
   files, 141; donors, 141; AI–H,
   129, 132, 140, 141, 143;
   by husband (AI–H), 129;
   donor, 380; insemination
   donor (AI–D), 129, 140–141;
   means, 31
ascribed status, 220
asexuality, 5
Ashanti, 201
Asia, Asians, 63, 83, 337, 375;
   South, 340; Southeast, 340;
   Southern, 386
assimilation, 381
asymptomatic, 348, 369, 376;
   symptom free, 366; HIV+
   individuals, 376
at-home; HIV test kits, 365;
   pregnancy tests, 124. *See* HIV,
   pregnancy
at risk, 87, 148, 195, 341, 344,
   361, 364, 370, 375

athletes, athletic, 148, 198, 345, 355, 360, 384; in training, 85; situations, 87
Atlanta, Georgia, xvi
atrophy, 94, 145
attitude(s), xiii, 5, 25, 26, 27, 30, 46, 110, 112, 116, 124, 131, 144, 155, 188, 199, 201, 202, 203, 208, 213, 218, 224, 225, 231, 322, 329, 333, 343, 364, 370, 379, 381, 383; change(s), xiv, 347, 375; change, 325, 337, 365; level, 82
attraction, 197, 312, 318, 319
attractive, 70, 107, 344, 380
attribution process, 34, 35
aunt(s), aunting, 138, 167, 169, 327
aural, 372
Australia, 120, 181, 199, 202, 383; aborigines, 202
Australopithecine, xvi, 47–48
Austria, 25
auto-stimulation, 203
autologous, 360
autonomy, 221, 228
average: life expectancy, 325; lifespan, 342; male, 85; marriage, 325; straights, 321
aversion, 193
avoidance rituals, 214; rules, 189. See taboo
avunculocality, 180
Aymara, 224

B

baboon(s), 42–43
baby(ies), 45, 49, 50, 68, 89, 113, 114, 119, 129, 130, 139, 141, 142, 143, 163, 164, 167, 168, 170, 171, 172, 173, 198, 237, 251, 267, 272, 328, 346, 362, 363; boom, 242; bottle feeding, 122; care, 197. See breastfeeding
bachelor, 215–216

bacteria, 119, 349, 350; pneumonias, 367; vaginosis, 348
Baja, 370
Baker, R. R. and Bellis, M. A., 60
baptism, 168, 291
bar/bat mitzvahs, 230
Barale, 28
bardaj, 296
Barker-Benfield, G. J., 93, 137
barren, 288. See infertile
barrier: contraceptives, 122, 258; method, 256, 264. See contraceptives
Bartholin's gland, 71, 95, 113
basal: body, 259, 261; body temperature, 74, 261; body temperature method, 259; metabolic method, 264
bath(ing), 125, 213, 239, 353
Bay Area, 342
beard growth, 69, 82, 83, 84, 86, 121
beauty, 109, 306–307
Becker and Coleman, 194
Becoming a Sexual Person, xiii
behavior: needle, 345; non-verbal, 26, 50, 62, 76, 221, 282, 284, 301, 328; risk free; 353, 364; risk taking, 258, 347, 354, 357, 364; self-destructive, 194
behavioral approach, 32; change(s), 282, 322, 323, 337, 365; definitions, 29, 32–33, 36
behavioral-physiological spectrum, 141
behaviorial-affective situations, 299
behaviorism(ist), 32, 33, 245
belief, xiii, 5, 16, 17, 23, 25, 26, 27, 36, 51, 54, 76, 93, 95, 98, 115, 124, 126, 128, 143, 155, 164, 184, 195, 213, 214, 220, 229, 235, 238, 255, 271, 273, 282, 298, 309, 313, 314, 317, 342, 347, 356, 357, 368, 373; system(s), 12, 131, 137, 151, 285, 302, 330

Belzer, E. G., et al., 247
Benedict, Ruth, 1, 7–8, 18, 286;
  anaylsis, 258
Bengali, 239
Benjamin, Dr. Harry, 152, 283
benzathine, 350
Berdache (two spirit), 31–32, 34,
  147, 150, 283–285, 279, 283–
  285, 287, 288, 296, 297, 298,
  310, 315, 322, 383. *See* two
  spirit
Bergalis, Kimberly, 361, 387
Bernarda, Doña, 167–169
Bernstein, R. J., 26
betwixt and between statuses,
  211, 230. *See Berdache*,
  transgender
Beyene, Y., 306
bilateral: descent, 183; descent
  societies, 54; kindred, 183. *See*
  descent, family, kin
bilineal: descent, 183; descent
  group, 112; descent societies,
  54; descent systems, 316
Billings, 261; method, 259, 264
bilocality, 180
bimodal: age, 325; distribution,
  323
bio-social, 165, 172
biobehavioral, 34, 41, 67, 68,
  104, 121, 129, 143, 170, 301,
  379, 381
biochemical: causes, 150; evidence, 384; explanations, 150;
  regulators, 76; responses, 55;
  sexuality, 72
bio-cultural, 1, 3–4, 153, 157,
  159–173, 177, 236, 251, 302–
  310; architecture, 17; 310;
  beings, 129; context, 24, 233;
  dimensions, xiii, 250; evolution, 14; experience, 18;
  framework, 277; perspective,
  14–18, 37, 279, 310; phenomenon, 53, 207–210, 277;
  system, 48; view, 22
biofeedback, 128

biogenic variables, 152
biological anthropology: approaches, 4, 21, 22, 34, 67,
  133, 285; baseline, 208, 303;
  clock, 136; continuum, 308;
  definitions, 29, 30–32, 33, 36;
  determinants, 30–32, 173;
  explanations, 11, 154, 190,
  298; father(s), 54, 58, 129, 141,
  194; perspective(s), 19, 21–22,
  24; predispositions, 153; sex, 4,
  30, 31, 34; structures, xv, 39,
  41, 64; system, 11; theories, 5,
  133, 154, 155, 316–317;
  triggering, 62
biologist(s), 47, 56
biology, 3–4, 14, 15, 22, 23, 33,
  35, 36–37, 42, 58, 77, 129,
  151, 177, 186, 192, 197, 235,
  266, 271, 284, 289, 310
biomedicine, 135, 139, 160, 166,
  167, 282, 363, 368
bipedal, xv, xvi, 39, 44, 46, 47–
  50, 57, 63, 64, 379, 381
birth, 31, 46, 53, 70, 89, 112,
  113, 130, 138, 140, 142, 145,
  159, 162–172, 173, 186, 187,
  197, 210, 216, 230, 251, 256,
  261, 268, 269, 270, 271, 272,
  275, 282, 286, 345, 346, 361,
  362, 380; attendant(s), 137,
  166, 167, 169; continuum, 157;
  control, xvi, 68, 95, 97, 155,
  160, 172, 224, 225, 227, 233,
  256–277, 305, 316, 381;
  measures, 257–258; methods,
  xvi; options, 73–74; pill, 262;
  practices, 233, 277, 374;
  premature, 328. *See* pregnancy
birthing: alternatives, 166;
  centers, 166; chairs, 163, 171;
  fetus, 113; models, 166–171;
  period, 166; position(s), 163,
  171; practices, 167; process,
  169, 170; rooms, 166, 171
birthrate(s), 256, 257
birthspacing, 256

bisexual, 155, 313, 314, 316,
318–319, 329, 331, 342, 344,
381; behavior(s), 4, 217, 220,
236, 296, 315; community, 387;
life-styles, 318; males, men,
320, 326, 341; orientation(s),
312, 315, 317, 318; phase, 216;
relationships, 320
Black 168, 241, 328, 340, 341;
community, 318
Blackwood, Evelyn, 27, 236, 296
bladder, 92, 94, 95, 110, 145,
186; disorders, 256
blended family(ies), 179, 327–
328, 331. See families
blended orgasm, 246, 247. See
orgasm
blended status, 296
blisters, 351
blood, bleed(ing), 89, 109, 112,
113, 119, 122, 124, 147, 159,
164, 169, 179, 189, 213, 262,
274, 338, 342, 351, 352, 354,
357, 361, 373, 376; brothers/
sisters, 345, 360; cell, 161;
contact, 357–361; flow, 96, 255;
fluids, 353; platelets, 345;
pressure, 161, 170, 255;
products, 336, 345, 347, 360,
374, 375; profile, 286; relatives, 180; sample(s), 74, 364,
365; sharing, 345; stream, 69,
74, 336, 344; tests, 346;
transmission, 360; types, 42;
vessels, 91, 97, 244
blood-borne disease, 347. See
HIV, sex
bloodied cloth, 219. See ritual,
virginity
Blue Collar Marriage, 22
boarder babies, 362. See HIV
Boas, Franz, 7
bodies, 46, 56, 81, 94, 95, 97,
106, 110, 114, 117, 120, 121,
123, 126, 127, 129, 137, 143,
150, 159, 160, 163, 173, 177,
201, 215, 230, 231, 244, 248,
249, 251, 254, 259, 264, 266,
270, 281, 283, 286, 288, 289,
299, 304, 309, 320, 336, 342,
347, 349, 350, 353, 363, 366,
374, 381; attitude, 199, 239,
354; body (man), 92; body
(woman), 105; cavity, 90, 355,
367; contours, 284; fat, 85, 86,
87, 105, 106–107, 122, 172,
208, 225, 272; functions, 69,
73–76, 104; hair, 45, 83, 85,
105, 208, 351; hair loss, 86;
language, 282; mutilation, 86;
odor, 89; piercing, 345, 360,
370; size, 47, 51; weight, 367;
zones, 187
Bogoras, W., 288, 296
Bolin, Anne, 283, 284, 285, 287
bond(ing), 13, 14, 24, 39, 45, 46,
52–55, 64, 67, 83, 125, 126,
164, 173, 178, 196, 200, 213,
216, 279, 300, 301, 310, 320,
325, 331, 345, 353, 379, 381
bone(s): development, 84; disease, 105; growth, 83, 84;
integrity, 104; size, 84
Borneman, 199
Boston Women's Health Collective, 306
Boswell, Holly, 287, 313, 314
Boverman, L., et al., 298
boy(s), 34, 35, 59, 68, 69, 84, 89,
112, 144, 145, 188, 189, 194,
197, 198, 199, 208, 209, 213,
214, 215, 216, 219, 220, 244,
250, 282, 285, 291, 303, 312,
313, 314, 319, 322, 380;
association, 212; babies, 139;
penis, 201; slave, 296
Boyd, R., and Richerson, P. J.,
25
boyfriend, 200, 248, 321
brain, 36, 45, 47, 48, 49, 73, 76,
153, 154, 282, 316, 379, 384;
centers, 22; complexity, 39, 44,
50–52; development, 160;
power, 51; size, 51; tissue, 51

Brandes, Stanley, 308
Brave New World, 131
Brazil, 53, 275, 337
breast, 47, 67–68, 149, 161, 217, 221, 237; bare, 68; cancer, 265, 306; coverage, 68; development, 30–31, 85, 104, 105–106, 107, 148, 149, 208; enlargement, 86, 147; self-exams (BSE), 91; shape, 105, 106, 305; tenderness, 116, 160, 262, 265; tissue, 106
breastfeed(ing), 68, 172, 346, 374; pattern, 169. *See* baby
breastmilk, 172, 336, 347, 375; infection, 361; tranmission, 346. *See* AIDS, baby, HIV
bride(s), 29–30, 112, 113; family, 11, 217; similarity, 217; wealth, 220
Britain, 269; attitudes, 93; views, 187
brother(s), 13, 53, 54, 180, 181, 182, 188, 189, 190, 191, 291, 292, 327, 345, 360
Broude, G. J., and Greene, S. J., 13
Brown, Judith K., and Kerns, Virgina, 308
Browner, C. H., 306
Brunei Malay, 137
bubonic plague, 335
Buddhism, 238
Burton, R. V., and Whiting, J. W. M., 212
Bushmen, 171, 172
Butler, R. W., et al., 306

C

Caesaerean, 346; birth, 361; rate, 166
caffeine, 128, 171
calcium, 171; intake, 105
calendrical method, 259. *See* contraceptives

caloric/calories, 92; intake, 225, 272, 276
Canada, 127, 128
cancer(s), 94, 105, 117, 256, 262, 265, 306, 336, 352, 365, 367, 384; pancreatic, 366; testicular cancer, 84, 90–91
Candida albicans, 348
candidiasis, 336, 348
capitalist ideology, 221
Caplan and Tripp, 253
carbohydrates, 128, 160
caregivers/caretakers, 50, 52, 178, 211, 329, 360, 371
Caribbean, 259, 300, 337, 375
Cartmill, Matt, 384
Cassell, C., 300
caste, 179, 181, 238
castrati, 84–85
castration, 84, 85, 116
catarrhinae, 41
Catholic, 14, 313, 314
Caucasian, 253, 268, 304, 342; American adolescent females, 225; females, 225; middle-class, 223; upper class, 223
Caucasus region, 309–310
caul, 163
Caulfield, M. D., 61
CC + H-P-G axis, 72, 79, 81, 98, 101, 131
CDC National AIDS Clearninghouse, 365
celibacy, 14, 238, 240, 252
Center for Population and Family Health, 169
Centers for Disease Control and Prevention (CDC), xvi, 335, 340, 342, 356, 362, 367, 387
cerebral cortex (CC), 43, 50, 51, 72, 76, 97, 103
ceremony(ies), 13, 121, 205, 210, 211, 212, 213, 216, 219, 230, 232, 288, 301, 309
cervix (cervical), 117–119, 163, 246, 247, 267, 365; abnormalcies, 117; cancer, 117, 265,

367; cap, 260, 264; mucous, 74, 119, 135, 136, 137, 140, 262, 263; normalcy, 117; plug, 119

cetamegolo virus (CMV), 367

Chagnon, Napoleon, 275–276

chalmydia, 117

Chambri, 35

change of life, 305. *See* menopause

Chappel, E., and Coon, C. S., 211

chaste, chastity, 11, 109, 112–113, 238, 371. *See* virginity

child(hood) 3, 23, 30, 33, 45, 46, 48, 49, 50, 53, 54, 58, 64, 112, 113, 119, 120, 138, 139, 141, 142, 143, 144, 152, 150, 159, 166, 167, 170, 171, 173, 207, 208, 209, 210, 211, 212, 216, 218, 219, 224, 225, 228, 229, 230, 231, 232, 242, 251, 265, 269, 271, 272, 273, 274, 275, 291, 299, 303, 308, 309, 320, 322, 324, 326, 328, 339, 340, 346, 361, 362, 363, 371, 372, 373, 380; abuse (sexual), 127, 128, 200, 344; care, 125, 197, 325, 327, 329; development, 175, 187, 197, 199; first, 216; masturbation, 199; molestation, 177; parent, 182, 317, 327; rearing, 59, 172, 258, 327; sexual behavior, xv, 175–203, 226, 317, 344, 381; support, 130, 323

childbirth, xiv, 22, 48, 53, 68, 106, 113, 137, 157, 159, 162–171, 173, 224, 268, 276, 328, 381. *See* birth

chimpanzees, 42–43, 44, 46, 47, 48, 52, 55, 63, 384

China, 105, 192, 193, 236, 259, 266, 269, 270, 380

chlamydia, 260, 261, 349, 350

Chodorow, Nancy, 211–212

chorion, 124, 131

Christianity, 182, 238, 240, 313, 314

chromosome, chromosomal, 42, 50, 317; characteristics, 145, 384; filtration, 129, 130, 132; male, 31; sex, 30, 34, 143, 144–145, 156, 380; variations, 131, 135–136, 147; X combination, 144; Y combination, 144

chronic villi sampling (CVS), 129, 131, 380

Chrysalis: The Journal of Transgressive Gender Identities, 289

Chuckchee, 32, 288, 296

circumcision, 89, 99, 110, 111, 132, 212

clan, 184–185, 191, 327. *See* descent, family, kin

climacteric, 303, 304, 305. *See* change of life, menopause

clinical: approach(es), 151, 254, 298; diagnosis, 128; entity, 127, 128; literature, 151, 298; methods, 243; perspective, 217, 298; pregnancy tests, 124; psychologists, 23; studies, 151, 255; syndromes, 385; theories, 217, 256

clitoral: glans, 111; hood, 71, 109; orgasm, 245; shaft, 71, 111; stimulation, 110, 246, 247; surgery, 110

clitoridectomy, 110, 111, 132, 213

clitoris, 70, 71, 87, 109–110, 110–111, 144, 145, 146, 148, 241, 242, 244, 245, 284

clotrimazole, 348

co-husbands, 182

co-wives, 53

cohabiting couples, 324

Cohen, C. B., 386

Cohen, C. B., and Mascia-Lees, F. F., 61

coitus, 28, 32, 60, 202, 215, 219, 222, 239, 241, 243, 245, 246, 270, 272, 303, 348, 350

coitus interruptus, 259, 270, 277
Collins, 384
colour vision, 46
Coming of Age in Samoa, 8, 9,
  207, 228, 383
communes, 323-324; hunting, 49;
  living, 323
communitas, 238
conceive, 6, 112, 114, 130, 137,
  138, 139, 141, 142, 143, 218,
  237, 257, 287, 288
conception, xiv, 53, 60, 62, 70,
  71, 95, 101, 120–121, 126, 133,
  135, 136–137, 138, 142–150,
  155, 208, 242, 256, 257, 259,
  265, 268, 269, 271, 275, 276,
  283, 304, 316. *See* pregnancy,
  impregnate
condom, 8, 93, 95, 141, 142, 223,
  259, 263, 264, 266, 270, 341,
  355, 356, 357, 371, 372, 374,
  387; prophylactic, 360, 364,
  376. *See* contraceptives
connective tissue, 115
constructionist theory, 151, 236,
  283
contraceptive(s), 74, 121, 198,
  223, 225, 227, 231, 251, 269,
  270, 271, 275, 277; behavior,
  226; effectiveness, 257; foams,
  119; method(s), 257–266;
  morning after pill, 265; pill,
  114, 241, 259, 262, 264, 265,
  270; research, 266; risk taking,
  258; use, 226, 258, 269. *See*
  birth control
Cook Island, 237, 250
Copper Eskimo, 202
Copper T380A, 263
copulins, 56
corpora cavernosa, 89, 95–97
corpus, 117; bodies, 95; luteum,
  123; spongiosum, 88, 89, 94,
  95, 97
cortisone, 148
Costa Rica, 270
Cote, James W., 383

courtship, 16, 221. *See* marriage
cousins: cross-cousin marriage,
  190–191; first, 190; parallel,
  190–191. *See* descent, family,
  kin, marriage
couvade, 166, 170
Cowper's glands, 71, 95, 113
crabs (pubic lice), 351
criminal offense, 202
cross-cultural: anthropological
  approach perspective, xiii, xiv,
  114, 131, 151, 153, 208, 210,
  236, 277, 299, 302–310, 313,
  314, 319, 320, 333, 369–374,
  381; approach, 138, 275;
  childhood sexuality, 203;
  correlation approach method,
  10, 11, 220, 297; documentation, 320; record, 11, 13, 14,
  22, 29, 31, 35, 154, 155, 180,
  203, 207, 236–240, 288, 296,
  303, 305, 309, 316
cross-dress, 150, 283, 287. *See*
  transvestite
cross-gender, 32, 151, 283, 296.
  See transgender
cross-sex identity, 212, 284
cross-species sexuality, 11
Crow, 219
crura, 87, 89, 98, 110
cuerttage, 267
cultural: activities, 46; analysis,
  308; anthorpology, 5, 19;
  approaches, 5; attitudes, 116,
  218; beliefs 5, 85, 99; conditioning, 253; configuration, 8;
  construct, 51, 128, 151, 208,
  236, 285, 289, 308; context, 4, 5,
  6, 18, 22, 24, 29, 213, 231, 233,
  254, 268–277, 297, 298;
  definitions, 4, 110, 120, 346;
  diversity, 27, 374; evolution, 6,
  271; holism, 8; ideals, 253;
  institutions, 151; meaning, 22,
  30, 178, 283–284, 298, 304;
  mores, 195, 302, 372; patterns,
  15, 107, 235, 328; patterning, 1,

24, 129, 256, 328; relativism, 24, 26, 27–28, 33, 85, 195, 198–203, 208, 215, 370, 374, 375, 376, 379, 381; specificity, 76, 283; system, 11, 22, 151, 235, 277; universal, 185, 189, 207
Culture and Human Sexuality, 386
Culture Element Distributions, 296
culture: bar, 307; car, 222; explicit, 25, 250; ideal, 13, 304; sex-positive, 226, 251; subcultures, 281, 370, 324; tacit, 26
culture-bound syndrome(s). xv, 283, 285, 298, 330
culture-free, 330
cunnilingus, 93, 236, 241, 344, 356
curandera, 137
Cutler, Winnifred, 56
cystitis, 110

D

Dahomeans, 201
Dani, 288
Darling, C. J., and Hicks, M., 226
Darwin, C., 5, 24–25
daughter(s), 11, 195, 226, 276, 291, 326, 371; first menstruation, 230
Davenport, William, 15, 270, 309–310
Davis, D. L., and Whitten, R. G., 11, 17
Davis, Dona, 229
day-care centers, 325
dayas, 170
death(s), 8, 73, 81, 82, 85, 87, 90, 104, 122, 136, 182, 210, 251, 268, 275, 305, 315, 325, 342, 350, 361, 362, 367, 368, 374, 376; certificate, 339; instinct, 23; rate(s), 256, 257, 343

deep kissing, 354, 355
deflower, defloration, 113, 202; ceremonies, 219. *See* rites of passage, virginity
degenerative, 104, 105, 115; bone disease, 84; process, 115
Delhi, India, 309
Democratic pro-choice platform, 267. *See* abortion
demography, demographic, 15, 271, 273, 277, 325, 366; patterns, 335, 337; transition, 257; trends, 256
Denny, Dallas, 287–288, 289
depopulation problem, 273, 274
depression, 120, 127, 157, 262, 304, 305
descent, 7, 10, 13, 14, 54, 112, 124, 143, 177, 178–185, 203, 208, 220; cognative, 183; groups, 138, 182, 184, 191; patricentered, 112; patterns, system(s), 138, 175, 183, 316; theory, 179. *See* family, kin groups
Descent of Man and Selection in Relation to Sex, The, 5
Desogestril, 264
developing countries, 217, 225, 256, 268, 270, 277
Devereux, George, 273, 275, 296
deviant, 155, 252, 283, 307, 317; behavior(s), 324, 370; text, 237
Devor, Holly, 152, 153, 284
diabetes, 255, 256, 262, 366
diagnostic: categories, 254; tests, 140; tool, 140
Diagnostic and Statistical Manual (DSM IV), 126–127, 317
diaphragm, 260, 264
diarrhea, 352
dichotomization, 285, 286
diet, 43, 51, 92, 105, 160, 171, 208, 257, 271, 276, 306, 367; factors, 251; fat, 107; habits, 170; restrictions, 213, 368
diethylstilbestrol (DES), 152

digging sticks, 49
dilate(s), dilated, 119, 163, 267, 268
dilation and evacuation (D and E), 268
Dildenafil, 255
dildoes, 344, 354
dioxide, 352
disease(s), 53, 105, 117, 172, 255, 256, 262, 263, 264, 273, 322, 333, 335, 337, 338, 339, 341, 342, 345, 347, 349, 361, 363, 364, 368, 369, 371, 373, 374, 376, 381; aspect, 336; category, 252; model, 151, 155, 254; of behavior, 375
diuretics, 256
diurnal, 46
Divale, W., and Harris, M., 275
diversity, xiv, 35, 57, 150, 151, 175, 193, 201, 218, 236, 253, 281, 285, 310, 318, 323, 330, 370, 381
Division of Sexually Transmitted Diseases, xvi
divorce, 130, 138, 139, 291, 309, 325, 327; rate(s), 193, 322, 323, 326, 380. *See* marriage
dl-norgestrel, 265
Dominican Republic Syndrome, 149
don, 288
donor (AI-D), 129, 130, 140, 141, 326, 360, 380. *See* artificial insemination
doña, 169
dorsal-lithotomy, 113
double descent, 183. *See* descent, family, kin
double standard, 112, 128, 194, 198, 201, 218, 219, 231, 251, 258, 282, 371
douches, douching, 114, 119, 259
doula, 169, 170
Down's Syndrome, 136
Downie, D. C., and Hally, D. J., 297

doxycycline, 349, 350
drug(s), xvi, 68, 97, 139, 165, 166, 256, 258, 270, 321, 343, 361, 363; abuse-addiction, 140, 352; combination anti-HIV therapy, 342–343, 368; intake, 127; recreational, 104, 128, 171; regimen, 86; therapy, 94, 116, 141, 362, 364, 368, 369; usage, 86, 370; use, 160, 360, 374; users, 345
dry humping, 248, 354
dual career couples, 327, 380. *See* marriage
dual gender system, 221
ductless, 69, 74
ducts, 95, 161, 266
dukun, 137
duolocaity, 180
dysmenorrhea, 120
dyspareunia, 120, 254
dysphoria, 150, 151, 152, 153, 254

E

ear, 299; piercing, 345, 360, 370
East Bay, 270
Easter Islanders, 202
eat(ing), 49, 125, 160, 171, 189, 213, 238; disorder, 107
ecological, 43, 182, 277, 324
econiche(s), 42, 43, 380
economic(s), 13, 15, 139, 151, 170, 179, 180, 181, 208, 209, 220, 230, 235, 243, 258, 269, 271, 281, 297, 313, 314, 315, 319, 320, 322, 323, 324, 325, 326, 328, 335, 336, 339, 346, 364, 365, 369, 371, 373, 374, 379
ecosystem, 16
ectasy, 239
educate, education, xiv, 9, 36, 141, 178, 198, 202, 205, 209, 216, 223, 223, 225, 232, 243,

257, 293, 294, 323, 326, 333, 344, 347, 357, 370, 371, 372, 373, 374, 375, 380
effeminate, 152
egalitarian, 36, 196
egg, 67, 71, 103, 116, 119, 121, 124, 136, 140, 142, 136, 140, 142, 168, 260, 264, 265; development, 69, 115; donors, 130; maturation, 74, 75, 122, 123; whites, 92
ego, 183; descent group, 191; mother's brother, 190; mother's sister, 190
Egypt, 169, 170, 271
Ehrenreich, Barbara, 245
Ehrenreich, B., et al., 241, 243
Eicher, 152
ejaculate(s), 59, 60, 71, 73, 84, 91, 92, 93, 94, 95, 97, 98, 99, 109, 112, 115, 129, 130, 135, 136, 137, 140, 141, 142, 214, 215, 222, 244, 247, 248, 253, 254, 255, 259, 261, 264, 304, 315; contact, 93; delayed, 254–255; duct(s), 91, 92; dysfunction, 253; process, 136; sex, 93; tract(s), 91, 92, 93, 94. See male, penis
elderly, 288, 303, 304, 308, 309, 310; age, 307; man, 305; sexuality, 305; woman, 305, 307
Electra complex, 188, 189
electrolysis, 289
ELISA, 364, 367
Ember, C., and Ember, M., 27
embryo, embryologist, 70, 116, 117, 119, 129, 130, 131, 135, 146, 151, 208, 252, 283; development, 124, 137, 144, 145; floating, 130; process, 143; transplants, 68, 129, 130–131, 141, 143, 380
embryo-sperm implantation, 141. See artificial insemination
emic, 24, 28; insiders view, 283
emotion, emotional, 4, 9, 23, 29, 35, 76, 111, 127, 128, 142, 152,

169, 195, 208, 249, 258, 286, 298, 301, 305, 318, 323, 325, 328, 330, 361, 369; bonding, 126, 196, 300, 310, 325; dependency, 197; intimacy, 148; love, 153–154; states, 5
enanthate, 265
enculturation, 26
endemic, 137, 224, 344, 373; warfare, 124
endocrine, 30; functions, 4; glands, 69, 74; hormones, 74; system, 153
endogamy, 12, 179, 181, 324. See descent, family, kin
endogenous, 140; hormone withdrawal, 172
endometrium, 117, 119–120, 121, 123–124, 140, 142; tissue, 120
enema, 171
English, 28, 187, 199
environment(s), xvi, 16, 17, 23, 25, 51, 152, 154, 163, 169, 235, 239, 276, 321, 324, 346, 371, 372, 379; changes, 190; context, 15; experiences, 153; feedback, 45; pollutants, 73; theories, 317
envy, 186, 197
enzyme, 136, 247
Eocene, 43
epicenter, 347
epididymis, 82, 89, 91, 145, 149, 349
epiphyses, 83
episiotomies, 113, 163, 166, 171. See childbirth, pregnancy
erectile: ability, 89, 94, 97; clitoris, 148; control, 386; dysfunction, 244, 253, 255; firmness, 97; problems, 68, 255; tissue, 244
erection(s), 60, 84, 89, 96, 97, 98, 99, 198, 201, 244, 251, 255, 304. See male, penis
erogenous zone, 77, 97, 106, 109. See sex

eroteogenic, 196
erotic, 28, 29, 47, 67, 68, 90,
    141, 192, 217, 236, 237, 252,
    353
erotocentricity, 203; zones, 187,
    196–197
erythromycin, 349, 350
essentialist/constructionist
    debate, 21, 151
estradiol, 74, 75, 122, 123, 265
estrogen, 31, 68, 73, 74, 75, 77,
    82, 84, 98, 103, 104, 107, 108,
    114, 115, 119, 122, 145. 152,
    155, 160, 172, 264, 289, 303,
    305, 317, 342; cream, 306;
    deficiency, 148; levels, 85, 105,
    123, 124; production, 69;
    relative, 105; unbound, 87
estrogen replacement therapy
    (ERT), 114, 306
estrogen-testosterone ratio, 106
estrus, 11, 47, 48, 55–58, 61, 67,
    316; loss, 39, 62, 63, 64
ethics, ethical, 339, 346, 365;
    relativism, 27
ethinyl, 265
ethnic: diversity, 253, 308;
    groups, 4, 139, 229, 230, 255,
    256, 257; identity, 368; issues,
    257, 258
ethnicity, 221, 229, 231, 256, 302,
    306, 324, 325, 338, 367, 374
ethno-theories, 270–271
ethnocentrism, 24, 26, 27, 30, 37,
    68, 203, 236, 254, 312, 373, 381
ethnography(ic), 10, 24, 297, 328;
    approaches, 1; method, 6;
    record, 236; reports, 296;
    spectrum, 207, 209, 235, 271;
    study, 9
ethnological, 24, 296
ethnomethodological, 285
ethos, 7
Ethos, 27
etic, 24, 28
etiology, 150–153, 298
eunuchs, 85

Europe, 6, 82, 89, 163, 189, 227,
    250, 265, 269, 270, 275, 310,
    325, 337, 338; Eastern, 269
eutheria, 41
Evans-Pritchard, E. E., 288, 296
evolution, xvi, 5, 15, 24–25, 37,
    43, 47, 48, 50, 52, 57, 58, 59,
    60, 62, 63, 64, 67, 69, 76, 129,
    190, 271, 276, 301, 314, 379,
    380, 385; adaptations, 16;
    anomaly, 122; distal, 50, 116;
    explanation, xiii, xv, 10, 11, 24,
    39, 41, 58–60, 106, 188, 313,
    314, 379; past, 18, 42, 45, 178;
    proximal, 50; social, 6, 301
*Evolution of Human Sexuality,
    The*, 57, 59
excision, 109, 111
excitement phase, 97, 243, 244,
    245, 253. *See* sex
exercise, 84, 98, 105, 106, 107,
    117, 126, 128, 160, 168
exogamy, 12, 179, 191–192, 193,
    217. *See* descent, family, kin,
    marriage
exotic, xiii, 187, 236, 269, 373;
    natives, 187; others, xiii
expressions, 126, 128, 131, 186,
    203, 219, 229, 283, 284, 298,
    302, 312, 328, 329; expression-
    repression, 253
expressive: roles, 35; orientation,
    307
expulsion, 92, 117, 164, 165, 246,
    263; phase, 98
extinction, 271
extramarital, 12, 13, 179, 193,
    217, 243, 250, 270. *See*
    marriage

F

face to face sex, 48
facial: expression, 299; hair, 82,
    83, 85, 105
failure rates, 259, 260–263

fallopian tube(s), 71, 91, 106,
113, 115, 116–117, 119, 120,
121, 123, 130, 135, 136, 137,
140, 141, 144, 145, 146, 148,
260, 266; blocked, 130; patent,
135; transplants, 143. *See*
female, sex
falsetto, 85. *See* puberty
Faludi, Susan, 387
family(ies) 10, 11, 41, 131, 138,
141, 170, 171, 173, 177–185,
194, 202, 203, 217, 229, 235,
258, 291, 298, 300, 319, 320,
326, 327, 331, 346, 357; audio,
353; dynamics, 7, 23, 24, 195,
197; forms, 175, 178–185;
health, 371, 372; immediate,
12, 186, 190; life, 383; members, 186, 191, 320, 339;
nuclear, 58, 175, 178, 179, 182,
192, 212, 325, 328; of orientation, 159, 166; of origin, 191;
of procreation, 159, 166;
planning, 257; primary, 191;
problems, 195; relations, 226;
relationship variables, 226;
roles, 193; size, 271, 324;
statuses, 190; structure(s), 63,
189, 190, 197, 243; two-income,
325–326
fantasy, 50, 131, 148, 353; audio,
353
Far East, 386
farm, farming, 55, 180, 183, 324
Farrell, W., 300
fat, 107, 109, 172, 208, 224, 225,
271; deposition, 85; hypothesis,
272; levels, 272; ratio(s), 106,
272
father(s) 7, 10–11, 54, 58, 59,
129, 138, 141, 152, 159, 170,
173, 177, 178, 182, 184, 188,
189, 194, 197, 212, 226, 228,
230, 241, 288, 291, 325, 326,
327, 328, 380; incest, 195;
brother, 190; sister, 190
fatherhood, 54, 210, 216

Fausto-Sterling, A., 61
favelas, 53
fecal, 351
fecundity, 62, 273
Federal Drug Administration
(FDA), 264, 265
feed(ing), 36, 155, 172, 215, 237,
353
fellatio, fellate, 92, 93, 98, 137,
214–216, 217, 236, 344, 356;
male-to-male, 150, 315, 373
female, 4, 22, 30, 34, 55, 56, 73,
76, 85, 87, 90, 93, 95, 97, 99,
141, 143, 147, 148, 149, 152,
153, 157, 160, 163, 167, 169,
172, 178, 188, 190, 195, 196,
201, 215, 216, 217, 218, 219,
221, 224, 225, 231, 237, 239,
248, 252, 253, 258, 260, 270,
272, 276, 279, 285, 286, 289,
294, 297, 298, 303, 313, 318,
322, 356, 357, 371, 380, 384,
386; adolescence, 9; adult
reproductive anatomy, 128;
anatomy, 67, 68, 101, 103, 208;
androgen deficiency syndrome,
254; arousal, 110; asexuality,
5; athlete(s), 74, 107; attractiveness, 106, 107; behavior(s),
53, 144, 281, 299, 302, 317,
328, 330; beings, 35; body fat,
132; breast development, 67,
69; castration, 116; centered,
63, 300; ceremonies, 212, 232;
condom, 264; cycle, 74; deity,
37; dysfunctions, 254; economic
contribution, 220; ejaculate,
114, 115, 247; embryos, 145;
evolution, 64; faithfulness, 54;
fertility, 61, 133, 135–136, 140;
fetuses, 130, 269; gender
crossing, 296; gender identity,
31; gender sexuality, 296;
genitalia(s), 31, 48; genital
area, 47; genital surgery, 109;
genotypic, 150; gonadotropins,
115; gonads, 115; heterosexuals,

female *(continued)*
314; hominid pattern, 122;
homosexual (lesbian), 154,
220, 316; hormones, 31, 103,
104, 131; hormone therapy,
284; household(s), 212, 331;
husband, 138; hygiene practices, 105; identity, 212, 294;
infanticide, 275, 277; infertility, 133, 136, 139; initiation
ceremonies, 213; insanity, 110,
116; kin group, 53; lineage, 54,
179, 183, 184; masturbation,
110, 222; nonhuman primates;
43, 59; nurturing, 58; orgasm,
11, 59, 60, 61, 117, 240, 242,
244, 245, 247; partners, 325;
phenotypic sex characteristics,
146; pheremone mixture, 56;
physiology, 101, 103, 128;
pollution avoidance rituals,
214; population, 194, 245;
premarital sexuality, 17;
primary sex characteristics,
75, 101, 104, 132, 282;
prostate, 115; reproduction,
101–131, 132, 145, 242;
role(s), 296, 327; secondary
sex characteristics, 31, 71, 86,
101, 104, 132, 147, 282; sex
cells, 31; sex drive, 246; sex
hormones, 69, 70, 74, 82, 104,
362; sexuality, 29, 54, 57, 59,
62, 76, 101–131, 132, 145,
198, 222, 240, 241–242, 243,
245, 246, 250, 305, 320;
socialization, 211; teen sexual
activity, 223; totems, 120–121,
137; vagina, 56; virginity, 220
Female of the Species, 33
female-child bonds, 53
female-female bonds, 53–54, 55,
64, 320
female-headed, 322, 327. *See*
descent, family, kin group
female-male bonds, 54

female-to-male transsexuals, 84,
152
Feminine Mystique, The, 242
feminist, 28, 59, 61–62, 128, 146,
242
Fernea, E. W., 300
fertile, fertility 17, 47, 57, 60,
61, 62, 70, 89, 112, 114, 120,
132, 133, 135–142, 147, 148,
149, 155, 178, 208, 213, 224–
225, 256–257, 271, 272, 277,
283, 302, 306, 371; capabilities, 36, 62; issues, xiv, 137;
patterns, 138; problem(s), 120,
138, 139, 140, 142; rate, 273;
window, 259. *See* infertile,
pregnancy, sex
fertilization, 58, 60, 71, 116, 119,
123, 124, 130, 136, 140, 141,
142, 143, 265, 380
fetal, 84, 153, 170, 346; choices,
131; development, 145, 151,
159, 160, 163, 173, 361;
heartbeat, 166; maturity, 163;
monitors, 166, 171;
movement(s), 160, 269; reduction, 131, 380; tissues, 267
fetishes, xv, 287
fetus(es), 31, 50, 106, 110, 115,
116, 119, 130, 131, 135, 142,
144, 159, 160, 161, 162, 163,
166, 266, 267, 269, 273, 316,
336, 345, 362, 363; viability, 268
fever, 336, 351, 352, 360
fief, 113
finger(s) 45, 117, 115, 146, 168,
202, 249, 344, 345, 349;
defloration, 202; insertion, 94;
vaginal exam, 163
Finkelhor, D., 195
Fiona, 28
Fisher, 226
fisting, 344, 355
fitness, 290
Flagyl, 348
flirting, 239

flora, 114
Florida, 339
folk: beliefs, 127, 143; community, 305; illness, 127; lore, 16, 93, 137, 250, 270; remedies, 142
follicle, 115, 121, 122, 123
follicular phase(s), 74, 103, 104, 122–123
follicular stimulating hormone (FSH), 73–75, 79, 90, 98, 103, 104, 115, 123, 124; activity, 81; level, 74, 122
food(s), 14, 43, 45, 47, 50, 51, 55, 85, 105, 161, 170, 173, 180, 300, 310, 373; cravings, 127; distribution, 137; gathering strategy, 49, 137, 178, 309; preparation, 125; resources, 106, 272; supply, 124
foragers, 17, 272, 273, 276; tropical, 68, 271
foraging: adaptations, 15; groups, 10, 213; population, 271; strategies, 51
Ford, Clellan S. and Beach, Frank A., 1, 9–10, 11, 13, 18, 45–46, 180, 199, 201, 202, 218–219, 224, 314–315
foreplay, 10, 46, 239, 114, 249
foreskin, 71, 88–89, 109
forest, xvi, 43–44, 276
Forty: The Age and the Symbol, 308
founding ancestor, 183, 184. See descent, family, kin
foundling homes, 275
four genders, 32
four-scheme model, 173, 244
France, 275. 338
Francoeur, Robert, xiii, 209, 227
Frankenstein, 291
fraternity(ies), 53, 55, 345
Frayser, Suzanne, 1, 11–12, 18, 70, 315
Freeman, Derek, 383
French kissing, 354, 355
frenulum, 89

frenum, 89
Freud, Sigmund, 7, 23, 33, 59, 109–110, 120, 144, 146, 175, 186, 187, 188–189, 190, 192, 196, 197, 198, 199, 200, 228, 240, 241, 242, 244, 245, 277, 317
Friday, Nancy, 241
Friedan, Betty, 242–243
friends, 3, 13, 54, 57, 170, 171, 200, 248, 291, 319, 320, 326, 327, 337, 345; platonic, 54
frigidity, 245. See sex
Fromm, Erich, 196
frontal lobe, 74
frustration hypothesis, 228
functionalist approach, 190, 192
fundus, 116, 117
fungus, 348
future: adults, 229; health risk, 268; life, 239

G

G Spot and Other Recent Discoveries About Human Sexuality, The, 247
Gagnon, J. H., 32–33
Gagnon, J. H., et al, 222
Gallup Poll, 266
gamete(s), 67, 121, 141
gamma benzene hexachloride (Kwell), 351
Gardnerella vaginalis, 348
gatherers, 49, 106, 180, 213, 271, 272, 276. See also foragers
gay(s), 154, 155, 312, 323, 331, 338, 341, 344, 347, 385, 386; behavior, 318; communities, 313, 314, 317, 339; disease, 345; female, 32; liberation, 321; men, 32, 317, 326, 342; parents, 326; relationships, 319, 320; youth, 321. See homosexuality, lesbian

Gay Related Immuno Deficiencey
Disease (GRID), 335, 345. *See*
HIV, AIDS
Gbaya, 111
Gebhard, 194
gender, 5, 8, 33, 37, 42, 61, 69,
81, 95, 103, 112, 114, 130, 135,
200, 210, 239, 244, 250, 251,
269, 271, 273, 274, 301, 312,
315, 317, 318, 319, 321, 322,
327, 342, 356, 367, 368, 372,
374; anomalies, 31; behavior,
116, 194; bias, 58, 60, 188,
198, 255, 296, 307; category,
199, 284; choice, 313, 314, 331;
crossing, 288, 297; differences,
23, 197, 211, 307; differentiation, 213; diversity, 285;
dysphoria, 150, 151, 152, 153,
156; identity(ies), 23, 31, 35,
133, 143, 144, 145, 147, 151,
152–153, 156, 279, 281–285,
289, 294, 298, 302, 310, 381;
identity conflict, 287; identity
variance, 152–153; ideo-
logy(ies), 221, 284, 286, 297,
298; inequity, 36, 151, 195,
276; reversal, 297; revolution,
286; role(s), 10, 15, 16, 31, 34–
35, 50, 133, 143, 144, 145,
147, 154, 156, 235, 240, 242,
253, 254, 279, 280, 281, 282,
285, 286, 287–288, 296, 298–
300, 302, 307, 310, 311, 316,
320, 323, 328–331, 381; role
behavior(s), 281, 298, 299, 311,
320, 328, 329, 330, 331; role
differences, 254, 297; role
expectations, 320, 323, 330;
role patterns, 298, 310, 323,
330; role socialization, 145;
role stereotype, 282; role
variation, 35; scheme, 34, 285,
286; segregated, 313, 314;
selection, 101, 129, 131, 380;
sex hormone production, 68;
sex predetermination, 68;

sexual behavior, 316; specific
restrictions, 219; status, 10,
34, 284, 288, 296; stereotypes,
62; stratification, 221; transformed status, 32, 296; variance, 29, 150, 151, 279, 285–
298
gene(s), genetic, 25, 30, 84, 144,
286, 316, 324; act, 61; avoidance, 190; basis, 83, 317;
cleansing, 190; development,
146; factors, 105, 310; females,
152; fitness, 24; influences,
208; male, 150, 296; mixture,
193; population, 310; predispositions, 317; problem, 31;
program, 14; screened, 141;
sex, 133, 143, 144–145, 146,
154, 156; sex characteristics,
147; studies, 42; tendencies,
83; trait, 384; variables, 83,
104; variation, 190; winner, 62
genealogies, 25. *See* descent,
family, kin groups
genital, 145, 175, 188, 221, 222,
222, 243, 297, 344; assessment, 286; contact, 123, 143,
316, 350, 353; differences, 284;
herpes, 351; herpes virus
(HSV–2), 351; manual practice,
203, 243, 351, 353, 355; mutilation, 212; practices, 203; sex
partner, 126; sexual, 301, 316,
348, 356; surgery, 89, 111, 132;
swellings, 60; tubercle, 145;
warts (veneral warts), 352. *See*
sex
genital-anal, 349, 350
genitalia, 28, 35, 48, 50, 57, 68,
90, 144, 146, 177, 188, 194,
197, 199, 202, 208, 216, 221,
250, 251, 284, 285, 286, 349,
351, 353, 355, 356, 370;
external, 93, 97, 108, 146;
female, 95. *See* sex
Genito-Urinary tract anoamalies,
148

genocide, 139
geriatric sexually breakdown
    syndrome, 307
Gertz, Alison, 357
gestalt, 8
gestation, 124. *See* pregnancy
Gibbons, 353
"giving head," 92. *See* sex
glands, 68, 69, 113, 367
glans, 87–88, 89, 94, 95, 97, 109,
    110, 111
glans clitoris, 71, 111. *See* also
    clitoris
glans penis, 71, 97. *See* also
    penis
Glascock and Feinman, 308
Glenn, 291
Godziehen-Shedlin, Michele, 167–
    169
Goldberg, H., 253, 297, 300
Goldberg, H., et al., 247
Goldman, R., and Goldman, J.,
    199
gonad(al) (G), 69, 72, 73–74, 115;
    sex, 30, 133, 144, 154, 156
gonadatropic releasing hormone
    (GnRH), 69, 70, 71, 73–76, 77,
    81, 87, 103, 104, 122
gonadectomy, 149
gonadotropins-androgens, 74
gonococcus, 349
gonorrhea, 117, 260, 261, 273;
    Neisseria, 349. *See* sexually
    transmitted disease
Gordon, S., and Gilgun, J. F.,
    223
gorilla(s), 42, 44, 63, 384
Grafenberg, Ernst (Dr.), 247
Grafenberg spot (G Spot), 114–
    115, 246–249
Granskog, Jane, 383
grasping hand, 39, 44, 45–46, 64,
    384
Great Britain, 127, 128, 130
Great Plains, 324
Greece, 112, 120, 169, 271, 299,
    306

Greeley, 304
Green, R., 152
Greenberg, 313, 314
Gregersen, Edgar, 10, 386
groin, 105
grooming, 44, 45–46, 64, 82
Guests of the Sheik, 300
Guinea, 6
Gussii, 213
Guttmacher, Alan Institute, 223
gynecologic: disorders, 339, 362,
    367; needs, 170
gynecologist, 110, 362, 386
gynecomastia, 85, 86, 147

H

H-P-G axis (hypothalamus-
    pituitary-gonads axis), 65, 69,
    73, 74, 76, 77, 101, 103, 119,
    121, 132, 172; functioning, 97,
    104; hormone release, 81, 121
H-Y antigen, 145, 146, 152, 385
hair, 45, 46, 83–84, 168, 171,
    208, 230; growth, 30, 56;
    patterning, 104
hairline, 105
Haiti, 201, 218
Hall, E., 26
hand(s), xv, 44, 45, 47, 49, 344,
    355
harem, 188
harrassment, 320
Harris, Marvin, 268, 275
Harvard Health Letter, 255
Hawaii royalty, 12
heads-of-households, 326. *See*
    descent, family, kin
healer, healing, 105, 137, 351
health, healthy, 45, 46, 104, 114,
    116, 117, 122, 129, 130, 137,
    139, 141, 157, 159, 160, 162,
    190, 238, 260, 262, 285, 285,
    299, 308, 309, 315, 316, 335,
    336, 343, 344, 354, 357, 363,
    364, 366, 371; care, 169, 172,

health, healthy *(continued)*
341, 342, 360, 361, 362, 366,
368, 372, 373; care providers,
336–337, 346, 360, 361; care
system, 362, 366, 373; ejaculate, 129; immune system, 336;
insurance, 339; practices, 357;
problems, 260–263, 265, 366,
373, 381; risk, 268; vagina, 114
heart, 5, 87; failure, 339, 350,
367; coronary heart disease
(CHD), 85, 106
hemophilia, 337, 339, 345, 360.
See HIV
Henderson, W. J., 84–85
hepatitis A, 351
hepatitis B, 87, 351
herbal remedies, 168, 171, 251,
256, 270, 271, 273, 369
Herdt, Gilbert, 27, 214, 217, 236,
315
heredity, 105, 106
Herman, J., and Hirschman, L.,
194
hermaphrodite, 31
herpes virus, 261, 367; HSV-1
(oral herpes virus), 351; HSV-2
(genital herpes virus), 351;
simplex 2 lesions, 354
heteroerotic, 220
heterosexism, 311, 312, 315, 331,
342
heterosexual, 126, 150, 154, 155,
156, 200, 216, 236, 238, 287,
312, 313, 314, 316, 317, 321–
322, 329, 331, 337, 341;
behavior(s), 4, 217, 220, 222–
223, 296, 315, 371, 385; bias,
321; bonding, 344; contact,
120, 123, 129, 319; couples,
320; genital contact, 95, 109,
143; life-style, 216–217; males,
255; marriage, 315, 323, 326;
nonmarital sexual behavior,
282; orientation(s), 310, 311,
315; parents, 326; penile-
vaginal intercourse, 215, 218,

219, 221, 222, 243; women,
307
Hidden Dimension, The, 26
high school students, 222, 225,
228, 328, 345, 357. *See*
adolescent, teenage
Hijra, 288
Hill, W. W., 296
Hindu, 202, 237, 238
hirsuteness, 31
Hispanics, 268, 328
Hite Report, The, 241
Hite, S., 241, 300
HIV (human immunodeficiency
virus), 335, 338, 340, 353, 370,
373; AZT, 352, 369; children,
346; continuum, 339, 367;
infant(s), 346, 362, 363; infec-
tion/AIDS, xiv, xvi, 54, 79, 87,
89, 95, 98, 220, 227, 258, 333,
335–376 387; infection discrimination, 339; infection positivity,
385; infection protection, 344;
infection rate, 342; mother(s),
346, 361, 362; Pattern I (HIV-1),
335, 337, 338; Pattern 2 (HIV-
2), 335, 337, 338; Pattern 3,
220, 337, 338; status, 172, 361,
364, 365; test, 333, 363, 365;
test antibody sensitive test/kits,
364, 365, 367, 376; test sites,
363; testing, 364; transmission,
92–93, 220, 345–346, 360, 363,
372; virus, 93, 223, 336, 366;
replication process, xvi; status,
342, 343, 346, 362, 365; women,
362
HMS Beagle, 24
Hochschild, A., 325
Hoebel, E. A., 296, 297
Hogopans, 44
holism, 34, 240, 273, 274, 336,
374, 379
homeopathy, 369
hominid/hominine, 47, 55, 57, 60,
122, 129, 178, 301, 353, 379;
ancestors, 44, 56; behavior, 43,

327; brain, 64; category, 384; characteristic, 53, 381; evolution, 39, 42, 45, 49, 51, 106; female genitalia, 48; females, 63; life cycle, 159; sexuality, 41, 48, 67, 377, 380; terrestrial adaptation, 44
hominidae, 41
hominoid, 51, 353; analogies, 42
hominoidea, 41
Homo erectus, 44, 63
Homo sapiens, 63, 256
homoerotic, 215, 216, 217, 220
homologous structures, 65, 70–71, 115, 121, 132
homonucleus, 138
homophobia, 155, 200, 217, 311, 313, 314, 315, 331. *See* homosexual
homosexual, 8, 27, 32, 135, 152, 153–155, 205, 215, 220, 232, 243, 270, 316, 319–320, 329, 331, 381; adolescents, 321; behavior(s), 4, 214, 216, 217, 222, 236, 296, 313, 314, 315, 317, 385; community(ies), 217, 318; couple(s), 320; intercourse, 222; men, 307, 326; orientation(s), 312, 313, 314, 315, 317, 319; parents, 326; prostitution, 297; women, 326; youth, 321
homosocial relations, 214
Hong Kong, 220, 270
Hopi, 202
horiticulturalists, 213
hormone(s)(al), 9, 22, 67, 69, 75, 117, 118, 128, 131, 141, 148, 151, 161, 163, 228, 255, 256, 264, 265, 266, 284, 287, 288, 289, 342, 362; abnormalities, 153; assays, 140; basis of human, 76; changes, 152, 159, 305; contraceptives, 258; differences, 153; fluctuations, 120, 155; functions, 77, 115; imbalances, 140, 153; im-

plants, 263; level(s), 70, 155, 303, 316; male, 31; men patterns, 81; reassignment, 31, 286; release pattern(s), 72, 73, 74, 83, 97, 103, 104, 124, 145, 146, 154, 172; response(s), 152; sex, 30, 34, 133, 156; sex characteristics, 147; studies, 155; system, 70, 77, 129, 153; therapy, 127
hormonal replacement therapy (HRT), 105, 254, 255, 306. *See* hormonal replacement therapy
hormonal-anatomical differentiation process, 144
Horney, Karen, 175, 196–197
horticulture(al), 10, 15, 49–50, 55, 111, 137, 140, 177, 180, 271, 315
hospital(s), 163, 164, 169, 346, 363, 364, 366; births, 166; staff, 362; stays, 171
Hostetler, J., and Huntington, G., 324
hot flashes, 305, 306. *See* menopause
household(s), 169, 170, 179, 181, 182, 212, 322, 325, 327, 331, 380; economy, 271; labor, 180; management, 320, 326. *See* descent, family, kin
housework, 325
how-to-advisors, 327
human(s), 41, 43, 69, 73, 116, 144, 153, 159, 160, 182, 186, 190, 198, 208, 224, 241, 246, 266, 299, 310, 333, 336, 345, 369, 370, 374, 376, 377, 380; arboreal adaptions, 39; behavior, 8, 23, 33, 121, 178, 197; biology, 14, 16, 18; body, 109, 114, 117, 153; brain, 47, 48, 50, 51, 64, 153; child, 52; cooperation, 52; couples, 62; cultures, 46, 49; embryo(s), 146; evolution, 22, 39, 42, 47, 59, 61, 62–63, 178; female(s),

human(s) *(continued)*
57, 59, 60, 62, 64; history, 22,
335, 384; life cycle, 32; males,
96; maturation, 64; nature, 8,
240; orgasm, 248; ovulation, 62;
parenting, 178; primate world,
353; reproduction, 64, 132;
sexuality, xiii, xv, 1, 3, 4, 5–18,
19, 21–22, 23, 24, 26, 28, 29,
30, 32–37, 39, 44, 45, 46, 48,
50, 51, 53, 55, 56, 57, 62, 64,
65, 67, 68, 72, 76, 79, 125, 137,
187, 201, 202, 233, 235–240,
248, 254, 310, 316, 317, 318,
319, 330, 331, 379, 381, 388;
textbooks, 309; species, xiv, 67.
See adolescent, baby, child,
female, infant, male
human chorionic gonadotropin
(HCG), 124
human physiological sexual
response function, 198
human reproductive cycles, 39.
See reproduction
human sexual response (HSR),
17, 32, 33, 37, 70, 97, 224,
233, 235–241, 243–244
Human Relations Area Files
(HRAF), 10, 11, 201, 218, 275
Human Sexual Inadequacy, 245
Human Sexual Response, 243
Hunt, M., 194
hunting, 120, 128, 179, 180, 213,
271, 272, 276, 384; big games,
49; cooperative, 58
husband, 58, 112, 129, 130, 137,
138, 141, 150, 167, 168, 180,
181, 182, 188, 193, 217, 274,
306, 309, 372. *See* descent,
family, kin, marriage
Hutterites, 139, 324
hwami, 32
Hyde, J. S., 266; study, 258
hygiene, 82, 89, 105, 114
hymen, 111–112
Hynie, 152
hypertension, 262

hypertestosterone levels, 86
hypothalamus (H), 69, 72, 73–76,
81, 97, 103, 122, 124, 317
hysterectomy, 120, 132. *See*
female

I

Iatmul, 288
identity(ies), 147, 151, 153, 154,
156, 211, 212, 229, 236, 279,
280, 282, 283, 287, 289, 294,
298, 302, 310, 315, 318, 324,
330, 342, 368, 381
ideology(ies), 16, 36, 59, 195,
221, 222, 237, 269, 284, 286,
297, 298, 304; systems, 16
Ifugao, 202
ill(ness), 114, 122, 140, 252, 274,
303, 333, 335, 336, 337, 343,
351, 361, 362, 363, 368, 369,
374; AIDS, 338–339; chronic,
335, 352, 366, 367
immigration, immigrant(s), 220,
257
immune system, 266, 336, 339,
342, 352, 361, 362, 365, 366,
369, 374. *See* AIDS, HIV
immunity, 173
immunological studies, 42
impotence(cy), 250, 251, 254,
255. *See* infertility
impregnate(ing), 11, 50, 60, 64,
129, 135, 136, 142, 270, 275,
347, 375. *See* pregnancy,
reproduction
in utero, 129, 133, 143, 144, 152,
155, 164, 198, 316, 317, 336,
345, 361, 374, 380
in vitro fertilization (IVF), 50,
68, 129, 130, 132, 141, 143,
380. *See* artificial insemination
*In Her Prime: A New View of
Middle-Aged Women*, 308
*In Search of Eve, Transsexual
Rites of Passage*, 283

Inca emperors, 12
incest, 28, 177, 185, 186, 191,
192, 193, 199, 203, 219, 344;
brother/sister, 189, 194;
inbreeding, 190; mating, 190;
regulations, 190; relationship,
196; statistics, 194; sexual
abuse, 194; sexual experience,
196; stepfather, 194, 195;
survivors, 194–196; taboo(s), 7,
12, 175, 179, 185–196, 224,
250; unions, 190; victims, 194
independent household(s), 179,
180, 285. See descent, families,
kin
India, 8, 131, 202, 237, 238, 259,
269, 288, 340, 386
Indian(s), 147, 167, 170, 181,
202, 275, 297, 322
Indonesia: societies, 202
industry, industrial, 282, 324;
adaptations, 15; capitalism, 36;
countries, 225, 257, 368; peoples,
179; revolution, 187; societies,
125, 180, 225; states, 15
inequality between the sexes, 151
Inez, Gloria, 167–168
infancy, 198, 199, 207, 329
infant, 31, 33, 45, 46, 63, 145,
164, 172, 173, 178, 186, 187,
202, 275, 336, 345, 346;
dependency, 39, 44, 52, 53, 64,
380; feeding, 50; formula, 50;
mortality rates, 380; sexuality,
188; survival, 52, 179. See
baby, breastfeeding
infant-mother bonding, 164–165,
171
infanticide, 53, 130, 269, 271,
272, 274–275, 276
infection, xiv, 93, 114, 173, 220,
227, 258, 260, 263, 274, 333,
336, 337, 338, 339, 340, 344,
347, 348, 349, 350, 352, 354,
356, 357, 360, 362, 363, 364,
365, 366, 367, 368, 375, 376;
rate, 342, 343

infertile, 17, 31, 133, 135,
138–142, 265, 268; couple(s),
138, 139, 140–142, 385;
male, 139; problems, 141,
155. See adopt, artificial
insemination, impotence,
marriage, pelvic inflammatory
disease, sterility
infibulation, 111, 132
inhibited sexual desire (ISD), 3,
252–253. See sex
Inis Beag, 303, 305
initiation(s), 97, 121, 205, 210–
214, 215, 216, 219, 230, 301.
See rites of passage
inseminate, 49, 59, 130, 140,
149, 214, 215, 216, 380
Instruction and Advice for the
Young Bride on the Conduct
and Procedures of the Intimate
and Personal Relationship of
the Marriage State, 29–30
interstitial cell stimulating
hormone (ICSH), 74–76, 79;
activity, 81
interbreed, 30, 42. See descent,
kin
intercourse, 10, 34, 28, 62, 64,
113, 114, 120, 143, 201, 202,
203, 214, 216, 219, 221, 222,
223, 224, 226, 238, 239, 243,
254, 257, 258, 259, 270, 306,
351, 380; ejaculatory-penile
vaginal, 140; first intercourse,
113, 223; intercourse-related
behaviors, 141; interfemoral
intercourse, 236, 354; penile-
anal (P/A), 93, 98, 344, 355;
penile-vaginal intercourse (P-V,
PV-I) 4, 70, 93, 98, 110–111,
112, 129, 140, 142, 150, 218,
264, 316, 344, 354, 355, 356,
385
intergender, 43, 283, 323
International Foundation for
Gender Education (IFGE), 294
intersex, 147, 281, 315

interstitial cell stimulating
hormone (ICSH), 74–76, 79,
81, 103
intimacy, 3, 52, 54, 148, 192,
221, 239, 253, 279, 298, 300–
302, 310, 316, 380. *See*
marriage, sex
intra-gender, 43
intraspecies, 155
intrauterine device(s) (IUD), 258,
259, 263, 265, 266. *See*
contraceptive
introitus, 109, 110–111, 112, 113,
163
in vitro fertilization, 68. *See*
artificial insemination
Ireland, 209, 269; folk community, 303, 305
Isoma fertility ritual, 29
Israel, 192
Ituri forest, 49
ius pramae noctis, 113
IV drugs, 166, 171, 352, 360. *See*
drugs

J

Jackson, Michael, 85
Jacobs, S. E, and Roberts, C.,
32, 34
Jacobs, Sue-Ellen, 383
Japan, 47, 220, 236, 282
jealousy, 53, 197
jergunda, 150, 214, 215
Jewish, 230
joint household, 180
Johnson, M., 244, 304
Jones, Ernest, 7
Joregensen, King and Torrey,
226
*Journal of the Society for
Psychological Anthropology*, 27
Joy of Sex, The, 241
Judeo-Christian, 237
"Just Say No" approach, 227

K

K-Y Jelly, 114
Katchadourian, H. A., 28, 29
Katchadourian, H. A., and
Lunde, D. T., 21–22
Kazak, 202
kegel exercise, 98
Kelly, G., xiii, 203
Kelso, Jack, 62, 63
Kessler, S. J., and McKenna, W.,
34
kibbutz, 192
kidney, 68, 87, 145
kin, 7, 15, 49, 52, 61, 137, 143,
151, 157, 170, 172, 190, 208,
229; classification, 6; diagram,
182; fictive, 53, 54; group(s),
relation(s), 12, 13, 14, 18, 54,
111, 138, 159, 169, 173, 175,
178–185; terminology, 53, 203.
See descent groups, family
kindred, 183. *See* descent, family,
kin
Kinsey, Alfred, 32–33, 222
Kinsey, A., et al., 199, 222
Kinsey Reports, 32
kissing, 220, 222
Klinefelter's Snydrome (XXY),
133, 146, 147, 148, 156. *See*
chromosome
Komarovsky, M., 22
Konker, C., 200
Korea: South, 269, 270
Kottak, C. P., 44
Kroeber, A. L., and Kluckhohn,
C., 25
Kubeo, 202
!Kung, 271–272, 276; women, 309
Kwoma, 201

L

labia, 148; majora, 71, 90, 109,
111; minora, 71, 109, 111, 145

labor contractions, 170. *See* birth
lactate(ing), 50, 53, 70, 75, 76,
    103, 122, 138, 161, 172, 225.
    See breast, female, pregnancy
Laetoli, 48
Lancaster, J. B., 225
laparoscopy, 120
larynx, 84
laser therapy, 94
Late Luteal Phase Disorder
    (LLPD), 126
latency stage, 186, 188, 199, 200
latex barriers, 93. *See* condoms,
    contraceptive
Latin America, 54, 105, 167,
    169–170
legal issues, 31, 86, 112, 126,
    127, 128, 130, 131, 141, 198,
    230, 231
Lepcha, 202
lesbian(s), 10, 27, 32, 129, 148,
    154, 155, 222. *See* gay, homo-
    sexuality
Lesu, 202
Levi-Strauss, C., 192
levirate, 182
Lewis, H. B., 226
leydig cells, 76, 81, 90
liaisons, 13. *See* descent, family,
    kin, marriage
libido, libidinal, 23, 69, 75, 79,
    82, 85, 86, 98, 147, 186, 187,
    197, 228; hormone, 104
life cycle, xiv–xv, 33, 63, 99, 105,
    121, 129, 159, 177, 178, 194,
    199, 207
life-giving attributes, 213
lifespan, 25, 178
life-style, 127–128, 154, 155, 159,
    216, 217
limbic system, 76, 97
liminal, 211, 229, 231
lineage, 12, 124, 179, 181, 184–
    185, 191. *See* descent, family,
    kin groups
Linnaean Society of London, 24

Longer, 52
love, lovemaking, lovers, 9, 55,
    141, 154, 186, 188, 195, 198,
    200, 221, 231
Lubbock, John, 6
lubricant(s), lubricate, 89, 71, 92,
    95, 109, 114. *See* contraceptive
luteal phase, 74, 103, 122, 123–
    124, 127
luteinizing hormone (LH), 73–76,
    79, 81, 90, 98, 103, 104, 115,
    122, 123–124, 155
luteotropic hormone (LTH), 73–
    76, 103, 117, 161; prolactin, 74

M

machismo, 54, 370, 371–372
macho women, 296
macrophages, 336
madonna, 54, 253
Mae Enga, 124, 140
Mahu, 288, 297
maithuna (sexual union), 238,
    239
Malaysia, 220
male(s)/man, 4, 13, 22, 25, 30,
    34, 35, 48, 50, 53, 54, 56, 57,
    61, 63, 64, 71, 77, 101, 105,
    106, 107, 110, 113, 125, 127,
    129, 135, 137, 139, 141, 142,
    143, 146, 148, 150, 153, 157,
    167, 169, 170, 180, 181, 182,
    184, 188, 190, 194, 195, 196,
    201, 202, 210, 215, 216, 217,
    221, 223, 224, 229, 231, 237,
    239, 243, 246, 247, 251, 252,
    253, 258, 259, 260, 266, 269,
    273, 274, 283, 284, 285, 286,
    288, 289, 291, 301, 306, 307,
    308, 309, 318, 319, 321, 322,
    323, 325, 326, 337, 340, 341,
    342, 344, 345, 348, 349, 350,
    356, 362, 363, 366, 368, 371,
    372, 373, 374, 376, 380;

male(s)/man *(continued)*
adolescents, 214; adult reproductive anatomy, 128; age, 90;
aggression, 297; anatomy, 67,
68, 79–99, 145, 208; behavior,
144, 281, 298, 299, 302, 328,
330; *Berdachism*, 32, 296; body
fat, 132; body hair, 230;
ceremonies, 232; child, 49, 189;
circumcision, 89; climacteric,
303, 305; culture, 147, 299,
300; deaths, 275; dominance,
36, 62, 112, 155, 177, 214, 299;
ejaculation, 94, 222; facial
hair, 82; fertility, 133, 136,
140; fetus, 130, 145; gender
role, 31; H-P-G axis, 103, 121;
hairline shape, 83; health, 238;
heterosexual, 314; homosexuality (gay), 154, 220, 313, 314,
316, 317; hormone production,
73, 303; hunting, 58; identity,
315; impotence, 250, 254, 255;
infertility, 133, 139–140;
initiation ceremonies, 212;
internal genitalia, 97; internal
reproductive physiology, 98;
kin, 138; libido, 98; life course,
303; lineage, 112, 124, 177,
179, 183; marriage, 218;
menopause, 304; middle-class
behavior, 298; orgasm, 59, 60,
98, 244; phenotype, 147;
primary sex characteristics, 76,
79, 82, 98, 282; reproductive,
59, 79, 99, 145; role(s), 149,
287, 296, 327; secondary sex
characteristics, 79, 98, 147,
282; sex cells, 31; sex
hormone(s), 69, 70, 74, 81, 82,
86, 87, 104; sexual reproductive anatomy, 76; sexual
response, 79, 97, 99; sexual
structures, 145; sexuality, 5,
58, 60, 62, 86, 93, 198, 222,
241, 242, 245, 250, 255, 320;
solidarity, 55, 212; sterliza-
tion,
92; strength, 315; supremacy
complex, 275, 276; tissues,
152; tonicity, 131; urethra, 94;
voice, 84
male-female 54, 373; bonds, 64;
fertility problems, 140; interaction, 301, 302; penetrative sex,
356; relations, 52, 143, 217,
258, 281, 298, 301, 315; sexual
relationships, 54, 310, 316
male-headed households, 327
male-male bonds, 55, 64
male-male competition, 55
male-to-female, 284, 293; post-
operative transsexual, 294;
pre-operative transsexual, 292;
ratio, 338; transsexuals, 31,
84, 152, 286, 287
Malinowski, Bronislaw, 1, 6–7, 9,
18, 186, 189–190; theory, 190
mammal(s), 41, 42, 45, 47, 51,
52, 311, 315, 330; first child,
216; level, 266; life, 210;
transition, 216; wife, 130
Mangaia, 109, 226, 237, 250,
251, 277, 303, 305
manhood, 82, 89, 214
"manly hearts," 296
Mantra, 239
Manus, 201
Maria Santisima, 168
marianisma complexes, 54
marijuana (grass, pot), 79, 86,
98, 361, 370. *See* drug
Mariposa Foundation Newsletter,
356
Marquesans, 181, 202
marriage, 7, 9, 11, 12–14, 15, 16,
30, 58, 61, 107, 112, 124, 137,
179–182, 185, 188, 189, 193,
197, 198, 203, 209, 210, 213,
214, 220, 228, 230, 231, 236,
237, 240, 241, 243, 250, 256,
269, 273, 288, 290, 291, 292,
296, 307, 315, 322, 323, 326,
328, 353, 380; arranged
marriages, 217; avoidance,

192; bonding, 344; coitus, 270; companionate, 252, couple(s), 17, 180, 253, 257, 304, 324; females, 5; first, 218; forms, 182, 379; in-group, 324; intercourse, 238; late, 217; medieval, 325; men, 216; open, 322; parallel, 190; pattern(s), 180, 309; prohibitions, 191; rate, 380; relationships, 53, 54; rules, 179; sexual behavior, 217, 224; systems, 13, 26, 33, 182; women, 270. See descent, family, kin
"marry out," 192
Marshall, D. S., and Suggs, R. C., 315
Marshall, Donald, 250, 251, 305
Martin, E., 104
Martin, M. K., and Voorhies, B., 1, 10–11, 17, 18, 32, 33, 218, 296
Mascia-Lees, 61
masculine, 34-35, 76, 82, 89, 129, 195, 212, 215, 255, 279, 281, 282, 283, 284, 285, 300, 301, 302, 328, 329, 371. See male
massage(s), 113, 163, 165, 167, 171, 273, 239, 353; prepartum, 170
Masters, William H., and Johnson, Virginia E., 3, 10, 17, 59, 110, 114, 240, 241, 242, 243–245, 246, 247, 277, 304
masturbation, 29, 32, 84, 89, 93, 141, 199, 201, 202, 203, 215, 220, 236, 242, 243, 270, 316, 353; mutual, 220, 236, 316; rates, 222. See childhood, female, sex
mate, mateship, 12, 13, 30, 55, 58, 59, 179, 180, 181, 186, 190, 380; instinct, 14
maternal, 170, 346; behavior, 116; mortality rates, 380; overprotection, 152

maternal-child care, 53
maternal-fetal transmission, 346
maternity suites, 163. See pregnancy
matriarchy, 53, 59, 63, 166. See descent, family, kin
matrilineal, 7, 143, 183; descent, 184, 220; horticulturists, 17; social organization, 17; societies, 10, 54. See descent, family, kin
matrilocal residence, 180, 220
mature, maturation, 52, 71, 115, 121, 160, 163, 187, 189, 202, 207, 208, 209, 212, 215, 222, 225, 231, 245, 267, 272, 303, 308; beauty, 307; egg(s), 74, 75, 90, 103, 104, 116, 122, 123, 130; orgasm, 242; process, 91; sexuality, 307; sperm, 91; woman, 224
Mauritius, 270
Maya, 137, 167, 171, 306
Mbuti, 49
McClintock, M. K., 57
McLennan, John, 6
Mead, Margaret, 1, 7, 8–9, 18, 35, 207, 228, 229, 237, 281, 329
Mead, Margaret and Samoa: The Making and Unmaking of an Anthropological Myth, 383
measles, 336, 368
meat, 37, 238, 239
media, 111, 200, 241, 252, 344, 369; mass, 226
medical, 104, 117, 163, 166, 169, 173, 252, 253, 259, 361, 365; anthropology, 4; classification, 128, 254; disclosure and confidentiality laws, 364; explanation, 89, 307; healthcare, 116, 141, 165, 254, 265, 268, 320, 336, 360, 373; perspective, 36, 89, 241, 242, 269, 274; profession, 127, 317; schools, 270; view(s), 29, 270

medication(s), 254, 255, 257, 270, 369

Mediterranean groups, 93

Melanesia, 140, 220, 337; groups, 315

menarche, 17, 121, 208, 213, 216, 218, 224, 272. *See* adolescents, female, teenage

Mendel, Gregor, 25

Mennonites, 324

menopause, 70, 104, 105, 114, 122, 136, 303, 304, 305, 306, 308. *See* female

menstruation, 31, 82, 98, 99, 116, 136, 148, 149, 208, 230, 235, 238, 264, 265, 272, 308, 310, 355, 357; blood, 110, 113, 119, 147, 213, 263, 336, 342, 344, 347, 354; cramps, 101, 104, 117, 126, 132; cycle(s), 55–57, 67, 72, 73, 74, 75, 101, 103, 104, 107, 115, 121–129, 132, 138, 172, 214, 262, 263, 305; first, 121; house, 125, 214; mental illness, 29, 252, 305; monthly cycle, 61; pain(s), 262, 263; periods, 126, 262, 263, 349; Premenstrual Tension Syndrome (PMS), 101, 104, 123, 126–128, 132, 262; phase, 121, 122; problems, 107, 120; synchrony, 57, 101, 104, 126, 132; taboos, 125. *See* adolescent, female, pregnancy, rites of passage, sex, teenage

Messenger, J. C., 303, 305

metronidazole, 348

Mexico, 137, 202

Meyer-Bahlburg, 154

Michael, George, 28

miconazole, 348

Micronesia, 220, 273, 337

mid-life crisis, 304. *See* adult

middle-class, xv, 29, 105, 107, 119, 122, 138, 142, 187, 209, 221, 223, 229, 231, 241, 253, 257, 281, 298, 300, 325, 327, 330, 338, 342, 368, 376

Middle Eastern, 191, 259, 297, 386

Mideast, 300

midwife(ves), 137, 169, 170, 171, 173, 309. *See* birth, pregnancy

Migeon, C. V., et al., 152

milk ducts, 106, 161. *See* breast, female, lactate

Miller, P. Y., and Simon, W., 231

Miocene, 47

Miracle, A. W., and Suggs, D. N., 10

miscarriage(s), 131, 265, 268, 273. *See* female, pregnancy

misogynist, 326

missionary position, 6, 48. *See* sex

mittelschmirtz, 123

Mohave, 32, 296

molestation, 177. *See* incest, rape

mon veneris, 109

Money, John, 55, 148, 151, 200

Money, J., and Ehrhardt, A., 30, 151–152

monkey(s), 10, 42–43, 46, 190

monogamy, 5, 6, 26, 30, 241, 246, 253, 255, 322, 324, 355, 375, 381; marriage, 112; relationship, 344, 353; serial, 246, 322, 353, 381; sex, 252; socieities, 353. *See* descent, family, kin, marriage

mons pubis, 109

Morgan, Louis Henry, 6, 194

Mormons, 324

morning sickness, 160–161, 170. *See* pregnancy

mortality, 225, 256, 362; rates, 380. *See* death

mother(s) 7, 45, 52, 53, 54, 63, 110, 124, 130, 159, 160, 162, 164, 165, 166, 168, 170, 171, 172, 173, 177, 178, 180, 188, 190, 197, 210, 212, 225, 226, 228, 268, 269, 270, 272, 273, 275, 299, 326, 336, 345, 361, 363, 371, 380; antibodies, 346; brother(s), 189, 327; defloration, 202; father, 327;

line, 184; role, 195. See child,
parents.
mother-daughter relationship, 226
mother-infant bonding, 173
mouth, 168, 186, 187, 274, 351,
354, 366
mucous, 114, 124, 136, 137, 140,
119, 262, 263, 264, 336, 356;
substance, 119, 163
Muecke, E. C., 144
Mullerian duct, 145
Mullerian Inhibiting Substance
(MIS), 145, 146
Mundugumor, 35
Munroe, R., 297
Murdock, George Peter, 12–13,
182, 192, 386
Murngin, 201
muscle(s), 4, 48, 74, 83, 87, 98,
107, 110, 113, 117, 147, 149,
160, 244, 247, 284; mass, 85,
87, 104, 106, 132; pectoral,
106, 109; pubococcygeal, 247
Muslim societies, 111, 300
My Secret Garden, 241
Myer, 195
myometrium, 117
myth(s), 137, 143, 146, 184, 186,
188, 221, 245, 369

N

nadle, 147, 322
Nama Hottentot, 203
Nanda, Serena, 288
natal groups, 190
National AIDS Clearinghouse,
xvi
National AIDS Hotline, xvi
National Lesbian And Gay
Heatlh Foundation, xvi
Native American, 230, 285, 288,
384; Berdache, 285, 288, 297;
two-spirit, 383
natural selection, 24, 48, 62, 190
nature/nurture, 151, 329

Nauruans, 219
Navaho, 147, 296, 322. See
Native Americans
Nayar, 181
Ndembu, 29
needle(s), 345, 357, 363; sharers
(IDU), 337, 338, 340, 366;
shares, 336; sharing, 352,
357–361, 370, 372; use, 338,
360, 375; use behaviors, 360;
use contact, 338; use risk
reduction behaviors, 363; 336,
338, 366; using, 342; using
behavior, 342
Nelson, J., 196
neocortex, neural, 64, 96, 97, 160
neonatal, 106, 171, 362; AIDS,
346, 361. See AIDS, birth, child
neotony, 48, 52
Nepal, 181
Netherlands, 269
neurophysiology, 23
New Guinea, 93, 137, 140, 147,
201, 213–216, 288, 315–316
New Woman Sex Report, 223
Newcomer, S. F., and Udry, J. R.,
222
newlywed, 216, 217. See marriage
Nigeria, 183
night sweats, 352, 367
nipple, 106
noble savage, 197
non-unilineald, 183. See descent
nongonococccal urethritis (NGU),
350
nonheterosexual, 311, 322. See
bisexual, gay, homosexual,
lesbian
nonhuman: 47, 49, 53, 55, 178,
327, 379; ancestors, 57
nonhuman primates, xiv, 41, 52,
178, 379; primate ancestors,
57; primate behavior, 64;
primate evolution, 39; primate
models, 43; primate record, 59;
primate sex, 56; primate
world, 353

nonmarital, nonmarried, 218, 282, 322, 344. *See* descent, family, kin, married
nonmedical support services, 362
nonobstertric: healthcare roles, 169
nonoperative transsexualism, 286. *See* transsexual
nonpregnant, 118, 119, 165; state, 172
nonreproductive: aspects, 313, 314; sexual behaviors, 15
nonsexual, 54, 347; relationship, 196
nonstigmatized life, 286
Norgestimate, 264
Norplant, 263, 265. *See* contraceptive
North America(n), 181, 198, 232, 284, 304, 309, 310, 337, 338, 340; socialization practices, 199
North Piegan, 296
"not man," 150. *See* Berdache, two spirit
Noyes, Robert, 95
Nuer, 138, 182, 288
nutrition(al), 47, 83, 84, 104, 122, 128, 161, 172, 208; omnivorous, 43, 51. *See* carbohydrate, fat
nystantin, 348

O

O'Connor, Dagmar, 200
obstetric, 170, 386. *See* female
Oceania, 386
Oedipal phase/complex, 7, 186, 188–189, 190, 197, 198, 242
Offir, C. W., 56
Ohno, S., 152
old age, xiv, 84, 94, 193, 200, 202, 214, 219, 244, 303, 307, 309, 310, 315, 323, 325, 372
olfactory system, 55, 56, 126. *See* smell

Olsen, 324
Omani, 288, 297
Oneida, 95, 182
oophrectomy, 116, 254
opportunistic infections (OIs), 333, 336, 339, 352, 357, 364, 365, 367, 368, 369, 374
oral, 150, 203, 222, 236, 237, 255, 344, 349, 353, 354, 355, 356; acyclovir (Zovirax), 351; contraceptive pill (OCP), 264–266, 270; herpes virus (HSV-1), 351; HIV test, 365; insemination, 214; stage, 186, 187; thrush, 367
oral genital/sexual practices, 92, 93, 95, 98–99, 137, 150, 203, 236, 316, 344; contact, 349, 350; interaction, 352; intercourse, 351; practices, 222; sex, 355
oral-anal; activities, 356; sexual contact, 351
orangutan, 44
orchidectomy, 84, 94
orgasm, 10, 32, 39, 57–60, 62, 64, 67, 84, 93, 97, 98, 99, 112, 117, 119, 126, 146, 198, 221, 222, 237, 240, 242, 243, 245, 248, 249, 250, 323; dysfunction, 115, 253, 254; function, 114; plateau, 243, 244, 245, 366; platform, 244, 246, 247; pre-, 254; responses, 59, 60, 94, 110, 114, 254
Origin of Species, The, 24
os, 117, 119, 121, 136, 137, 163
osteoporosis, 75, 84, 105
Oswalt, W., 207, 308
Otomi, 167
ovarian, ova, 5, 69, 71, 73, 75, 90, 106, 107, 113, 115–116, 120, 121, 122, 123, 130, 144, 146, 254, 305; cancer, 262; cells, follicles, 74, 148; cysts, 262; hormones, 103, 104, 148; sacs, 74

Ovral, 265
ovulate, ovulation, 47, 55, 58, 60,
  63, 64, 75, 116, 119, 121, 123,
  129, 135, 140, 142, 172, 224,
  261, 262, 263, 264, 265, 272;
  concealed, 57; irregularities,
  107, 136, 141; patterns, 123;
  phase, 122. See female,
  menustration, pregnancy
oxytocin, 163

P

pairbonding, 246. See marriage
Panchattattva, 239
PAP smears, 117, 366. See
  female, menustration, sex
Papua, 214
parametrium, 117
paraphilias, xv
parent(s), parenting 3, 9, 12, 17,
  45, 46, 52, 60, 129, 130, 131,
  142, 159, 166, 172, 175, 177,
  179, 180, 183, 186, 191, 193,
  194, 197, 198, 199, 200, 202,
  203, 208, 209, 224, 226, 228,
  230, 257, 276, 291, 320, 321,
  322, 380; roles, 178; single,
  258, 322, 325, 327; styles,
  325–328, 331. See child,
  mother, father
participant observation, 6, 26,
  297
parturient, 160, 166, 169
passage ceremony, 210. See
  adolescent, initiation, rites of
  passage, teenage
passion, 141, 240
passive-aggressiveness, 298
pastoralists, 180
paternity, 54, 58, 112, 130, 141,
  177, 178. See child, parent
pathogen(s), 336, 346, 363
pathology, 8, 23, 367; states, 162,
  283
patrescence, 53, 166

patriarchal, 6, 7, 36; culture,
  304; privilege, 195; societies,
  190; system, 195. See descent,
  family, kin
patrilineal, 10, 124, 138; descent,
  112, 183–184, 316; households,
  212; kinship system, 143;
  organizations, 17; socieities,
  54, 215. See descent, family,
  kin
patrilocal residence, 180
Patterning of Human Sexuality,
  The, 14
Patterns of Culture, 7
Patterns of Sexual Behavior, 9–
  10, 13, 46
Patton, C., 373
Paul, J., 319
peasant(s), 105, 112, 306, 309,
  310
pediatric AIDS, 342, 346, 362,
  375
pediatrician, 175, 327
peer(s), 9, 52, 125, 200, 370;
  groups 198, 226, 258, 321, 371;
  ritual, 111
pelvis, 48, 91, 123, 244, 262,
  263, 379; bone, 117; exam(s),
  115, 116, 117, 119, 361;
  hernias, 115
pelvic inflammatory disease
  (PID), 117, 140, 262, 263, 265,
  349, 367. See infertility, sex
penicillin, 350
penile, 137, 140, 198, 242, 380;
  implants, 68, 255; injections,
  255; shaft, 71; skin, 71, 88;
  stimulation, 351. See male,
  penis
penis, 69, 70, 71, 87–89, 93, 94,
  97, 98, 109, 110, 114, 144, 145,
  146, 148, 149, 201, 236, 243,
  244, 246, 250, 251, 255, 264,
  284, 304, 349, 350, 354; bone,
  96; cervix contact, 246; envy,
  197; external, 82, 87; size, 89,
  221, 237; sleeping, 251

People with AIDS (PWAs), 336, 362, 366, 368, 369, 371. *See* AIDS

perimenopause, 305. *See* female, menopause, menustration

perinatal: AIDS, 346, 361–362, 375; care, 170; infection, 361; transmission, 361–363, 374. *See* birth, child, pregnancy

perineum, perineal, 90, 149, 171, 356; stimulation, 94; tearing, 113, 163. *See* birth, childbirth, pregnancy

Perper, Timothy, 47

Perry, J. D., and Whipple, B., 247

personality(ies), 4, 7, 8, 186; disorders, 317

Peru, 300

petting, 4, 32, 220, 222, 228. *See* intercourse

Pfafflin, 152

Pfizer, 255. *See* Viagra

pH, 92; balance, 114, 136–137, 140; incompatibility, 141; levels, 114

phallic stage, 186, 188

Phelan, 386

phenotype, phenotypic: expression, 143; female, 150; male, 146, 150, 299; matching, 141; sex, 133, 144, 147, 154, 156, 282

phenylethylamine, 385

pheremone(s), 55–56, 60, 73, 77, 126

Phillipines, 49, 220

phratries, 185

Phthirus pubis, 351

physical: ability (boy), 82; aggression, 128, 147, 371; anthorpology, 19, 178; body, 177; defects, 289; development, 4, 33, 45, 69, 146, 161, 208, 315; evolution, 129; female, 31, 140, 146, 284, 286; maturation, 76; strength, 85, 215

physician(s), xvi, 36, 115, 126, 127, 250, 258, 267, 270, 365

physiology: of puberty, 208; of sexual arousal, 32

Pima, 296

pituitary, 69, 72, 73–76, 81, 122; gland, 76, 90; hormone(s), 74, 76, 103, 115, 161, 163; LH, 123

placenta, 124, 161, 164–165, 168, 169, 361. *See* childbirth, pregnancy

Plains Indians, 147, 296, 322

Planned Parenthood Federation of America, xvi

pleasure principle, 198

Pneumocistis carinii pneumonia (PCP), 336, 339, 357, 366, 367

pneumonia, 365

podophyllin, 352

political: alliances, 55, 137; biology, 36; conservatism, 251; economy/economic interpretaiton, 36, 273; organization(s), 15, 235; system, xiv, 235, 366

polyandry, 181

polygamy, 180–181. *See* marriage

polygyny, 53, 124, 126, 138, 180–182, 212, 276, 324, 373; sororal, 181

Polynesia, 89, 226

Ponapeans, 202

population: control, 256, 277; decline, 274; densities, 218; growth, 256, 257, 271; pressure(s), 138, 139, 276, 381

post-reproductive, 178. *See* female, sex

postnatal: care, 167, 170; development, 156; hormonal evidence, 155; hormonal patterns, 154; life, 160; phenotypic expression, 143

postpartum: care, 170; depression, 157, 172–173; infant-mother contact, 171; massage, 170; period, 157, 166, 169,

172–173. *See* female, pregnancy, sex, reproduction
pregnancy(ies) xiv, 17, 49, 50, 53,
59, 68, 70, 107, 118, 119, 121,
122, 123, 131, 138, 142, 157,
159, 160–161, 162, 163, 165,
166, 167, 169, 170, 171, 172,
173, 182, 223, 227, 256, 257,
258, 259, 264, 265, 266, 267,
268, 269, 271, 272, 274, 275,
306, 323, 344, 346, 347, 353,
354, 361, 362, 363, 375, 380,
381; complication rate, 136,
328; delayed, 120; ectopic,
116–117; first, 325; hormone,
124; multiple fetus, 131; rate,
224, 225, 282, 380. *See*
childbirth, female, prenatal,
postpartum, sex
prejaculatory fluid, 95
premarital, 179, 231, 251–252;
abstinence, 227; restrictions,
11, 17, 218, sexual behavior, 6,
10, 13, 17, 18, 217, 218, 219,
220, 224, 226, 241, 243, 250,
252; sexual permissiveness, 17,
218, 220. *See* contraceptives,
intercourse, marriage, taboos
prenatal, 120, 150, 151, 152,
153, 156, 316, 317; care, 170;
development, 145; differentiation, 143, 148; hormone, 151–
152, 154, 155; pelvic exam,
119; theories, 133, 154. *See*
birth, child, pregnancy
prepubescent, 93; age, 214;
boy(s), 60, 84, 93; girl(s), 107,
218. *See* adolescent, child,
teenage
prepuce, 71, 109, 110–111
preputial glands, 89
Preti, George, 56
primate(a), 10, 24, 44, 47, 49, 56,
178, 190, 301, 330, 331, 353,
379, 384; adaptation, 51; analogies, 43; ancestors, 42, 43, 46;
arboreal adaptation, 45, 64;

behavior, 43; bonding, 53;
dependency, 52; evolution, xiii,
xiv, 41, 48, 53, 55, 67, 83, 311,
315; grooming, 45; group, 42, 52
pro-choice, 267. *See* abortion
procreation, 159, 166, 227, 238,
240, 305, 315. *See* birth,
fertility, pregnancy
progestasert, 263
progesterone(s), 31, 68, 69, 73,
74, 75, 76, 77, 82, 85, 86, 98,
103, 104, 107, 108, 115, 116,
117, 119, 123–124, 127, 160,
172, 264, 265, 303, 305, 306
progestin, 86, 148, 262, 264, 265,
306
promiscuity, 6, 7, 59, 215. *See*
marriage, premarital
prosimians, 43
prostaglandin, 117, 126, 268, 385
prostate, 99, 144, 145, 149, 247;
cancer(s), 84, 94; enlargement,
94; fluid, 115; gland, 90; PSA
blood test, 94
prostitute(s), 92, 297, 344, 371,
373
protease inhibitors, xvi, 369. *See*
HIV
psycho-sexual maturation, 187
psychoanalytic perspective/
theories, 7, 21, 33, 154, 175,
186, 197, 203, 212, 255, 316,
317
psychological: adjustment, 266;
anthropologists, 23; development, 267; disorder, 147, 252;
factors, 11, 152, 195, 298;
femininity, 279, 300; masculinity, 279, 300; perspective(s), 19,
21, 23–24, 285; problems, 194,
266, 321, 366; responses, 335;
social learning theories, 23;
therapy, 255
psychology, 23, 24, 37, 250, 292
psychosexual, 23; development,
186, 188, 242, 315, 317, 319;
orientation, 149

psychosocial, 5, 173; maturation, 209; qualities, 34
psychosomatic, 255
psychotherapy, 195, 255
puberty, 9, 31, 70, 73, 76, 77, 81, 82, 84, 85, 97, 90, 104, 106, 107, 122, 129, 136, 143, 149, 186, 188, 189, 200, 203, 205, 207–214, 219, 221, 224, 228, 230, 232, 244, 276, 383; age groups, 217; boys, 111, 220; girls, 111, 220; physiology, 208; rites, 213; rituals, 210–214; sexual arousal, 232; sexuality, 217, 218. *See* adolescent, teens
pubic hair, 69, 71, 83, 105, 107, 109, 171, 208; men's, 84; lice, 351
pubis, 91, 105, 109, 114, 161
Pukapukans, 202, 203
Purdah, 300
"put down," 212

Q

quickening, 160, 269

R

racism, 373
Raiders of the Lost Ark, 3
ramage, 183
Ramey, E., 73
rape, 28, 86; societies, 36–37
rectum, 94, 344, 355
Reiss, I., 231
relationship(s): consanguineal, 179, 180, 181, 185, 191; sexual relationships, 230, 354
religion(s), 15, 53, 186, 237–239, 271, 294, 313, 314; ecstasy, 239; experience, 238; guardian spirits, 120; life, 36; power, 238; sect, 95; system, 208, 238, 316; view(s), 240, 255

remarriage, 138, 182, 322. *See* marriage
remedies, 171, 369
Renshaw, Domeena, 46
reproduction 11, 21, 22, 25, 29, 33, 34, 39, 44, 52, 93, 106, 107, 129, 142, 143, 145, 154, 159, 163, 179, 207, 208, 213, 226, 227, 235, 241, 242, 285, 305, 310, 344, 385; auxiliary structures, 82; age, 63, 106; anatomy, 67, 70; behavior(s), 15, 28, 30, 55, 56, 61, 62, 67, 76, 82, 115, 121, 122, 127, 131; biology, 32; choices, 131; cycle(s), 65, 70, 77, 81; development, 69; efficiency, 63; embryologist, 130; fertility, 178; functions, 67, 68, 69, 70, 73–76, 82, 87, 115, 116, 117, 121, 125; hormones, 101; male sex characterstics, 84; maturity, 224; organ(s), 36, 114, 120, 122; problems, 274; strategies, 57, 61; structures, 70, 77, 97, 132; success, 49, 50, 53, 58, 59, 60, 62, 64, 90, 112, 126, 316, 319, 380; technology, 50, 129–131, 270, 331, 377; women, 49, 221
research method(s): methodological 6, 10, 11, 26, 33, 61, 130, 242, 243, 256, 257, 258, 259, 260, 264, 265, 266, 267, 268, 269, 270, 271, 272, 273, 274, 275, 276, 277, 297, 303, 306, 309
residence rules, 175, 177, 180, 182, 203, 220, 271
restrictive societies, 201, 218, 219
rete testes, 145
retrovirus, 335. *See* AIDS, HIV
rhythm methods, 259. *See* contraceptive
Riddle, J. M., 271
Rienzo and Marron, 306
Riportella-Muller, R., 303

risk, 89, 105, 252, 258, 262,
265, 268, 333, 342, 346, 347,
353, 354, 355, 356, 357, 360,
361, 362, 363, 365, 368, 370,
373, 375, 376; of infection,
93; reduction behaviors, 335,
343, 345, 369. *See* HIV,
premarital
Rite of the Five Essences, 239
rite(s), 205, 210, 211, 228, 229,
230
rites of passage, 111, 121, 205,
210–216, 226, 228, 229, 230,
232, 284. *See* female, initiation, male, taboos
ritual(s), 16, 29, 93, 109, 111,
113, 120, 121, 137, 170, 189,
202, 203, 205, 210, 213, 214,
232, 239, 315, 320, 345, 370;
homosexual initiation, 27, 215;
marriage, 181; occasions, 26,
138, 288; of transition, 231;
ordeals, 230; place, 112;
pollution, 238; pollution
avoidance, 214. *See* avoidance,
female, homosexuality, peer,
puberty, sex, virginity
*Ritualized Homosexuality in
Melanesia*, 315
Roe v. Wade, 267, 270. *See*
abortion
Romania, 269
Romans, 271
Ross, E., and Rapp, R., 36
RU 486, 265. *See* contraceptives
Rubin, L. B., 22
rural, 208, 370, 373
rural-urban, 337
Ruth, Dr., 3

S

sacred, 12, 214, 237, 238, 239,
240, 297, 309. *See* religion
Sado-masochsim (S/M) activities,
342. *See* sex

safer sex practices, 93, 335, 343,
344, 347, 353–357, 370, 371, 375
saliva, 351, 354
Sambia, 27, 93, 137, 140, 147,
150, 154, 205, 214–217, 232,
315–316; heterosexual behaviors, 217; homosexual behaviors, 217; initiation ceremony,
216; male(s), 138, 214, 321;
sexual beliefs, 137–138;
women, 215
*Sambia: Ritual and Gender in
New Guinea, The*, 214
Samoa, 9, 202, 228, 237
San Cristobal Huixchochitlan, 167
sanction(s), 201, 219, 252, 269,
273, 305, 309, 310
Sanday, 36
Sarvis, B., and Rodmon, H., 273
Saudi Arabia, 292
Scheper-Hughes, N., 53
Schlegel, A. and Barry, H., 11,
213, 218, 219–221
Schneider, David M., 273
scrotum, 109, 284; sac, 71, 82,
89–90, 91, 141, 145, 160; skin,
349
sebaceous gland, 85
Second Shift, The, 325
secrete(d), secretions, 85, 87, 89,
90, 95, 123, 124, 136, 148,
160, 351
sects, 238
sedentism, 107, 225, 271, 272, 276
self-esteem, 23, 194, 196, 321,
347, 354, 357, 371
self-pleasuring, 188, 203; masturbation, 28. *See* masturbation,
sex
semen, 29, 73, 84, 92, 94, 95,
114, 130, 135, 136, 137, 147,
150, 214, 215, 216, 217, 238,
240, 270, 304, 315, 336, 344,
347, 351, 352, 354, 356, 374,
375; analysis, 140; fluids, 93,
247, 353. *See* male, penis,
reproduction, sex

semi-restrictive societies, 201, 219

seminal vesicles, 82, 91, 92, 145, 149

seminiferous tubules, 74, 81, 90, 91

sensuous, 97, 357; behaviors, 141, 222, 347, 353; dance, 250; feeding, 353; sensuous-sexual alternative, 354

Sensuous Woman, The, 241

serfs, 113

seroconversion, 342, 346, 362; rate, 343

seronegativity, 364

sex, xiii, 3–5, 6, 12–14, 15, 28, 34–35, 36–37, 41, 45, 50, 57, 58, 61, 62, 125, 140, 146, 150, 152, 159, 163, 179, 181, 186, 188, 189, 191, 199, 201, 208, 212, 215, 217, 219, 221, 222, 224, 225, 228, 231, 235, 236, 241, 244, 249, 251, 252, 258, 259, 261, 265, 268, 288, 296, 301, 303, 305, 310, 324, 335, 341, 356, 366, 370, 371, 375, 377; abuse, 194, 195, 344; abusive father, 195; addiction(s), 3, 252, 254; adult(s), 82, 107, 121, 143, 196, 207; aging, 279, 281; anatomy, 70; arousal, xv, 4, 11, 35, 46, 63, 86, 91, 94, 96, 109, 114, 119, 136, 240, 252, 254; attraction, 64, 154, 197, 248, 318, 319; aversion, 193; behavior (acceptable), 380; behavior (risky), 93, 115, 342, 356–357, 376; beliefs, 137; choice(s), 131, 155; coital, 61, 239; compulsion, 287; conservatism, 200, 251; contact, xiv, 195, 196, 198, 200, 203, 316, 348, 349, 351, 352, 357, 372; contraceptive behavior, 226; counseling, 344; crimes, 86; curiosity, 199; cycle, 22, 65, 70, 77; cycle

hormone, 81; desire(s), 33, 192, 231, 240, 251, 254; development, 69, 199, 242; dichotomization, 285; differences, 245; differentiation, 129, 133, 135, 143, 144, 145, 155, 160, 317; dreams, 32; drive, 23, 82, 84, 85, 86, 104, 246; dysfunction(s), 5, 23, 24, 194, 233, 245, 250, 252, 253, 254, 256, 277, 317; education, 17, 201, 202, 205, 216, 218, 223, 225–227, 232, 237, 257; exclusiveness, 54, 324; experimentation, 175, 200, 201, 224; fantasy, 29; folklore, 270; gender-system, 33; gratification, 13; hormones, 55, 69, 77, 83, 104, 151, 362, 384; hypoactive sex desire, 252; identity(ies), 199, 315, 330; ideologies, 12; innocence, 198; insatible woman, 59; intercourse, 202, 216, 223; latency, 199; lives, 84; love, 238, 312; manuals, 237; maturation, 52, 207; morphological, 34, 151; negative aspects, 199, 200, 201, 251, 252; of assignment, 31; of internal reproductive structures, 30; offender, 86; orientation, 70, 89, 133, 135, 153, 154, 155, 156, 216, 289, 311, 312, 313–331, 345, 374, 381, 387; organ(s), 22, 120, 199, 250, 307; partner(s), 13, 46, 54, 112, 126, 142, 196, 200, 223, 309, 313, 314, 337, 338, 353, 373; permissiveness, 10, 220; play, 199; pleasure, 34, 71, 84, 109, 115, 187, 199, 242, 347; positions, 10, 48; primary characteristics, 69, 75, 76, 85, 86, 87–98, 107, 108–121, 147; promiscuity, 7; ratio, 130, 131, 307, 380; receptivity, 48, 55, 64, 126; recreational model,

253; repressive cultural atmosphere, 29; repressive sexual attitudes, 228, 305; research, 22, 26, 33, 254, 304, 309; researchers, 13, 380; response, 14, 23, 35, 56, 68, 70, 97, 98, 104, 110, 115, 116, 132, 175, 198, 199, 224, 233, 240, 241, 243, 250–256, 277, 381; restrictions, 11, 224, 231; revolution, 155, 241, 253, 254, 281–282, 302, 311, 322, 323, 327, 331, 344; rights, 12, 13, 231; risk, 343, 363; ritual(s), 238, 239; role(s), 8, 34, 131, 195, 197, 285, 297, 326; satiation, 59; scent signals, 73, 126; secondary characteristic(s), 30, 31, 67, 69, 75, 82, 84, 85, 87, 104–107, 121, 147, 286; secondary characteristic development, 69, 86; secondary female characteristics, 85, 132; secondary male characteristics, 82–87; selection, 380; sensitive, 71, 105; skin, 55; stimulation, 84, 88, 89, 109, 113, 243; structure, 11, 62, 70–76, 77, 87, 97, 107, 113–121, 145; styles, 229; swelling, 48, 55; symbolism, 29, 187; system, 12; taboo(s), 170, 171, 172, 253, 256, 272; therapy, 46, 200, 277, 386; toys, 342, 344, 354, 355, 357; transmitted diseases (STD), 117, 263, 264, 322, 336, 347, 348–352; violence, 86; with animals, 32; workers, 344. *See* adolescent, child, female, gender, male, safer sex practices, sexuality
Sex and Repression in Savage Society, 7, 189
Sex and Temperament in Three Primitive Societies, 35
Sex Information and Education Council of the United States, The (SIECUS), 226

Sex in Primitive Society, 8
sex change/sex reassignment surgery, 31, 286, 289, 291. *See* transsexual
sex reassignment, 286, 289; surgery, 287
sexologists, 13, 46, 114, 200, 240, 242, 244, 253
sexology, 8, 19, 36, 236–237, 240, 241, 243, 248, 277; research, 11, 22; terms, 21; theorizing, 19–37, 59, 242
Sexual Behavior in the Human Male, 32
Sexual Behavior in the Human Female, 32
Sexual Behavior on Mangaia, 250
Sexual Disorders in the Diagnostic and Statistical Manual of Mental Disorders IV, The, 256
Sexual Interactions, xiii
The Sexual Life of Savages in North-Western Melanesia: An Ethnographic Account of Courtship, Marriage and Family Life among the Natives of the Trobriand Islands, British New Guinea, 6
sexual-reproductive functioning, 159. *See* reproduction
sexuality, xiii, xiv, 5, 6, 8, 9, 10, 12, 13, 14, 15, 17, 18, 21, 23, 27, 28, 29, 30, 32, 33, 36, 43, 46, 47, 50, 52, 54, 57, 62, 64, 68, 70, 76, 77, 89, 97, 99, 110, 114, 115, 120, 124, 129, 131, 137, 143, 175, 177, 185, 186, 187, 188, 187, 197, 198, 199, 200, 201, 202, 203, 215, 216, 217, 218, 219, 220, 221, 222, 224, 227, 231, 232, 233, 235, 237, 240, 241, 243, 245, 246, 248, 251, 252, 254, 255, 277, 279, 296, 301, 302, 303, 304, 305, 306, 307, 308, 309, 310, 312, 315, 317, 318, 320, 322,

sexuality *(continued)*
323, 331, 344, 347, 357, 369,
370, 372, 374, 377, 379, 380,
381, 383
*Sexuality Today: The Human
Perspective*, xiii
sexually-confused daughters/sons,
326
sexually transmitted disease
(STDs), 114, 139, 140, 142,
224, 227, 252, 302, 336, 344,
347–352, 353, 354, 357, 370,
375. *See* AIDS, Gardnerella
vaginalis, gonorrhea, herpes,
HIV, safer sex practices,
syphillis, veneral warts
sexus, 28
Seyler, et al., 152
Shakti, 239
shaman, 309. *See* Native Americans, sacred
Shepher, 190
Sherfey, 59, 64
Shiva, 239
sibling(s), 12, 46, 54, 182, 186,
191, 192, 193, 194, 226. *See*
brother, sister
significant others, 362
Silber, S., 86, 128
Sildenafil (Viagra), 254. *See*
Pfizer, Viagra
Silphium, 271
Simon and Gagnon, 222
Singapore, 217
Singer, J., and Singer, I., 110,
246, 277
single-parent(s), 322, 326;
adoptions, 138; fathers, 380;
households, 326–327, 331, 380
Sinhalese, 181
Sipova, 152
Siriono, 202, 219
sister(s), 13, 53, 182, 188, 189,
190, 191, 292, 327, 345, 360,
371; children, 54, 181
skeletal: changes, 48; development, 160; information, 42

skene, 115
skin, 90, 109, 146, 265, 306, 336,
349, 355; aging, 104, 105;
cells, 105; condoms, 356;
rash(es), 350, 352; surgically
removed, 89; tone, 75
"the slims," 367
smegma, 89, 109
smell, 46, 55, 56, 347, 348
smoke(ing), 97, 251, 262
Smythers, Ruth, 29–30
social service: agencies, 327;
people, 172; system, 328
Social Structure, 12
societies: complex, 11, 179, 180,
203, 226, 229
socio/environmental arguments,
316
socio-cultural, 141, 316; complexity, 17; construct, 33; context,
377; definitions, 33–34, 36;
interpretation, 33–34, 93, 178,
212, 258; perspective, 38, 224,
253, 298
socio-economic, 5, 324, 327, 374,
380; classes, 257; complications, 325, 366; factors, 258;
status, 229, 374; structures,
366
socio-psychological, 169, 302,
323, 328; make-up, 318; needs,
167; theories, 166, 317
socio-sexual, 324; roles, xv; rules,
302
sociobiology, 57, 61–62
sociology, 35, 37, 151, 252;
perspectives, 19, 21, 22–23, 24,
33, 129, 138, 170
Softman, 288
sonograms, 269
sorority sisters, 53
South America, 181, 269, 337
South Pacific, 270, 375
southern states, 27
species, xiv, xv, 10, 22, 24, 29,
30, 41, 43, 48, 49, 55, 61, 63,
64, 67, 68, 69, 76, 129, 301,

312, 316, 335, 336, 337, 376, 381
speculum, 117
sperm, 60, 67, 71, 73, 84, 91, 93, 95, 107, 113, 114, 119, 121, 122, 135, 137, 260, 261, 264; banks, 50, 68, 129–130, 380, 385; count, 136, 139, 140, 142; deformed, 139; donors, 130; ducts, 266; ejaculate, 139; motility, 92, 136, 139–140; production, 74, 81, 90, 136, 303; supply, 140, 141; transport, 92. See male, penis, reproduction, sex
sperm-cervical mucous, 141
spermatic: cords, 89, 91; economy(ies), 93, 137, 140, 315. See sperm
spermatogenesis, 74, 75, 81, 82, 90, 98, 101, 104, 107, 115, 121–122, 125, 142
spermicide(al), 258, 260, 261, 263, 356; condom, 355; cream, 264; jelly, 264. See contraceptives
Speroff, L., 122
sphincter, 95, 96, 186
spirit, spirituality, 137, 143, 239, 294, 297, 315. See religion, sacred
Spirit and the Flesh: Sexual Diversity in American Indian Culture, The, 297
Spiro, 192
spirochetes, 350
Spitz, 45
Spock, Benjamin (Dr.), 175, 197–198, 203, 228–229; Dr. Spock's Baby and Child Care, 197
sports, 355; amateur, 55
spouse(s), 180, 181, 182, 195, 214, 216, 252, 309, 325, 372. See marriage, partner
Sri Lanka, 181, 220
St. Augustine, 240
Stark, 194

Starka, L., Sipova, I, and Hynie, J., 152
status, 12, 31, 43, 55, 57, 61, 139, 150, 159, 172, 181, 190, 209, 210, 211, 213, 214, 216, 218, 220, 226, 228, 229, 230, 231, 236, 258, 269, 274, 284, 287, 288, 296, 297, 302, 304, 305, 306, 307, 308, 309, 342, 346, 362, 365, 374; change(s), 89, 166, 224; considerations, 11; group, 22; marker, 224
stereoscopic vision, 39, 44, 45, 46–47, 64
steriods, 74, 87, 106, 256, 345, 360, 370, 384
sterility, 107, 133, 205, 218, 232, 260, 266, 272, 373, 374. See infertility
Steward, 296
stimga, 287, 298, 307, 331, 337, 339, 341, 345
Stoller, R. J., 151–152
stratification, 15, 17, 208, 220, 221
stress(es), 5, 6, 7, 21, 22, 91, 114, 123, 127, 140, 151, 211, 238, 250, 268, 321, 336; reduction techniques, 128
structure, 15, 23, 39, 69, 70, 82, 87, 89, 91, 92, 96, 97, 113, 114, 115, 119, 120, 121, 129, 144, 145, 148, 175, 189, 197, 208, 214, 218, 219, 235, 247, 317, 325, 366; kin groups, 179; premarital sex rules, 6
sub-Saharan, 111, 333; Africa, 372; coastal areas, 337
subincision, 89, 97, 99
substance abuse patterns, 368. See drugs
Sudan, 109, 111
Suggs, David N., and Miracle, Andrew W., 386
suicide, 127, 128, 321
superincision, 89, 99

Supermale Syndrome (XYY), 133, 146, 147, 148, 156. *See Berdache*, sex change, two spirit, XYY chromosomes
suppositories, 255, 264, 348
Supreme Court, 3, 267
surgery, 31, 115, 122, 141, 171, 255, 260, 284, 288, 289, 291, 352; equipment, 373; health-care, 360; incision, 113; procedure, 120, 256, 258, 266, 286–287
surrogate mothers, 130, 326, 380. *See* artificial insem-ination, family, infertile, reproduction
survival: fitness, 50; of the fittest, 25
Sussman, 384
sweat, 56, 352, 367
Swedish, 199
Symons, Donald, 57–58, 59–60, 64
synchrony, 57, 62
syphillis, 335, 350

T

T cell(s), 336, 365, 367
taboo(s), 6, 132, 171, 175, 185, 188, 189, 192, 237, 253, 256, 272, 308, 316. *See* menstrua-tion, rites of passage
tactile, 45, 105
Tahitian, 288, 297
Taiwan, 193, 220, 270
Tanner, N., and Zihlman, A., 42–43, 61
Tantric, 47, 233, 238–239, 240, 277
Tanzania, 48
tattooing, 245, 360
Taylor, J. M., and Ward, V., 257
Tchambuli, 35
technology, xiv, 15, 50, 131, 132,

142, 180, 208, 220, 270, 286, 286, 289, 331, 377
teenage, 209, 210, 214, 216, 218, 229, 230, 231, 237, 291, 323, 342, 344; boys, 244; girls, 223; mothers, 225; pregnancy, 224, 226, 227, 282, 325, 328, 331, 380; premarital sex, 219; sexuality, 223, 228, 232; trauma, 207. *See* adolescent
terrestrial, 42–46, 64; adapta-tion(s), xv, 43–44, 46; environment(s), 44, 64
testerone enanthate (TE), 265–266
testes, 69, 71, 73, 74, 75, 76, 81, 82, 90, 91, 115, 121, 145, 148, 149, 244. *See* male, reproduction
testicles, 74, 81, 84, 89, 90, 91, 94, 144, 147, 160, 349; atrophied, 147. *See* male, reproduction
testicular: feminizing syndrome, 31; function, 81; self-exam (TSE), 90
testosterone, 31, 68, 69, 75, 77, 81–82, 83, 84, 85, 87, 90, 98, 104, 105, 128, 146, 147, 148, 152, 155, 246, 266, 303, 306, 317; levels, 73, 79, 86, 255; ocryptorchid, 149; production, 73, 76; release, 145; replace-ment therapy, 254
tetracycline, 349, 350
Thailand, 220, 270
Thayer, 297
therapy(ies), 86, 141, 200, 251, 254, 255, 277, 284, 351, 352, 362, 364, 368, 381
thigh(s), 106, 236, 354
third gender, 297, 315
Third World, 139, 232, 256, 257, 268, 270, 339, 373
Three Essays on the Theory of Sexuality, 187

threshold rites, 82, 85, 104, 211
throat, 168, 352
thrush, 336, 362, 366, 367
Tibet, 181
Tiffany Club, 294
Tikopia, 202
tira, 250–251
Tiwi, 120–121, 137, 143, 181,
    217
Tlingit, 213
tobacco, 97. *See* smoke
toilet training, 186
tomboy(s), 148, 296
Tongans, 219
tongue, 249, 352
tonicity, 73, 77, 79, 81, 82, 98,
    103, 121
Tontonac, 202
totem(s), 120, 137, 143, 189
Totem and Tabu, 188
touch, 26, 44, 45–46, 47, 64, 200,
    347
tranquilizers, 95
transgender, 286, 288–295, 385,
    386; community, 294; identity,
    150, 153, 288; issues, 289;
    leadership development, 294;
    persons, 279; populations, 151
Transgender Alternative, The,
    287
transociptase inhibitors, 369
transsexual (TS), xv, 84, 135,
    150–153, 154, 155, 279,
    283–298, 385; etiology, 151,
    152; hormonal responses,
    152; identity, 283–285, 298,
    310; issues, 289; people, 152,
    283, 284, 287, 385; preopera-
    tive, 283. *See Berdache*, two-
    spirit
transvestite (TV), 135, 150, 151,
    279, 283, 286, 287, 291, 294,
    385. *See* crossdressing
Treponema pallidum, 350
tribe(s), 15, 179, 191, 275, 276
Trichomonas vaginalis, 348, 350

trimester(s), 116, 119, 131, 145,
    159, 160, 173, 267, 361, 362.
    See pregnancy
trimethoprim-sufamethoxazole,
    349
Trisomic X, 148
Trobriand Islands, 6–7, 189, 203,
    224
Troiden, Richard, 252
Truke, 201
tubal, 116, 130, 149, 267, 350;
    blockage, 120, 140, 141;
    ligation, 259, 266; patency,
    136, 140; sterilization, 260
tuberculosis (TB), 336, 365, 367
tumors, 352
Turner, Victor, 29, 211
Turner's Syndrome, 133, 146,
    148, 156
TV/TS Tapestry Journal, 294
two spirit, 31–32, 279, 283–285,
    287, 288, 296, 297, 384. *See
    Berdache*
Tylor, Sir Edward Burnett, 25,
    192

U

United States, xv, xvi, 9, 47, 50,
    85, 88, 89, 98, 107, 109, 116,
    119, 120, 127, 128, 129, 130,
    132, 133, 137, 138, 139, 140,
    142, 156, 160, 164, 165, 167,
    170, 171, 172, 175, 179, 187,
    193, 194, 196, 197, 200, 201,
    203, 205, 207, 208, 209, 221,
    225–227, 228, 230, 232;
    adolescent(s), 229, 231; births,
    113, 166; childbirth practices,
    113; college students, 112;
    college coaches, 93; culture(s),
    22, 55, 67–68, 82, 95, 103,
    105, 125, 126, 135, 144, 163,
    169, 173; folk beliefs, 143;
    male-female relationships, 52;

United States *(continued)*
men, 73; population, 224;
pregnancy rate, 225; sexology,
23; sexual behavior, xiii, xiv;
view, 162; women, 110, 114
U. S. Center for Disease Control,
223
umbilical cord, 161, 164, 168
unilineal descent(s), 138, 178,
183, 184–185, 185, 190. *See*
descent, family, kin
United Nations, 111
upper class(es), 209, 223
urban, 179, 180, 208
urethra, 88, 92, 93–94, 97, 110,
115, 145; acidity, 71; opening,
149; sphincter, 95
urinary: meatus, 71, 88, 94, 95,
97, 110, 115, 145; stress
incontinence (USI), 115; tract
infections (UTIs), 110
urine 71, 74, 94, 95, 97, 109,
110, 115; specimen, 124; tests,
87
uterus, uterine 31, 59, 106, 113,
115, 117–121, 130, 136, 138,
145, 146, 148, 149, 161, 162,
163, 165, 172; cavity, 116, 119,
123; contractions, 117, 126;
lining, 124; tone, 69, 75, 76,
103, 116
utopian, 182

V

vaccine(s), 266, 347, 361, 366,
368, 369, 375
vacuum curettage, 255, 267–268
vagina(al) 56, 59, 70, 111, 113–
114, 117, 119, 136, 137, 140,
144, 145, 149, 163, 168, 202,
237, 242, 244, 246, 250, 254,
261, 284, 286, 307, 346, 350,
361, 367, 380; blind vagina,
31, 146; childbirth, 113; dam

barrier, 93, 355, 356, 357;
fluids (lubrication), 93, 114,
306, 336, 342, 344, 347, 351,
353, 354, 374, 375; frigidity,
245; intercourse, 222, 236, 351,
354, 355, 357; mucosa, 114,
116, 264; musculature, 247;
orgasm(s), 110, 146, 245;
suppositories, 348
vaginalis, 348
vaginismus, 254
vaginosis, 348
Van Gennep, Arnold, 210
Vance, Carol, 321
Vance, C. S, and Pollis, C. A., 236
Varieties of Sexual Experience, 11
vas deferens, 70, 71, 82, 91–92,
93, 116, 121, 141, 144, 145,
149, 260, 266
vasectomy, 92, 259, 260, 266
vaso-congestion, 59, 244
vasomotor, 306
Vatuk, S., 309
Vedas, 238
vehicle(s), 46, 59, 179, 192, 193,
209, 216, 238, 272, 352
vein, 141, 345
venereal warts (genital warts),
352
Venezuela, 275
vernix, 164
Viagra, 254, 255. *See* Pfizer,
Sildenafil
vibrators, 344, 354
Vietnam, 194, 257–258, 292
violence, 36, 86, 127, 128, 147,
371
virus, 87, 223, 336, 337, 338,
340, 342, 345, 347, 352, 354,
356, 360, 361, 363, 364, 365,
366, 368, 369, 373, 375, 376;
heptatitis, 351; pneumonias,
367
Virginia, 290, 324, 326
virginity, 109, 112–113, 167, 219,
220, 241, 385

virility, 112, 142, 303, 371
visual, 105, 144, 286, 344,
372; centers, 45, 56; cortices,
46–47; fantasies, 353; predation theory, 384; sex cues, 47,
82, 107
voice, 52, 84, 149, 208, 230,
270, 288, 299, 320; changes,
121
vulva, 93, 108, 109, 236, 264,
348, 356; orgasm, 246

W

Wallace, Alfred Russell, 24
Wallis, 223
Walters, W., 193
Ward, Martha, 195
warfare, 49, 50, 55, 124, 137,
151, 214, 217, 275, 276, 297;
warrior, 150, 215, 288; warrior
woman, 150
warts, 352
Washburn, S. L., and DeVore, I.,
42
Waxenberg et al., 254
Webster decision, 267
wedding, 29, 30, 112–113. See
bride, marriage
weight, 92, 127, 265, 328, 362,
367; gain, 160, 161, 262; loss,
352
Westermarck, Edward, 192, 296;
effect, 192, 193
west(ern), 105, 125, 138, 139,
142, 147, 177, 186, 189, 190,
199, 200, 203, 205, 207, 208,
209, 215, 221, 222, 223, 228,
231, 232, 233, 238, 239, 242,
256, 259, 268, 270, 277, 281,
282, 287, 304, 308, 310, 315,
321, 328, 341, 373, 375, 380,
383; adolescence, 8, 221–223,
226, 230; adult sexuality,
202; child (children), 175,

230; culture, 9, 23, 27, 126,
127, 132, 155, 283, 299, 313,
314, 316; ethnocentrism, 236;
European, 6, 325; family, 7,
178; femininity, 34; gender
ideology (paradigm), 34, 35,
194, 285–286, 297; homophobic perspective, 217, 220;
manuals, 237; perspective(s),
143, 159, 201, 286, 381;
sexology, 240, 241; sexuality,
22; tomboy, 296; transsexual
identity, 153; women, 122,
160, 254, 305, 306
Western Blot test, 364, 367. See
AIDS, HIV
Western Uttar Paradesh, 309
wet dreams, 121, 125, 215, 216,
222
wet kissing, 354, 355
wet nurses, 275. See breastfeeding, infant
Whelehan, Patricia, 196, 300
When Harry Met Sally, 54
Whipple, B., and Perry, J. D.,
240, 246–249, 277
White, 386
White, Leslie, 192
Whiting, 297
widowhood, 309
wife, 54, 55, 110, 124, 130,
137, 138, 180, 181, 182,
193, 216, 217, 218, 276, 288,
324, 371, 372; adoption,
193. See descent, family,
kin, marriage
Wikan, 297
Williams, 27, 285, 288, 297, 315,
322
Williams, Gertrude, 200
Williams, Walter, 236, 296
Williamsburg, 290
Wilson, 144
Wilson, Robert (Dr.), 306
Winch and Blumberg, 180
Wogeo, 202

Wolf, Arthur, 192–193
Wolffian ducts, 145, 148, 385
woman, 25, 28, 31, 32, 34, 54,
  68, 93, 101–131, 136, 137, 138,
  139, 140, 142, 143, 144, 150,
  159, 161, 162, 166, 167, 169,
  170, 171, 172, 173, 181, 188,
  201, 210, 224, 229, 238, 239,
  245, 250, 257, 258, 264, 266,
  268, 269, 272, 282, 286, 287,
  288, 289, 292, 295, 307, 308,
  309, 318, 322, 326, 327, 336,
  344, 346, 350, 362, 363;
  marriage, 182; menstrual
  cycle, 259, 305; social, 286. *See*
  female, reproductive, sex
woman-husband, 182
woman-marriage, 288
womb, 117, 120; envy, 197
women, 12, 13, 30, 34, 35, 46,
  47, 49, 50, 53, 54, 55, 56, 57,
  58, 59, 60, 62, 68, 73, 77, 81,
  82, 83, 84, 85, 87, 90, 91, 93,
  98, 101–131, 137, 138, 142,
  147, 153, 160, 163, 165, 167,
  168, 169, 170, 171, 172, 182,
  188, 189, 190, 194, 196, 197,
  211, 212, 213, 214, 215, 216,
  217, 218, 220, 223, 224, 231,
  237, 238, 242, 243, 244, 245,
  247, 250, 251, 252, 253, 254,
  255, 257, 258, 261, 262, 264,
  265, 266, 267, 268, 269, 270,
  272, 273, 275, 276, 283, 284,
  285, 286, 288, 296, 297, 302,
  306, 308, 310, 315–316, 318,
  319, 321, 323, 324, 325, 326,
  327, 337, 338, 339, 340 341,
  342, 344, 346, 348, 349, 355,
  362, 363, 366, 367, 368, 371,
  372, 373, 375; athletes, 107;
  attitudes, 274; clothing, 287;
  culture, 299, 300; desire, 241;
  fertility, 136; groups, 301;
  movement, 241; passivity, 240;
  reproductive structures, 110;

role, 36; self-concepts, 307;
  sexual life, 303; sexuality, 241;
  status, 305, 309
world, 4, 8, 9, 15, 26, 47, 54,
  138, 178, 180, 183, 187, 191,
  207, 209, 211, 212, 215, 237,
  238, 248, 257, 258, 268, 270,
  276, 282, 286, 294, 300, 311,
  315, 324, 346, 347, 353, 354,
  380; birthrate, 256; population,
  256, 269
World Health Organization
  (WHO), 265, 338
World of Human Sexuality, The,
  386
World of Human Sexuality:
  Behaviors, Customs and
  Beliefs, The, 10
Worlds of Pain, 22

X

X chromosome, 130, 145, 146,
  148, 317
/Xai/xai, 309
Xanith, 288, 297
XO chromosomes: female, 146
XX chromosomes, 144, 146, 148,
  149, 156
XXX chromosomes, 148
XXY chromosomes (Klinefelter's
  Syndrome), 146, 147, 148
XY chromosomes, 145, 144, 146,
  149, 156
XYY chromosomes, 146, 147, 148

Y

Y chromosomes, 146, 147;
  filtration, 130
YO chromosome male, 146
Yako, 183
Yanamami, 124, 275–276
Yap, 202, 224, 273–274

yeast infection, 336, 348, 366, 367

Young, F. W., 212

youth, 195, 207, 211, 216, 218, 220, 223, 224, 226, 228, 230, 237, 250, 274, 304, 306, 307, 321; lives, 209

Yucantan, 171

Yuppies, 302

Z

Zaire, 372

Zambia, 340

Zande, 270

Zelnik and Kanter, 223

Zihlman, A., 42–43

Zilbergeld, B., 300

Zovirax (oral acyclovir), 351